MATERIALS SCIENCE AND TECHNOLOGIES

BOROPHOSPHOSILICATE GLASS THIN FILMS IN ELECTRONICS

Materials Science and Technologies

Additional books in this series can be found on Nova's website under the Series tab.

Additional e-books in this series can be found on Nova's website under the e-book tab.

Electronics and Telecommunications Research

Additional books in this series can be found on Nova's website under the Series tab.

Additional e-books in this series can be found on Nova's website under the e-book tab.

MATERIALS SCIENCE AND TECHNOLOGIES

BOROPHOSPHOSILICATE GLASS THIN FILMS IN ELECTRONICS

VLADISLAV YU. VASILYEV

nova publishers
New York

For permission to use material from this book please contact us:
Telephone 631-231-7269; Fax 631-231-8175
Web Site: http://www.novapublishers.com

Library of Congress Cataloging-in-Publication Data

Tonkie sloi borofosforosilikatnogo stekla v tekhnologii kremnievoi mikroelektroniki. English.
Borophosphosilicate glass thin films in electronics / editor, Vladislav Yu. Vasilyev (Semiconductor Devices and Microelectronics Department, Radio Engineering, Electronics, and Physics Faculty, Novosibirsk State Technical University, Novosibirsk, Russia).
pages cm
Translation of: Tonkie sloi borofosforosilikatnogo stekla v tekhnologii kremnievoi mikroelektroniki.
Includes bibliographical references and index.
ISBN: 978-1-62417-959-4 (hardcover)
1. Electronic apparatus and appliances--Materials. 2. Glass coatings. 3. Boron compounds. 4. Silicon compounds. 5. Glass manufacture--Chemistry. I. Vasil_ev, V. IU. (Vladislav IUr_evich) II. Title.
TK7871.15.S55T6613 2013
621.3815'31--dc23 2013002527

Published by Nova Science Publishers, Inc. † New York

"To my wife Lidia"

CONTENTS

PREFACE

This comprehensive monograph summarizes the 30-year studies of borophosphosilicate glass (BPSG) thin film used in electronic technologies, including the author's personal experience with film deposition, characterization, and implementation in microelectronic technology. The main focus of the monograph is the interrelation of chemical vapor deposition (CVD) kinetic features, thin film material properties, and electronic device technology aspects. Part 1 of the monograph is devoted to the analysis of thin film synthesis, such as: CVD methodology and BPSG film processes, silicon dioxide and glass film growth kinetics, CVD step coverage and gap-fill features. Part 2 of the book is a description of BPSG film properties, film structure, glass flow capability, BPSG film-moisture interaction and the film defect formation phenomenon. A number of experimental data are presented and discussed in detail.

In Part 1, Chapters 1 to 5 focus on the different aspects of BPSG thin film synthesis. Among them are CVD techniques, methodology and processes parameters, silicon dioxide and glass thin film growth kinetics, CVD step coverage and gap-fill features, and the analysis of commonalities and differences of CVD processes. Chapter 6 is devoted to the description of the most important industrial production processes used for the film deposition in electronics. As a result of these considerations, the reader can obtain a complete view of the CVD thin film processes, including both silicon dioxide and silicate glass thin films. This will help to understand the consolidated material in Part 2 of the monograph that is devoted to the analysis of the film composition, structure, properties, and their interrelations with BPSG thin film CVD process features.

Part 2 (Chapters 7 to 14) is devoted to a description of BPSG thin film composition, structure, and basic properties. As long as the CVD thin film technique is not the traditional glass forming technique, special attention is given to find the links between the BPSG film structures and properties with definitions of the classical glass chemistry. This allows to see the relationships between glass film composition, structure, and properties with BPSG thin film CVD process features.

Part 3 of the monograph (Chapters 15 to 20) is devoted to the description of BPSG thin film implementation aspects. As long as BPSG films were introduced in electronic technology as flow-able dielectric, the most common use of this film is to smooth sharp edges of device structures. Chapter 15 is devoted to the detailed analysis of the flow aspects of BPSG films on the real IC device structures. It is important to note that data presented in Chapter 15 were obtained experimentally and explained quantitatively. We also showed that the approaches presented can be applied to explain the flow of other types of flow-able glass films such as BPSG films with germanium oxide additive. Chapter 16 shows approaches for obtaining the optimized BPSG composition used in electronic technology. Chapter 17 is devoted to the aspects which are rarely discussed in the related literature, namely BPSG film durability. Some approaches to improve BPSG film post deposition thermal treatment are proposed. Finally, we generalize some data regarding BPSG film application in electronics in Chapter 18.

List of Abbreviations (in Order of Appearance)

IC	integrated circuits
CVD	chemical vapor deposition
BSG	borosilicate glass
PSG	phosphosilicate glass
BPSG	borophosphosilicate glass
APCVD	atmospheric pressure chemical vapor deposition
SACVD	subatmospheric pressure chemical vapor deposition
LPCVD	low pressure chemical vapor deposition
PECVD	plasma enhanced chemical vapor deposition
HDP-CVD	high-density plasma chemical vapor deposition
VLSI	very large-scale integrated (circuit)
ULSI	ultra large-scale integrated (circuit)
PMD	pre-metal dielectric
SEM	scanning electron microscopy
CMP	chemical-mechanical polishing
IMD	inter-metal dielectric
LT	low temperature
MT	middle temperature
HT	high temperature
TEOS	tetraethylorthosilicate
[B], [P]	boron- and phosphorus content (concentration) in the film
IMP	intermediate reaction product
AR	aspect ratio
SC	structure complexity
RF	radio frequency
HMG	homogeneous (route)
HTG	heterogeneous (route)
FA	furnace anneal
RTA	rapid thermal anneal
RI	refractive index
ICP	inductively-coupled plasma
SIMS	secondary-ion mass spectrometry

NMR	nuclear magnetic resonance
HF	hydrogen fluoride
XRF	X-ray fluorescence
AFM	atomic force microscopy
SPV	surface photo voltage
SCA	surface charge analysis
WER	wet etch rate
WERR	wet etch rate ratio
DD	defect density, cm-2
OM	optical microscopy
LS	laser scanning
TEM	transmission electron microscopy
AES	Auger electron spectroscopy
EDS	energy-dispersive spectroscopy
GFC	glass flow capability, dimensionless

LIST OF IMPORTANT SYMBOLS (IN ORDER OF APPEARANCE)

T_d	deposition temperature, °C
C_P (or [P])	elemental phosphorus content (concentration), wt %
C_B (or [B])	elemental boron content (concentration), wt %
d	thin film thickness, μm, Å
P_d, p_i	total and compound partial deposition pressure, torr
V	reactor volume, l, cm^3
G_0, g_i	total and compound gas flows, l/min, sccm, lpm, cm^3/s
x_i	spacing between gas injection and semiconductor wafer, cm, mm, mls
S	reactor surface, cm^2
W	deposition rate, Å/min, nm/min, μm/hour
τ	residence time, s
υ	gas flow speed, or gas velocity, m/s, cm/s
s_R	reactor cross-square, cm^2
T_r	room temperature, °C
$[C_A]$, $[C_B]$	concentrations of compounds A and B
a,b	stoichiometric coefficients, or reaction orders for A and B compounds
k_f, k_0	coefficients of proportionality, or reaction constants
E_a	activation energy, kJ/mol
R	gas constant
k_{eff}	effective constant of the deposition rate, cm/s
$[C_{Si}]$	silicon compound concentration, mol/l
χ	coefficient of proportionality
n_i	number of compound moles in reaction gas phase
A_0	initial thin film angle on the device step after deposition, degrees
H	depth of the gap, μm
G	width of the gap, μm
d_2	film thickness on the step side wall, μm, Å
d_1	film thickness on the step top, μm, Å
SC	structure complexity, μm^{-1}
T_a	annealing temperature, °C
T_g	glass transition temperature, °C

μ	glass viscosity, Poise
T_f	flow temperature, °C
t_e	exposure time to ambient air for BPSG films, minutes, hours
t_m	defect monitoring time, minutes, hours
t_a	thermal anneal time (minutes for FA, seconds for RTA conditions)
U	activation energy for glass flow
A_t	thin film angle on the device step after glass flow process, degrees
T_{bf}	temperature of the beginning of the glass flow, °C

INTRODUCTION

Thin film materials have become basic construction elements in modern solid-state technologies, such as microelectronics, micro-electro mechanical device technology, nanoelectronics, etc. Traditionally, integrated circuit (IC) technology is the most advanced field of solid-state electronic technology that has always motivated thin film electronic materials research and development. Silicon dioxide-based thin film materials have been found to be excellent inorganic insulators with some specifically important material properties compatible with mono- and polycrystalline silicon. Boron-, phosphorus-, arsenic-, germanium-contained silicon dioxide thin film materials (in electronic technology, sometimes erroneously called B-, P-, As-, Ge-"doped" silicon dioxide, respectively) were described to have so-called "vitreous" (or "glassy") structures. This is because they consist of SiO_2, P_2O_5, B_2O_3, GeO_2, As_2O_3 and other types of oxides linked to each other in the matrix. Chemical bonding and structures of vitreous thin films are expected to be similar to those of bulk silicate glasses with similar glass composition. Therefore, knowledge of classical chemistry of the vitreous materials can be applied to the explanation of structural features and the differences in thin glass film properties and structures.

Traditional bulk glasses are considered to be a kind of supercooled liquid, which is formed from mixed oxides by the melt cooling method. In contrast, thin glass films are formed using nontraditional glass-forming techniques, such as chemical vapor deposition (CVD). This is a material synthesis method in which chemical compounds (also named reactants, reagents) in vapor or gaseous form enter into a chemical reaction either in a heated reaction volume or on a heated substrate surface resulting in thin solid film formation. For electronic applications, thin films are created using gaseous or liquid compounds and a variety of CVD techniques are realized in flow-type reactors. CVD methods allow depositing at a relatively low temperature, a large variety of materials with well-controllable stoichiometric composition, as well as vitreous and crystalline films having a high degree of purity, perfection, and thin film structures that are impossible to attain by any other techniques.

The most common feature of the CVD methods is that the formation of the glass structure (formation, for instance, $Si - O$, $B - O$, $P = O$ bonds in the films) occurs directly during thin film synthesis. Therefore, the glass film composition and properties are strongly defined by the compounds used, as well as by CVD conditions. The film deposition temperatures (T_d) for microelectronic technology applications are normally high enough ($T_d > 350$ °C), excluding the formation of amorphous or polymeric compounds. Subsequent heat treatment of thin silicate glass films, which in a number of cases is dictated by technological requirements, is performed at temperatures < ~ 900 °C that is lower than that of the glass crystallization temperature. Thus, as-deposited glass film structures can be considered to be fixed and to have already relatively stable structures with Si – O, B – O, P = O, etc. bonds. Post deposition glass film anneal at the above-mentioned temperatures therefore causing relatively slight structural improvement. As long as practically used so-called "binary" (phosphosilicate glass, PSG, and borosilicate glass, BSG) and "ternary" (borophosphosilicate glass, BPSG) films are significantly enriched with a SiO_2 component, these complex vitreous materials can be reasonably analyzed using a silica structural network base where the other oxides (P_2O_5, B_2O_3, GeO_2, etc) are assumed to be additives.

The author studied CVD thin film process technology and thin film materials used in electronics for approximately 30 years. Experimental data have been obtained personally by the author for polysilicon, silicon nitride, silicon dioxide and silicate glasses, silicides, as well as ruthenium metal thin films, synthesized by different CVD techniques with a number of compound combinations. The results have been generalized in the thesis and reviews [1-4].

When considering an idea to write a monograph on CVD thin film dielectric materials for electronic applications, the author realized that borophosphosilicate glass thin film could serve as an "ideal object". In fact, up to date BPSG thin film is the most complicated glass material that has been used in electronics for more than 30 years. At the same time, during this period the film has been a permanently modified commercial product. Analysis of BPSG thin film technology and material properties allows reviewing many CVD and material related topics, such as: tools and methodology of CVD thin film synthesis, growth kinetics in a variety of chemical compound combinations and CVD process conditions, ultra-low gap-fill methodology, characterization of glass film composition, structure, and film properties, quantitative analysis of film features, film application issues in electronics, etc.

This book is devoted to the BPSG thin film analysis for electronic applications. The main core of this monograph is the interrelation of CVD kinetic features, thin film material properties, and technology aspects. The large body of the reviewed information has been divided into three parts.

Part 1 (Chapters 1 – 6) of the monograph is devoted to the different aspects of thin film synthesis, such as CVD methodology and BPSG thin film CVD processes, silicon dioxide and glass film growth kinetics, CVD step coverage and gap-fill features, and analysis of the commonalities and differences of CVD processes. All of these topics together allow a complete view of CVD thin film processes. This helps to understand the observed differences of the film properties and their interrelation with CVD process features.

Part 2 (Chapters 7 – 14) is a description of BPSG film properties, film structure, quantitative glass flow capability and gap-fill analysis, BPSG defect formation phenomenon, etc. As long as the CVD technique is not the traditional glass forming technique, special attention is given to show the links between the BPSG film structures and the film properties with definitions of the classical glass chemistry. This allows seeing the relationship between glass film structure and CVD thin film peculiarities.

Part 3 (Chapters 15 – 18) covers different aspects of BPSG film applications in electronic technology, optics, nanotechnology.

The author would like to reiterate that all data presented in this book have been experimentally obtained by a number of research groups. Most of these data have been double-checked and confirmed. Some data can be considered to be non-repeatable because they were obtained using some specific deposition processes. Nevertheless, the author would like to stress that he considers this book an attempt to create a complete view of the BPSG thin film material for electronic applications. In this regard, the author tried to generalize a number of experimental data selecting the most common features of the film synthesis, composition, structure, and properties.

This book is intended for scientists, researchers, material and electronic engineers, working in solid-state electronics, as well as for students. It is supposed to be helpful for engineers, which are involved in the everyday solution of practical issues in electronic technology.

PART 1. BPSG THIN FILM DEPOSITION

In Part 1, Chapters 1 to 5 focus on the different aspects of BPSG thin film synthesis. Among them are CVD techniques, methodology and processes parameters, silicon dioxide and glass thin film growth kinetics, CVD step coverage and gap-fill features, and analysis of commonalities and differences of CVD processes. Chapter 6 is devoted to the description of the most important industrial production processes used for film deposition in electronics. As a result of these considerations, the reader can obtain a complete view of CVD thin film processes including both silicon dioxide and silicate glass thin films. This will help to understand the consolidated material in Part 2 of the monograph that is devoted to the analysis of the film composition, structure, properties, and their interrelations with BPSG thin film CVD process features.

Chapter 1

BRIEF HISTORICAL OVERVIEW OF BPSG THIN FILM RESEARCH

Inorganic materials are used in solid-state technologies in the form of thin films with thickness in a range from a few tens of angstroms to a few micrometers. These films are created on the surface of a semiconductor or insulating substrates using a variety of CVD techniques. The first most comprehensive monograph covering many aspects of thin film CVD was published by Powell, Oxley, and Blocher in 1966 [5].

Generally, silicon-based films such as silicon dioxide and silicate glasses, polycrystalline silicon, silicon nitride, as well as metals and metal silicides, have become the basic thin film materials in IC technology [6-14]. A variety of tools, chemical compounds and deposition conditions have been investigated for thin film deposition over the last 40 years. Some aspects of film deposition techniques, thin film properties and application issues in microelectronic technology have been summarized in a number of monographs, reviews and surveys published in the last part of the 20th Century [6-47].

Silicon dioxide-based thin film materials revealed some specifically important material properties such as insulating properties and gettering properties with respect to alkali ions (phosphorus-contained silicon dioxide). Phosphorus- and boron-contained silicon dioxide thin films also revealed a capability to soften at enhanced post-deposition temperatures. These features presented a reason why the silicon dioxide-based thin films have been chosen as a base for silicon IC technology. These dielectric films are widely used in microelectronic technology for shallow trench isolation, spacer isolation, dielectric isolation between metallization layers, final device passivation, etc. These dielectric films are deposited using a variety of chemical compound combinations and CVD methods, such as atmospheric, subatmospheric, or low-pressure chemical vapor deposition (APCVD, SACVD, and LPCVD, accordingly), and plasma-enhanced (PECVD) and high-density plasma (HDP-CVD) deposition.

Intensive studies of thin P_2O_5-containing insulating phosphosilicate glass films were stimulated by the rapid development of microelectronic technology, and the advent of very large-scale integrated (VLSI) circuits followed by ultra large-scale integrated (ULSI) circuits [7-10]. These films were considered for use in electronic technology for two basic purposes: (a) as an IC device passivation dielectric, and (b) as a "Pre-Metal Dielectric, PMD" for the separation of IC transistor level from the first metallization level [15-20]. Dielectrics like PSG thin films with elemental phosphorus content within a few percents revealed gettering properties against undesirable alkali metal diffusion from ambient into the transistor structures. Being thermally treated at high temperatures caused the film flow and PSG films were found to provide a smoothening (planarization) of the dielectric surface and rounding of the contact hole top edges. Planarization was found to be necessary in IC device technology due to the non-conformal metal sputtering on the sharp device steps (see Figure 1.1(a)) and, therefore, metal line imperfection on the side walls of IC device structures (see Figure 1.1(b)). Thus, flow-able dielectrics provide a reliable multilevel metallization system formation.

In the mid-1970s, PSG films were used as PMD with about 1 μm thickness and an elemental phosphorus content (C_p) up to 13 wt % (about 30 wt % P_2O_5). PSG films were planarized by viscous flow during a thermal heating of the semiconductor structures covered with glass films. The PSG flow temperature in microelectronic technology was as high as above 1000 °C. This temperature was not applicable for some types of VLSI devices, which required a significant reduction of thermal treatment conditions (frequently called "thermal budget") in order to minimize lateral and vertical dopant diffusion in silicon substrate. Furthermore, the phosphorus oxide in the film was found to react with the ambient moisture to produce orthophosphoric acid (H_3PO_4). This acid attacks the aluminum metallization causing severe corrosion of the latter [8-10]. As a result, highly P_2O_5 concentrated PSG thin film application in microelectronic technology as PMD material was restricted by the revealed film disadvantages. At the same time, BSG films were rarely used in semiconductor technology.

In the early 1980s, borophosphosilicate glass (BPSG) was proposed as an alternative PMD film. BPSG thin film was introduced in VLSI circuit technology in 1982, when Ramiller and Yau [49] showed some BPSG film advantages over the traditionally used PSG thin films. The same year, the first detailed paper by Kern and Schnable [50]) appeared that was devoted to BPSG film synthesis, composition, properties, and applications in VLSI devices. The proposed BPSG films contained up to about 30 wt % B_2O_3 and P_2O_5 taken together or expressed in elemental content 3 – 6 wt % B and 3 – 7 wt % P, or expressed in oxide content 9.6 – 19.2 wt % B_2O_3 and 6.9 – 13.8 wt % P_2O_5. With about two-thirds of the P_2O_5 replaced by B_2O_3, the problem of VLSI aluminum metallization corrosion was overcome.

In addition, the flow temperature of BPSG films was greatly (by a few hundreds Celsius degrees) reduced compared to PSG films, providing good film planarity and smoothing the contact hole top sharp edges (see Scanning Electron Microscope (SEM) example photos in Figure 1.1(c-f)). The latter feature made BPSG suitable for use with short-channel and shallow-junction transistors and with radiation-hardened VLSI.

Note that despite the fact that BPSG films were first employed in IC with minimum feature size over 1μm, they are also compatible with IC technologies well below that level, such as ULSI technology.

However, due to ULSI device complexity, additional Chemical-Mechanical Polishing (CMP) is used in technology to planarize BPSG PMD films. A simplified sketch of the device structure in Figure 1.2 shows the implementation of glass thin films in ULSI technology.

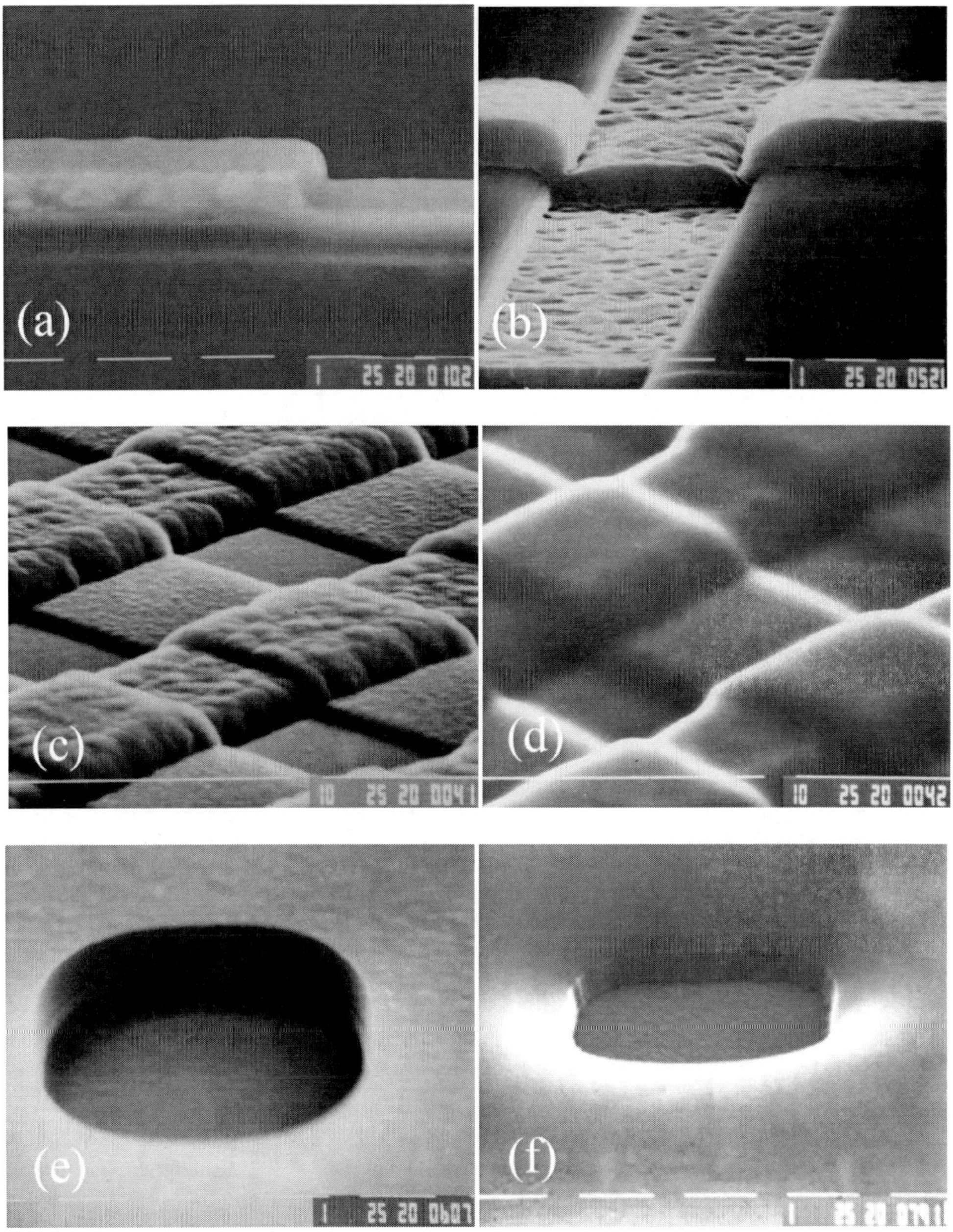

Figure 1.1. Scanning Electron Microscopy images of polysilicon step covered with glass film (a), metal line imperfection (severe thinning) on the bottom corners of the vertical glass side walls (b), crossing polysilicon lines covered with as-deposited (c) and thermally-flowed (d) BPSG films, and contact holes in as-deposited (e) and thermally-flowed (f) BPSG films on silicon dioxide underlayer that does not flow.

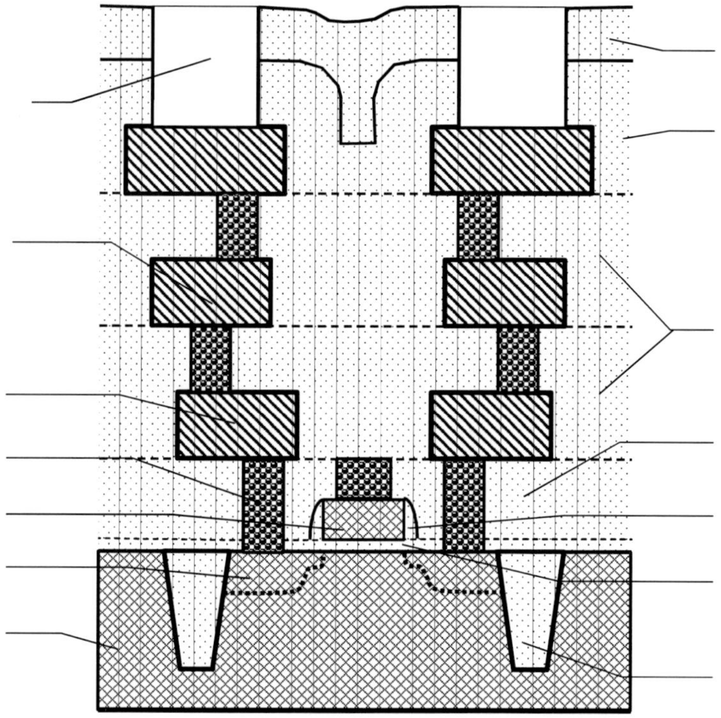

Figure 1.2. Simplified cross-sectional sketch of ULSI structure (not in the scale). Definitions: 1 – silicon substrate; 2 – shallow trench isolation areas; 3 – source-drain doped areas in substrate; 4 – thermally grown gate dielectric; 5 – gate; 6 – spacer; 7 – tungsten plug; 8 – Pre-Metal Dielectric (PMD), made of phosphosilicate glass (PSG) or borophosphosilicate glass (BPSG); 9 – 1st metal layer; 10 – Inter Metal Dielectric (IMD); 11 – 2nd metal layer; 12 – PSG passivation layer; 13 – silicon nitride passivation layer; 14 – contact area.

The first promising results enhanced world-wide BPSG-related research works followed by intensive patent and paper publications during the1983 – 1989 period. Historically, this time frame could be defined as the first period of BPSG investigation. The related papers are listed in references [51-98] in chronological order. The maximum publication activity took place between1985 – 1987. During the first years of the film study, some important BPSG-related topics were covered while searching for the optimized glass compositions applicable in the microelectronic technology. They included: a variety of the film deposition aspects (such as type and characterization of the deposition tools, CVD methodology, chemicals characterization and applications, CVD process parameters, etc.), film property characterization (including flow capability at post-deposition anneal temperatures higher than 900 °C, film devitrification and defect formation issues), precise boron and phosphorus concentration analysis, etc. As a basic result of these studies, a suitable BPSG film composition for VLSI device technology was described as a composition with less than about 8 – 10 wt % of elements (or 25 – 30 wt % of P_2O_5 and B_2O_3 totally). The first detailed survey of BPSG film publications was performed by Dukhanova, Vassiliev and Veretenin in 1988 [93].

The new research impetus for BPSG film study was stimulated by a number of principal changes in integrated circuit technology and manufacturing. First, it was due to the trend of the wafer diameter increase from 100 to 150, followed by an increase to 200 mm. This trend pushed CVD deposition techniques from a multi-wafer CVD tool design to different single-wafer CVD tool designs. Secondly, the appearance of VLSI and ULSI circuits with sub-micron technology nodes enhanced issues of CVD film step coverage and gap-fill.

A second period of BPSG investigation started in 1990 (with the maximum publication activity corresponding to the years between 1993 and 1997). BPSG-related research and technology publications were significantly wider and deeper. They are shown in [99-241] in chronological order until 1999. The second survey of BPSG thin film publications was published by Vassiliev in 1998 [207]. The highest patent activity in BPSG-related areas (deposition methods, tools, compounds, film application, etc.), was found to correspond to the years between 1990 and 1997, when about 50 US patents were published (as compared to about 10 patents published before 1990) [207].

The publications of the previous decade were devoted to the deep analysis of BPSG film deposition, properties, and quantified modeling of gap-fill and flow on the device structures, analysis of specific properties of PECVD films, details analysis of production processes, new applications of BPSG films, etc. These publications are listed as references [242-291] in chronological order. A couple of brief BPSG thin film reviews were published by Vasilyev in the middle of the first decade of the 21st Century [276,281].

When analyzing the BPSG publication history, it can be seen that the research activity was followed by the appearance of new BPSG deposition methods. CVD methods of BPSG thin film deposition were the initial push for appearance of new data. The most important milestones of BPSG thin film CVD methods are plotted in the diagram in Figure 1.3 in chronological order.

In Figure 1.3, it can be seen that the most CVD methods were realized at a low deposition temperature (around 400 °C). Important changes in the BPSG film development can be definitely linked to the changes in the chemical compounds used for BPSG thin film CVD: (Silane + O_2) → (TEOS + O_2) → (TEOS + ozone). This was mostly due to the inherent ability of hydrides to form particles during oxidation reactions because of the chain-brain reaction mechanism. Obviously, particle formation phenomenon during thin film formation was in contrast with the IC device-scaling trend. Therefore, the necessity to search for proper compounds to reduce (or exclude) the defect formation processes during the film deposition, become clear. This difficult task was solved by using TEOS-chemistry: first by oxygen oxidation at the middle-temperature range followed by ozone oxidation reaction at low temperatures.

After a detailed analysis of publications, the author has selected nine important BPSG CVD methods, which were realized earlier or later as the industrial deposition processes using the designated industrial tools. These methods will be detailed in the following chapters. However, it is important to emphasize that the most important issue of this monograph is not only analysis of published BPSG-related publications, but also the generalization of the data that are obtained in order to establish a strong interrelation of CVD methods, BPSG material film composition, structure, and properties.

In the diagram in Figure 1.3, we also show two BPSG formation methods which are not CVD-related. The first one is by boron diffusion from BN wafers at high temperature [65] and the second one is by sol-gel method from organic ethers spin-on followed by the film baking [119]. Both these methods will not be analyzed below.

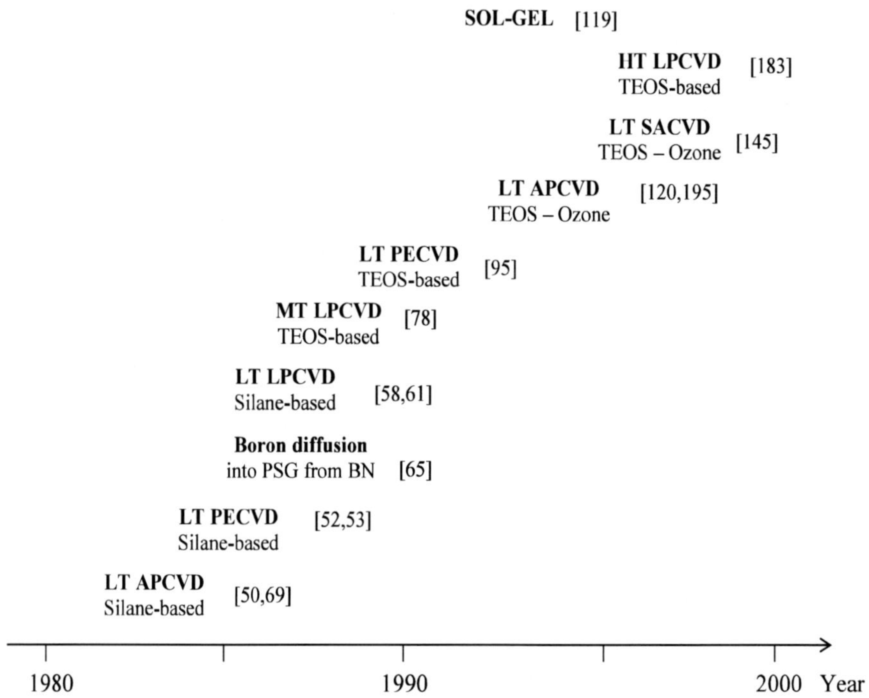

Figure 1.3. A chronology of BPSG thin film CVD method appearance. Definitions: LT - low temperature, MT – middle temperature, HT – high temperature. Silane (SiH_4), represents hydride-based chemistry, TEOS – tetraethylorthosilicate, $Si(C_2H_5O)_4$, represents organic ethers-based chemistry. References in the diagram indicate the first publications regarding the method.

Chapter 2

BASIC BPSG FILM CVD TOOLS AND METHODOLOGY

CVD thin film processes are characterized in terms of process chemistry, deposition tool design, process conditions such as temperature, compound pressures and concentrations, gas flow rates, etc. Before depositing the film, the substrates (or wafers) are usually treated in standard chemical solutions followed by washing in deionized water. The thickness (d) of glass films on the silicon wafers is normally equal to 0.2 – 1.0 μm. The content of the glass, i.e. the concentrations of boron C_B([B]) and phosphorus C_P([P]) in the film were expressed in terms of percentage weight of the elements (in order to obtain weight percent of oxides, B_2O_3 and P_2O_5, corresponded elemental content values should be multiplied by ~ 3.2 for boron and ~ 2.3 for phosphorus).

The general objective of thin film research is to define the best thin film deposition conditions in order to obtain the film with composition, structure and a set of film properties sufficient for a particular thin film application in IC technology.

Normally, this research requires extensive expensive experimental works on the film synthesis and characterization followed by an evaluation of the deposited thin film in an electronic circuit. It is obvious that a variety of compounds, deposition apparatus, and CVD process variables significantly complicate the understanding and comparison of the thin film deposition methods.

An additional complication is due to the progress in microelectronic technology that permanently enhances the requirements for thin film CVD processes. Detailed analyses of silicon dioxide thin film CVD processes development history can be found, for instance, in reference [122]. It is necessary to note that apparatus, CVD process conditions, silicon compounds and oxidizers are similar for silicon dioxide, binary and ternary silicate glass thin film synthesis. In fact, to obtain the necessary boron- or phosphorus content in silicon dioxide films, respective B- and P-compounds, frequently erroneously named “dopants”, need to be added to the mixture of silicon compound and oxidizer. This commonality simplifies the analysis of silicon dioxide based thin film CVD processes.

Table 1.1. Some selected properties of BPSG thin film related compounds

Compound name, abbreviation	Chemical formula	Molecular weight	Physical state at normal conditions	Boiling point (°C)	Density: g/l (gas) or g/ml (liquid)
Silane (monosilane)	SiH_4	32.12	Gas	-112	1.44
Tetraethylorthosilicate, TEOS	$Si(OC_2H_5)_4$	208.3	Liquid	169	0.946
Dimethyldichlorosilane, DMDCS	$Si(CH_3)_2Cl_2$	129.06	Liquid	70.3	1.06
Phosphine	PH_3	34.0	Gas	-87.4	1.53
Phosphorus chloride	PCl_3	137.33	Liquid	75.1	1.56
Trimethylphosphite, TMPI	$P(OCH_3)_3$	124.1	Liquid	111	1.05
Trimethylphosphate, TMPO	$PO(CH_3O)_3$	140.1	Liquid	193	1.21
Triethylphosphate, TEPO	$PO(C_2H_5O)_3$	182.16	Liquid	215	1.07
Diborane	B_2H_6	26.77	Gas	-92.5	1.24
Boron chloride	BCl_3	117.17	Liquid	12.4	1.43
Trymethylborate, TMB	$B(OCH_3)_3$	103.9	Liquid	68.7	0.92
Triethylborate, TEB	$B(C_2H_5O)_3$	146	Liquid	117.5	0.86
Tri-n-propylborate, TPB	$B(OC_3H_7)_3$	188.08	Liquid	175-178	0.86
Tri-iso-propylborate, TiPB	$B(OC_3H_7)_3$	188.08	Liquid	138-140	0.82

2.1. CVD Process Compounds

BPSG films can be deposited in a flow regime by a number of irreversible chemical oxidation reactions using either silane (SiH_4) or tetraethylorthosilicate (TEOS) chemistry (some other studied options are not of interest). The input compounds can be jointly used in a various combinations of silane or TEOS with PH_3, B_2H_6, esters of boric, orthophosphorous, or orthophosphoric acids; boron or phosphorus chlorides, etc. In addition, oxygen (O_2), ozone-oxygen (O_3/O_2) mixture, and nitrous oxide (N_2O) with plasma activation have been used as oxidizers depending on the CVD technique, silicon compound and CVD process conditions. Some data for compounds used to deposit BPSG films can be found in Table 1.1.

All of the CVD processes occur with more or less complicated reaction mechanisms. They can include a variety of intermediate reactions and, intermediate reaction products (IMP). For instance, typical CVD process includes the gas-phase interaction of chemical compounds resulting in different types of IMPs, the transport of source compounds and IMPs to the substrate surface followed by their adsorption on the substrate surface and heterogeneous reactions, etc.. It is important to note that depending on CVD process conditions, the basic CVD oxidation reaction products can exist in a form of a solid film, an aerosol (fine particles, frequently named also aerosil, silica dust, silica micro-particles, etc.), and gaseous reaction by-products. Most authors of referred publications used the simplest general reactions schemes, see example equations (1.1) – (1.3) for hydride compounds oxidation:

$$SiH_4 + O_2 \rightarrow SiO_2 + 2\,H_2, \tag{1.1}$$

$$4\,PH_3 + 5O_2 \rightarrow 2\,P_2O_5 + 6\,H_2, \tag{1.2}$$

$$2\,B_2H_6 + 3O_2 \rightarrow 2\,B_2O_3 + 6\,H_2 \tag{1.3}$$

The aerosol formation phenomenon during CVD processes is the feature of fast oxidation reactions and, especially, the feature of glass formation processes. This is due to the specific chain-brain oxidation reaction mechanism that will be discussed below. The goal of CVD thin film process optimization is to find a set of process conditions where the basic reaction product is the only thin solid film (i.e. aerosol formation process is excluded). In addition, CVD reaction effectiveness (chemical reaction yield, i.e. a share of input compounds transformed into the film, or compound conversion) should be as high as possible. The typical value of CVD reaction effectiveness lies in the range of about 30 – 50 %. Due to all of these circumstances, it is not possible to indicate the precise reaction schemes for chemical oxidation reactions without detailed analysis of the reaction gas phase, for instance, with an in-situ mass-spectrometry method.

It is necessary to highlight that those reaction schemes (1.1) – (1.3) are very general and they cannot be true for CVD process description. Moreover, reaction by-products can be sometimes embedded (trapped) into the growing film that was shown using profiling analysis techniques (see the Part 2 of the monograph). For instance, the CVD oxidation reaction of hydrides in the excess of oxygen (normally the latter concentration for about 20 times exceed the hydride concentrations) can lead to the formation of either H_2, or H_2O, or their mixture depending on CVD process conditions (APCVD, LPCVD, or PECVD), and deposition temperature. It is clear that the compositions of those CVD thin films are different.

In the case of organic compounds like ethers (see Table 1.1), or Plasma Enhanced CVD processes, the reaction schemes are simply unknown even in general form. In addition to hydrogen or water vapor presence in the films, the use of organic compounds revealed the issue of the carbon contamination of the films due to an insufficient oxidizer concentration.

2.2. Basic CVD Tool Design For Silicon Dioxide and Glass Film Deposition

At the beginning of their appearance in microelectronic technology in 1970, silicon dioxide and glass thin films were deposited in a few types of cold-wall single- or multi-wafer APCVD reactors (see example sketches in Figure 1.4(a,b)). Deposition was performed at the relatively low temperatures of 350 – 450 °C using hydrides oxidation reactions [15,16,50,88]. We will refer to this classical hydride BPSG APCVD method as Method 1. Due to the low productivity of APCVD reactors, this method had a limited industrial potential and has become a convenient research method.

In addition, a continuous APCVD apparatus with a moving belt [26] was proposed by Watkins-Johnson (US), referred to below as WJ, with the use of the same hydride chemistry, see Figure 1.4(c). The basic tool design feature is the use of separated nozzles for hydrides and oxidizer injection separated by high nitrogen flow that provided the mixing of compounds on the heated substrate surface. Because of this design, the size and the shape of flat substrates were not limited.

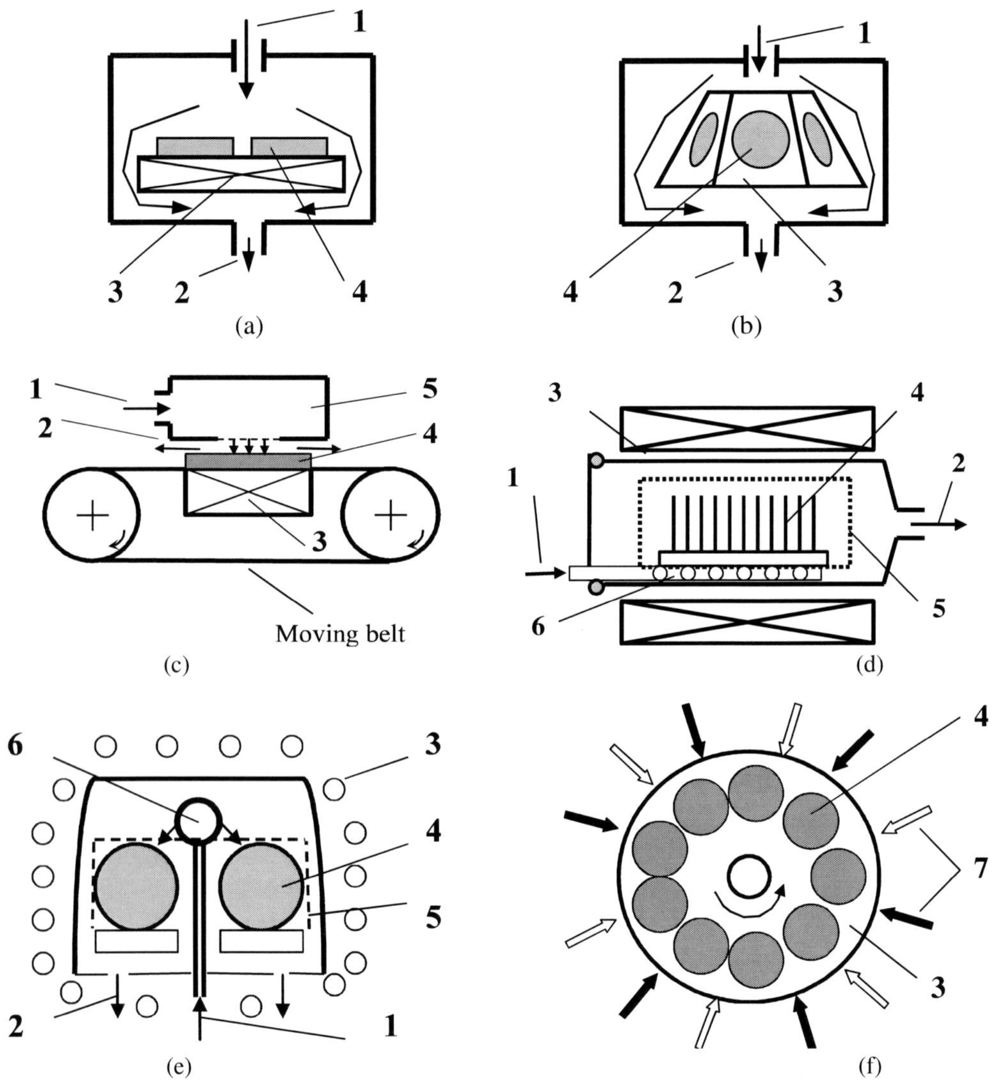

Figure 1.4. Simplified sketches of basic multi-wafer CVD reactor chambers in order of their appearance in microelectronic technology: cold-wall (a,b,c); hot-wall (d,e,f); atmospheric pressure (APCVD) (a,b,c), low pressure (LPCVD) (d,e,f). Definitions: 1 – compound mixture "in"; 2 – reaction by-products "out" or vacuum pumping; 3 – heater; 4 – substrate (wafer); 5 – wafer cage; 6 – distributed-feed; 7 – separated injection of TEOS-TEB-O_2-N_2 and PH_3-O_2-N_2 mixtures.

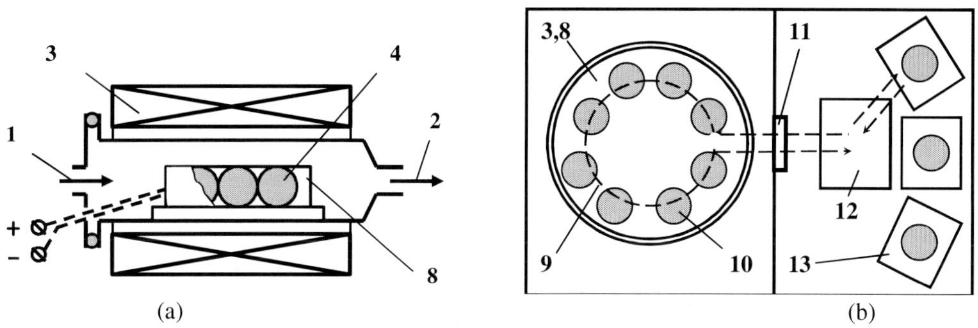

Figure 1.5. Simplified sketches of multi-wafer PECVD reactor chambers: hot-wall multi-wafer tubular PECVD chamber (a); cold-wall PECVD (b). Definitions: 1 – 4 are the same as in Figure 1.4; 8 – RF electrode; 9 – wafer moving direction through six deposition stations with wafers; 10 – wafer on the station; 11 – vacuum loadlock; 12 – robot system; 13 – cassettes with wafers.

Due to undesirable APCVD inherent film thickness non-uniformity, it was commonly accepted that decreasing the working pressure in CVD systems would help to decrease the process productivity and thickness uniformity. As a result, so-called multi-wafer hot-wall LPCVD reactors with either common tubular design, well-developed by many manufacturers like Thermco (US) [58], or special bell jar shape like the Full Spectrum CVD System (below FS CVD) developed by Anicon (US) [59-61] were developed to be used for BPSG film deposition, see Figure 1.4(d,e), accordingly.

The advantage of Plasma Enhanced CVD (or PECVD) processes where a Radio Frequency (RF) plasma is used to activate the chemical reactions, were also demonstrated [14,22,24,28-32]. A plasma (or glow discharge) is a partially ionized gas containing an equal number of positive (ions) and negative (electrons and ions) charges, as well as some other non-ionized gas-particles of electrons, ions and neutral species. For RF plasma activated CVD processes, the deposition rate is normally linearly proportional to the plasma density increase followed by saturation.

BPSG thin film PECVD processes were developed at low temperatures using hydrides and nitrous oxide (sometimes with a small addition of oxygen) either in a hot-wall P-CVD reactor by ASM America (US) [24,51,52] (see Figure 1.5(a)) or in a continuous diode-type cold-wall reactor developed by Novellus Systems (US), below Novellus, with a so-called "Lazy Susan" reactor plate presented in [9], see Figure 1.5(b), or in a single-wafer reactor [53].More details are presented below.

Most of the CVD PECVD multi-wafer reactors cited above allowed simultaneous deposition on from tens to hundreds of silicon wafers 4 inches in diameter. In addition, due to the growth of the substrate diameter from 4 to 6 and up to 8 inches, as well as because of the inherent particle formation issues of hydride-based chemistry, the general trend of CVD processes development for microelectronic applications shifted from hydride-based to organic ether-based deposition methods, either CVD, or PECVD.

The TEOS-based hot-wall tubular LPCVD process was developed at a so-called "mid-temperature range" of about 700 °C [78]. The tubular LPCVD design presented in Figure 1.4(d) was realized as a horizontal reactor first followed by a vertical reactor (not shown). The latest in-time multi-wafer rotating LPCVD "Integrity reactor" was developed by LamResearch (US) [183] (see Figure 1.4(f)) to deposit BPSG films using TEOS-based chemistry at temperatures close to the BPSG film flow temperature (800 °C). The primary purpose of high-temperature BPSG CVD [76] was to provide simultaneous film growth/film softening that could help, for instance, to improve the film gap-fill. Among the tool schematics presented in Figure 1.4(a-f), the only multi-wafer CVD reactors shown in Figures1.4(c,d,e) and PECVD reactors shown in Figures 1.5(a,b) allowed to get significantly higher BPSG thin film deposition productivity and convenient design as compared to the others. As a result, these tools were well-adopted by electronic industry. We will refer to the production BPSG CVD methods realized on these tools as Method 2, Method 3, Method 6, Method 7, accordingly. At the end of 1980, the PECVD cold-wall single-wafer approach (see Figure 1.6(a)) with TEOS-based chemistry was developed [95]. This new approach allowed a very high BPSG deposition rate and productivity. Because of oxygen use as an oxidizer, this new approach allowed a solution to some issues caused by incomplete compound oxidation by nitrous oxide (see details in the next chapters). This method, frequently mentioned as "BPTEOS", is referred to as Method 8.

However, above cited BPSG film deposition methods were found to be not effective for the half-micron device technology node due to the poor film step coverage and gap-fill capability. To solve these issues, ozone-based oxidation organic ethers chemistry was proposed. TEOS-TMB(TEB)-TMPO-O_3/O_2) mixtures either in a APCVD single wafer chamber or TEOS-TEB-TEPO-O_3/O_2 in a SACVD single wafer chamber were developed at low temperatures. To realize this new technology, at the end of 1980, new single-wafer cold-wall CVD production tool concepts appeared(see Figure 1.7(a,b)). Reactor design in Figure 1.7(a) was proposed by Japanese Semiconductor Process Laboratory intentionally for SiO_2 and silicate glass film deposition with the use of a TEOS-based chemistry and ozone/oxygen oxidizer [88]. In addition, this type of reactor design was realized on production tools by Quester Technology (US) (see below Quester [122]). The unique feature of this APCVD reactor design is a wafer heater placed on the top of reactor. Therefore, the wafer has a "face-down" position in this reactor. To provide this unusual wafer placement, a special lifting system was created to place the substrate to the heater. Another feature of this reactor is a showerhead with parallel slots for compound vapor injection into the space between the substrate and the showerhead. This TEOS-ozone CVD production method below will be cited as Method 4.

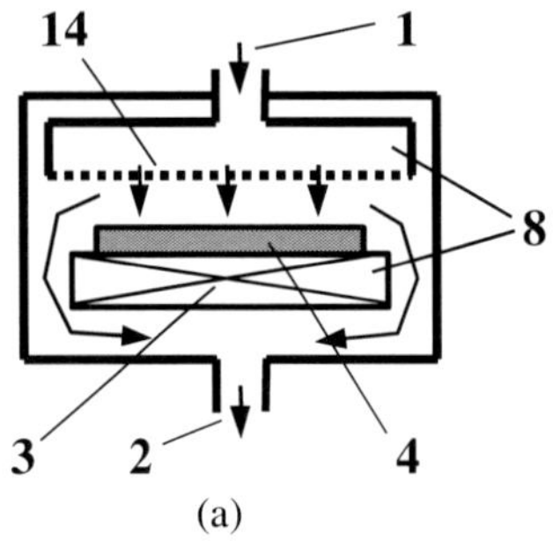

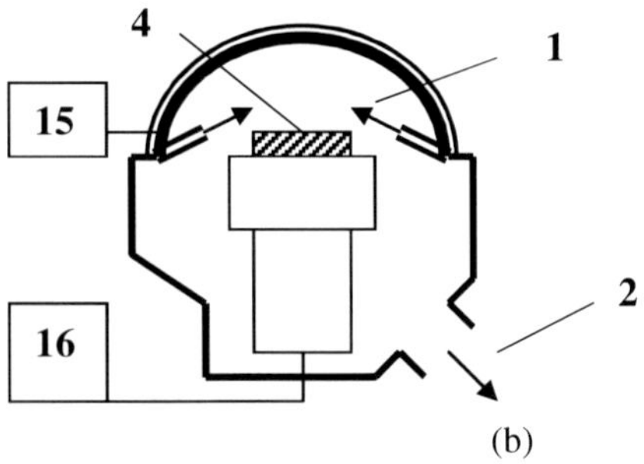

Figure 1.6. Simplified sketches of single-wafer PECVD reactor chambers: cold-wall single-wafer PECVD chamber (a); cold-wall High Density Plasma chamber (HDP-CVD) (b). Definitions 1 – 4 are the same as in Figure 1.4 and Figure 1.5; 14 – gas mixture assembly or showerhead; 15 RF generator, connected to the coil; 16 – RF-generator connected to the electrostatic chuck.

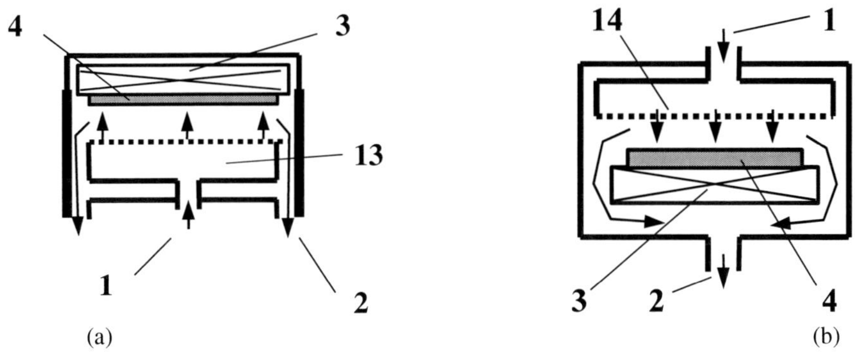

Figure 1.7. Simplified sketches of basic single-wafer cold-wall CVD reactor chambers in order of appearance in microelectronic technology: atmospheric pressure (APCVD) (a), subatmospheric pressure (SACVD) (b). Definitions are the same as in Figure 1.4 and Figure 1.6.

The single-wafer chamber created by Applied Materials (US), below AMAT, known also as Precision 5000 (below P-5000) is presented in Figure 1.7(b). This design includes also a perforated showerhead and heater with the usual "face-up" substrate placement. The feature of this chamber is availability to operate in a wide range of CVD process pressures: from low pressure to atmospheric pressure. This basic CVD chamber design was very successful and it was used for LPCVD, SACVD, APCVD processes. With slight modification of the reaction chamber, this design was also adopted for PECVD processes (see, for instance, Figure 1.6(f)). The history of the P-5000 chamber was detailed in [206]. This TEOS-ozone SACVD production method, frequently known as SABPSG, is cited below as Method 5.

In the mid 90's, the sub-quarter micron ULSI circuit technology required a new CVD technique which would be able to fill the ultra-narrow gaps in IC devices with the high quality insulators. Single-wafer High Density Plasma (HDP-CVD) reactors (see, for instance, an example reactor sketch in Figure 1.6(b)), with in-situ sputtering of deposited materials were introduced for silicon dioxide and PSG film deposition [292-301]. HDP-CVD tools were developed, for instance, by Novellus and AMAT. So-called "biased" conditions were used to allow ion-induced argon sputtering (frequently called "etch") to occur simultaneously with the film deposition. This provides a very specific thin film profile on the device elements and a void-free gap-fill (see for instance [302]). This deposition method is designated below as Method 9. HDP-CVD tools are also applicable for BPSG film deposition [303].

Based on the information presented above, all cited types of reactor design, therefore, can be classified in a number of ways. For instance, it can be done as follows:

- based on deposition pressure as: atmospheric pressure (APCVD), low pressure (LPCVD) or subatmospheric pressure (SACVD) reactors. Tubular LPCVD reactors can be either horizontally or vertically oriented;
- based on chemical reaction activation method as: thermally activated (all of the above mentioned), or plasma activated, namely, plasma-enhanced (PECVD) or high-density plasma (HDP-CVD);
- based on the quantity of wafers placed in a chamber as: multi-wafer and single-wafer reaction chambers;

- based on the wafer heating method such as: cold-wall and hot-wall (or isothermal) reactors, and so on.

Despite the differences in the deposition reactor design, the main assemblies of tools are similar. They normally include: a deposition chamber itself with a substrate heater, compound distribution and mixing systems, a pump system (optional), pressure sensors and plasma generators (optional). In this monograph, we have no intention of analyzing the advantages or disadvantages of existing reactors that overloads readers with unnecessary details. In contrast, we will pay an attention to the features of reactor design and determine the basic deposition parameters, which are common for any type of reactor.

2.3. Some Basic Thin Film CVD Process Parameters

To deposit glass films, the following CVD parameters need to be well-controlled:

- deposition temperature T_d [°C], (it is necessary to know the real semiconductor wafer temperature that can be lower than the set temperature, especially at very low pressures [4]);
- total deposition pressure P_0 [torr] (needs to be measured with high accuracy, especially in a low pressure range of about 0.1 – 1 torr);
- reactor volume V [l, cm^3];
- all compound flows g_i, which allow the calculation of total gas flow G_0 and compound partial pressures p_i ($G_0 = \Sigma\ g_i$ [l/min, sccm, lpm, cm^3/s, etc.], $P_0 = \Sigma\ p_i$).
- spacing between the gas injection and semiconductor wafer x [cm, mm, mls];
- ratio of reactor surface S [cm^2] to reactor volume V, S/V [cm^{-1}];
- RF power [W] and power density with respect to wafer square [W/cm^2].

Knowing these CVD process data, it is possible to make calculations of some useful process characteristics, such as temperature dependence, compound concentration, etc.(see for instance, reviews [8-10]). More details and some additional important CVD process parameters are discussed below.

Chapter 3

BASIC BPSG CVD KINETIC FEATURES

As we mentioned in Chapter 2, different chemistries (chemical compounds) can be used for BPSG thin film deposition. In this chapter, we briefly summarize basic $SiO_2$and silicate glass film deposition kinetic features.

This is necessary for understanding the differences in thin film structure, composition and properties. Depending on the basic reaction chemistry used, silicon dioxide deposition kinetics (and silicate glass thin film) differed. The differences were found to be more pronounced for SiO_2 thin films, and they become less pronounced for silicate glass CVD kinetics.

The main difficulty with the adequate kinetic analysis was linked to a variety of the CVD reaction chambers that were used. To perform proper comparative analysis, it was necessary to use a valid methodology that could allow the definition of some commonalities and differences in CVD thin film kinetics.

In this Chapter 3, we briefly analyze the typical common thin film kinetic CVD features and summarize a methodology for thin film CVD kinetic analysis. The analysis combines a number of experimental data summarized previously for polysilicon, silicon nitride, silicon dioxide and glass thin film deposition. We then analyze basic silicon dioxide and silicate glass deposition features. This sequence will allow us to understand the data presented in Chapters 4 to 5.

3.1. BASIC THIN FILM CVD KINETIC FEATURES

In order to demonstrate the CVD kinetic basics regardless of the thin film nature, the following important CVD thin film features need to be highlighted for wafer samples with a flat surface [1-4,11,35-38,42-48].

(1) An analysis of thin film deposition kinetics is better if performed in dedicated tubular reaction chambers like hot-wall LPCVD, Figure 1.8(a), with relatively long and flat temperature distribution within the reactor, see Figure 1.8(b). This allows you to clearly see the basic features of thin film CVD kinetics - typical deposition rate (or kinetic) curves like presented in Figure 1.8(c) in general form. The shape of such curves typically revealed an increase, a maximum and a decrease area. The shape of

the curves and, especially, the position of the maximum on the curves, differ depending on the deposition chemistry and CVD process conditions. The analysis of maximum position gives the opportunity to establish its numerical characteristics for fast and simple CVD processes characterization.

(2) It is important to note that the squares under the kinetic curves, such as presented in Figure 1.8(c), are proportional generally to the deposited material quantities. According to the general chemical approach, the chemical synthesis should be performed at optimized reaction conditions, i.e. in the area close to the maximum, because the maximum position corresponds to the maximum of the CVD reaction yield. Therefore, the characterization of the thin film CVD process should be done after preliminary optimization of the reaction kinetics, i.e. after obtaining some kinetic curves and the determination of the maximum position on the curves. It is necessary to note that the determination of kinetic curves and their maximums is not an obvious procedure, especially because of the variety of CVD reactors used. We highlight the main methodology aspects for this procedure. General approaches about the theory and methodology of CVD reactions can be found in [1-4].

(3) The shape of the reaction curve and the position of the maximum on the curve with respect to the CVD reactor length always depend on the process variables (temperature, pressure, etc.). It is very important to highlight that at the constant CVD temperature and CVD pressure it also depends on the total gas flow (G_0). The higher total flow value corresponds to the longer maximum position in the tubular reactor, as shown in Figure 1.8(c).This observation plays the crucial role in the understanding of CVD thin film kinetics.

(4) An analysis of silicon nitride experimental data in hot-wall tubular LPCVD reactors allows us to make a very important conclusion. After special re-calculation, typical kinetic curves presented in Figure 1.8(c) coincided with the so-called "time curve"(see Figure 1.9 [42]). This kinetic time curve was derived from experimental data. Instead of the wafer coordinate in the reactor, a so-called "residence time" τ coordinate was used. The latter was calculated as follows: $\tau = (x_i - x_0)/\upsilon$, where $x_i$is the substrate coordinate in the reactor (cm, mm), and υ is the gas flow speed (or gas velocity), for instance, expressed in m/s or cm/s. Here the point x_0 corresponds to a coordinate of the beginning of the film growth in the reactor. This can be found only experimentally, analyzing the beginning of the film appearance on the quartz tubular reactor side walls. It is obvious that the gas flow speed is proportional to the total gas flow. Therefore, the curves presented in Figure 1.8(c) are the same in shape as the curves presented in Figure 1.9 (top right) because $G_{0(1)} \sim \upsilon_1$, $G_{0(2)} \sim \upsilon_2$, $G_{0(3)} \sim \upsilon_3$ (under the same other CVD conditions).

Gas flow speed values differ by a few orders of magnitude depending on the CVD process conditions. The largest υ values, up to tens of meters per second, are typical for LPCVD systems. The smallest υ values, like a few tens of cm/s, are typical for APCVD conditions. Using the examples of tubular APCVD and LPCVD reactors with cross-square s_R,[cm^2], and the gas flow speed can be easily calculated using formulas (1.4) and (1.5), accordingly:

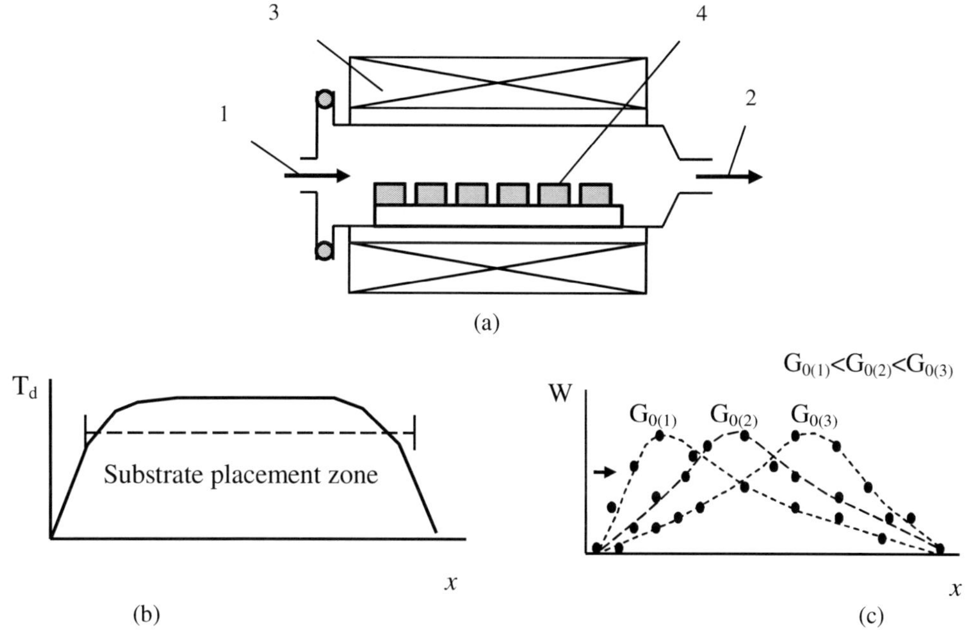

Figure 1.8. Sketch of hot-wall tubular LPCVD with horizontal substrates placement (a), typical shape of deposition temperature distribution along with reactor length (b) and typical shapes of kinetic curves obtained at different process flows G0 and some other CVD process conditions (c). Definitions are the same as in Figure 1.4.

$$\upsilon_{APCVD} = G_0 \times T_d / s_R \times T_r \,, \tag{1.4}$$

$$\upsilon_{LPCVD} = G_0 \times T_d \times P_a / s_R \times T_r \times P_d \,, \tag{1.5}$$

where T_d and T_r are the deposition and room temperature, K, accordingly, and P_d and P_a are the deposition and atmospheric pressure, torr, accordingly.

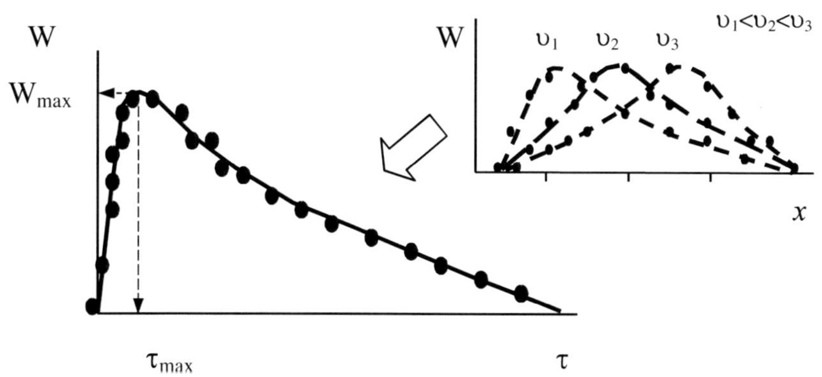

Figure 1.9. Kinetic time curve calculation example (see details in the text).

Thus, the time curve for a pre-selected set of CVD thin film process conditions can be obtained experimentally by changing only the total gas flow rates while keeping constant the other process conditions.

(5) The time curve analysis allows the definition of two basic kinetic parameters (see Figure 1.9): the absolute value of the deposition rate in maximum (W_{max}) and corresponding residence time value τ_{max}. These two characteristics always change in shape and values depending on the process conditions, as shown in Figure 1.10 in general form. In this graph, one can see that an increase of deposition temperature and deposition pressure cause the increase of W_{max} and its shift to the shorter τ_{max} value, and vice versa. This graph clearly shows that proper CVD thin film process analysis in any reactor should be performed taking into account real kinetic time curves. For instance, by placing the only one control substrate just at the same place in the tubular reactor that is indicated as A, B, C, D in the graph, a researcher will get the wrong kinetic trends, as detailed in the top right corner of the graph.

(6) According to classical chemistry, the coincidence of kinetic curves presented in Figure 1.8(c) into the time curve presented in Figure 1.9 and Figure 1.10 indicates that CVD process conditions in flow reactor used corresponded to the so-called "plug reactor" conditions. The main important feature of the plug reactor is that the residence time τ is equal to the chemical reaction time. Thus, the kinetic time curves in Figure 1.9 and Figure 1.10 represent the features of studied chemical reaction. This is a very important conclusion because it provides the basis for accurate proposals regarding the process schemes of CVD thin film reactions.

(7) The kinetic curves presented in Figure 1.9 and Figure 1.10 using the example of the tubular reactor can be in part obtained in any other flow-type CVD reactors. However, it is necessary to highlight that for single-wafer chambers with showerheads, the parameter x_i *is* equal to the spacing between the showerhead and the wafer surface. In this case, the re-calculations give a good coincidence of the kinetic curves [224].

(8) The analysis presented above allows us to find the proper experimental methodology for CVD thin film process characterization. For instance, the right way to analyze CVD kinetics is to register the kinetic curves in order to find the values of W_{max} and τ_{max}. We accept that in many cases the experimental work cannot be done in a wide range of process parameters. Nevertheless, even one completed kinetic time curve that can be roughly calculated based on small volume experiments, can give the basic understanding of the studied CVD process and CVD tool conditions.

(9) The other CVD process characterization includes, for instance, a search for an order of CVD reaction with respect to the used compounds, an activation energy calculation, etc.

A simplified example of chemical reaction of two compounds (A and B) with formation of thin solid film (D) and a gaseous by-product (E) can be presented in general form as follows (1.6):

$$aA + bB \rightarrow dD + eE \tag{1.6}$$

According to the classical chemistry, the rate of chemical reaction W can be written as follows (1.7):

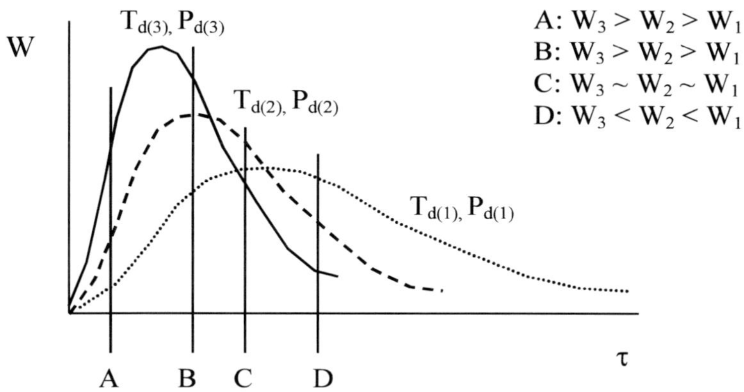

Figure 1.10. Typical changes in kinetic curves along with the changes of Td and Pd and basic kinetic trends in case of fixed positions of the control wafer in the reactor (right top).

$$W = k_f \times [C_A]^a \times [C_B]^b, \tag{1.7}$$

where, k_f – coefficient of proportionality, or reaction constant; $[C_A]$, $[C_B]$ – concentrations of A and B compounds; a and b – stoichiometric coefficients, which are equal to the reaction orders with respect to compounds A and B. It was observed that the studied thin film CVD reactions revealed mostly the first order with respect to the silicon compounds used for CVD [1-3]. At the same time, CVD thin film reactions were conducted typically at considerable excess of the second compound (oxidizer or ammonia). This indicates the pseudo zero reaction order with respect to the second compounds that is shown in general form in Figure 1.11(a) as concentration dependencies W = f($[C_i]$). This conclusion was reached for all studied silicon-base thin film CVD thin film processes [1-3]. This conclusion allowed the simplification of the CVD thin film characterization. As for the activation energy, E_a, its value is normally calculated using equation (1.8) and temperature dependence like presented in Figure 1.11(b):

$$k_f = k_0 \times \exp(-E_a / RT_d), \tag{1.8}$$

where k_0 is a coefficient of proportionality and R is a gaseous constant. For CVD thin film processes, E_a was variously reported to be 62 – 209 kJ/mol except for PECVD, in which E_a is usually lower by an order of magnitude.

(10) Summarizing thin film CVD kinetic information, it was concluded that the film deposition rate W_{max} can be considered as the only important gas-phase deposition process parameter. In order to exclude any influence of the deposition process parameters, a generalized parameter named an "effective constant of the deposition rate, k_{eff}" [cm/s] was established in [1,2]. The lower k_{eff} value corresponds to the slower CVD process and vice versa.

The parameter k_{eff} can be easily calculated based on the experimental kinetic curves using the deposition rate values in the maximum (W_{max}), and silicon compound concentration $[C_{Si}]$ as follows:

$$k_{eff} = \chi \frac{W_{max}}{[C_{Si}]} \tag{1.9}$$

where χ is the coefficient of proportionality and $[C_{Si}]$ is calculated using the well-known equation (1.10) according to equation (1.11):

$$p_i \times V = n_i \times R \times T_d \tag{1.10}$$

$$[c_{Si}] = \frac{n_{Si}}{V} = \frac{p_{Si}}{R \times T_d}, \tag{1.11}$$

where p_{Si} is a partial pressure of silicon compound, n_{Si} is a number of silicon compound moles. Below we show that k_{eff} value is a main kinetic parameter that allows the description of the most of obtained CVD thin film experimental data.

In conclusion, it follows from the above discussion that basic measurable CVD process parameters (T_d, P_d, etc) are insufficient for proper CVD kinetic characterization. The analysis must include the residence time determination.

3.2. Basic Silicon Dioxide Thin Film CVD Kinetic Features

The kinetic analysis presented above is applicable for all types of CVD thin film processes. In this paragraph we briefly focus on the silicon dioxide thin film kinetic features. Based on accumulated experience, we would like to highlight two important features: basic reaction chemistry and the oxidation compounds used.

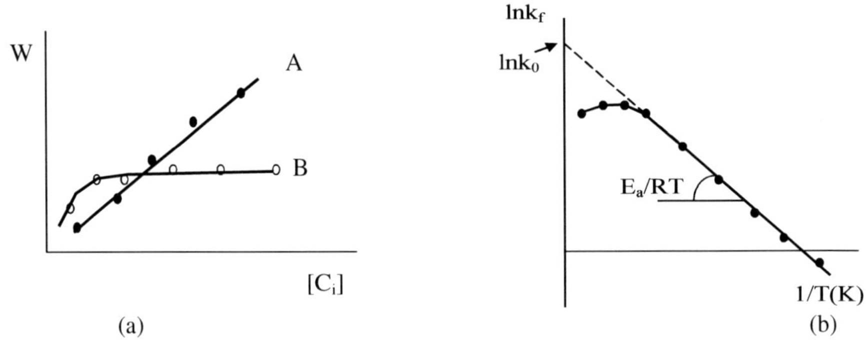

Figure 1.11. Typical concentration dependencies (a) and reaction temperature dependence (b) (in general form).

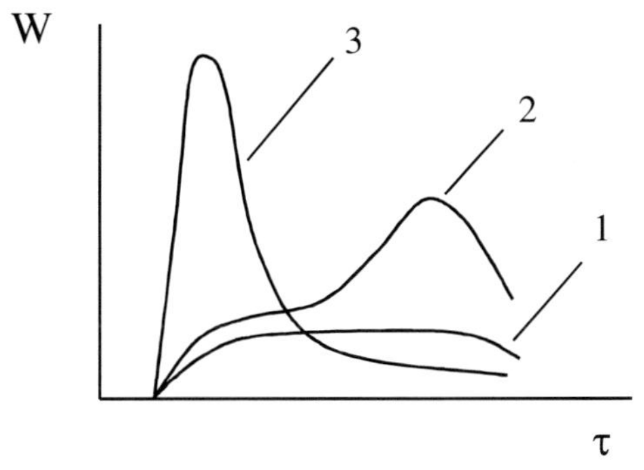

Figure 1.12. Typical silicon dioxide thin film kinetic curves obtained in tubular hot-wall CVD reactor with flat temperature profile, see Figure 1.8(a,b). Explanations are in the text.

(1) As we have already mentioned above, silane-based and TEOS-based chemistries have been used for silicon dioxide and silicate glass film deposition. These chemistries revealed totally different CVD kinetic peculiarities.
The main difference can be illustrated by typical kinetic curves obtained in tubular hot-wall CVD reactors and presented in Figure 1.12 in general form. The curve 1 in Figure 1.12 is typical for relatively low deposition rate TEOS-based processes. The curve 3 is typical for high deposition rate silane-based processes. Severe differences in the shapes of these types of kinetic curves will be explained in detail in the next chapter by the difference in CVD reaction schemes and mechanisms.
Another feature of silicon dioxide CVD processes is the capability of aerosol formation during the film deposition. It is necessary to emphasize that according to accumulated experimental data, silane-based chemistry is much more prone to form particles in the gas-phase of CVD reactors [3]. This is also explained by the difference in the reaction mechanism. It is important to highlight that aerosol formation effects were found to appear at enhanced CVD working temperatures and pressures, as well as at longer residence times. The latter can be explained by the appearance and development of side gas-phase reactions. These observations indicated that areas of the deposition rate decrease in kinetic curves presented in Figure 1.12 are significantly less applicable for the high-quality silicon dioxide thin film deposition. It is necessary to mention that this conclusion, in general, is applicable for the other types of CVD thin films, but silicon dioxide thin film CVD serves as the best example.
As for the curve 2, it is typical for CVD processes with acceleration effects. This process starts with a low deposition rate followed by sharp acceleration up to a high deposition rate. This curve was found by the author to be representative for two studied chemical systems: SiH_4-N_2O and DMDCS-O_2[1,2]. Both reactions were studied in a mid-temperature range in tubular hot-wall LPCVD apparatus.

(2) The type of second reaction component, i.e. oxidant, was found to be also very important for CVD. Historically, oxygen was considered to be the main oxidation agent. However, "soft" oxidizers like CO_2 and N_2O were also examined in the reaction with silane that was reviewed in [3,42]. It was found that their main disadvantage is the relatively low deposition rates as those provided by oxygen. As a

result, these two oxidizers could be used at a significantly higher CVD temperature (above ~ 600 °C) that is in contrast with many IC technology requirements. Another negative feature of these soft oxidizers is the impossibility to use these compounds to deposit silicate glasses during the thermal activation of the gas-phase. Phosphorus additives were found to suppress the CVD processes with N_2O as the oxidizer [1-3].As a result, these oxidizers, preferably N_2O, were used for the only PECVD processes.

The last attempt to use a different oxidizer was done at the end of 1980 [96]. It was found that ozone can play a very important role in the reaction with TEOS, significantly enhancing the deposition rate of thermally-activated CVD processes at low temperatures. In Figure 1.13(a), some temperature dependences for TEOS-based silicon dioxide CVD are presented using data [304] obtained in APCVD at the end of 1960. Data shows that TEOS thermally-activated pyrolysis either in the presence of Ar, or H_2, started at about 500 °C. However, the curve for TEOS-O_2 system shows very unusual behavior at a low deposition temperature. However, the deposition rate was found to be very low for practical use in microelectronic technology.

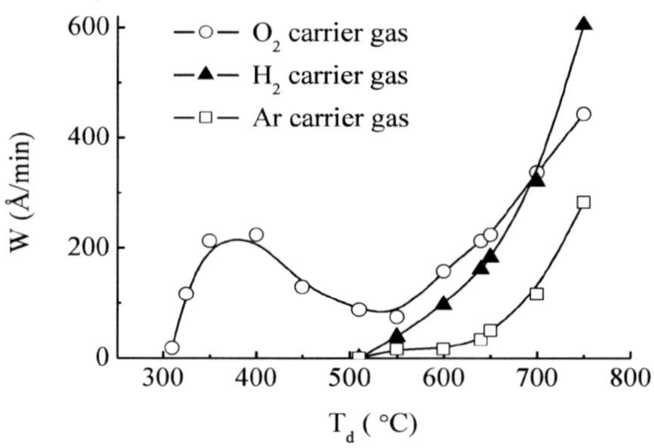

(a)

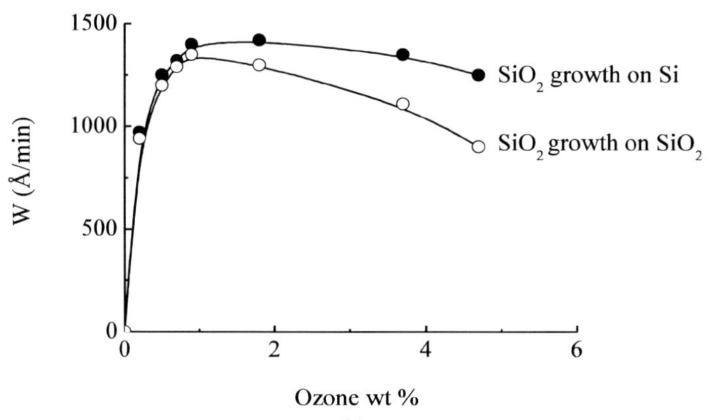

(b)

Figure 1.13. Ability of TEOS to react with oxygen at very low temperature as compared to hydrogen and argon (a) (re-drawn using data [304]), and acceleration of TEOS-O_2 reaction with ozone and substrate dependence of deposition rate ("surface sensitivity" effect) (re-drawn from [122]).

However, ozone addition to oxygen [96] was found to accelerate the deposition rate significantly, making this low-temperature deposition applicable for industrial silicon dioxide and silicate glass thin film deposition [106,114,120,122,161]. Moreover, the TEOS-ozone CVD process was found to depend on the substrate material. This feature is illustrated in Figure 1.13(b) for silicon and silicon nitride substrate surfaces. It can be seen that depending on the ozone concentration in oxygen, the difference between the deposition rates on silicon and silicon nitride can reach about 30%, and the deposition rate is faster on silicon surfaces. In addition, this ozone-based deposition process was found to provide much better step coverage on stepped IC device structures.

The low-temperature thermally-activated TEOS-ozone CVD process was found to be very promising for the semiconductor industry because it was considered as a good replacement of the silane-based chemistry, especially because of the potential absence of the aerosol-related problems.

3.3. Basic Silicate Glass Thin Film CVD Kinetic Features

It was discovered that the phosphorus and boron additives have an important part in the silicate glass film deposition [2,3]. Phosphorus compounds are known to accelerate the oxidation of silicon compounds, reducing the activation energy by a factor of 1.3 to 3. The acceleration is accompanied by a proportional increase in the phosphorus concentration for both PSG and BPSG thin films. This acceleration also degrades the film uniformity and enhances the formation of aerosol by-products. Maximal acceleration (by a factor of 3 to 7) was found with TEOS-based chemistry, namely with TEOS-PH_3 combinations. However, it is minimal with silane-based chemistry in any combination with phosphorus compounds. It has also been found that boron compounds tend to inhibit CVD involving silane and phosphine. This feature of silane-based oxidation in the presence of phosphorus and boron compounds is presented in Figure 1.14 using the kinetic time curves of Figure 1.10 in general form. In Figure 1.14, one can see that the introduction of phosphorus compounds into the CVD reactor to deposit PSG thin films causes the change of the SiO_2 kinetic curve. The maximum shifts to the shorter τ_{max} value and W_{max} increases. In contrast, in the presence of boron compounds one can see a change of the SiO_2 curve reflecting a suppression of oxidation reaction. In this case, maximum shifts to the longer τ_{max} value with respect to the SiO_2 kinetic curve and W_{max} decreases.

The author would like to emphasize that similar to the graphs in Figure 1.10, the curves in Figure 1.14 clearly confirm that silicate glass CVD thin film process analysis in a reactor without taking into account real kinetic curves is meaningless. In fact, by placing the only one control substrates just at the same place in the tubular reactor that is indicated as A, B, C, D in the graph will give the wrong kinetic trends, as shown in the top right corner of the graph.

We believe that one of the most important silicate glass CVD features is the differences of the silane-based and TEOS-based kinetic curve behavior after the CVD acceleration by phosphorus compounds. It was found that acceleration of the TEOS-based thin film CVD (whether oxygen or ozone is used as oxidizer) is much more pronounced as compared to silane-base CVD. As an example, about a 5.5 times acceleration effect is clearly demonstrated in Figure 1.15 [184]. In addition, accelerated reaction clearly reveals the presence of reaction maximum.

Summarizing all of the above, it has been found that phosphorus compounds exert a major influence on deposition kinetics:

- phosphorus compounds accelerate the reaction and shift the time curve toward shorter residence times;
- phosphorus compounds enhance the incorporation of boron into the film, whereas boron compounds inhibit the incorporation of phosphorus. When the oxidation reactions are inhibited by boron compounds, the time curve is shifted toward longer residence times and aerosol by-products are not observed;
- phosphorus compounds also enhance the formation of aerosol by-products (see below).

As for the boron compounds effect, it is significantly less and, basically, can be ignored for major analysis.

Experimental comparison of acceleration effects in different CVD silicate glass chemistry was done by Vassiliev [207,224]. Direct comparison of experimental CVD kinetic time curves with and without acceleration is shown in Figure 1.16. Here the silicon dioxide time curves (presented in Figure 1.12 in general form) were taken as the bases. It is important to highlight that the PSG and BPSG CVD processes taken for comparison provided close enough comparative values of phosphorus content in the films; the latter gave the hard base for comparison.

Data in Figure 1.16(a) show that silane-based acceleration was found to be about 30 %, while TEOS-based acceleration was about 5.5 times (see Figure 1.16(b)) re-calculated using the graph in Figure 1.15(a). In addition, a clear maximum close to low residence time values appeared on the accelerated time curve for TEOS-ozone reaction. At the same time, DMDCS-oxygen system, taken as an example of CVD reaction with in-situ acceleration (see Figure 1.12, curve 2) showed a middle-acceleration behavior, of about 3 – 4 times. This comparison clearly demonstrates that the higher acceleration value is a feature of relatively slow SiO_2 CVD using TEOS-oxygen, or TEOS-ozone chemical compounds. Another important feature is that accelerated CVD reactions demonstrate clear maximums on the time curves and the positions of these maximums are close to each other, being near 0.3 – 0.5 s.

A final comparison of additive effects on the CVD silicate glass kinetics is presented in Figure 1.17 [224,276]. These results show that TEOS-based chemistry revealed significantly higher acceleration effects, especially for phosphorus compounds. At the same time, boron compounds were found to accelerate TEOS-based oxidation, but inhibited the silane-based oxidation CVD.

The acceleration effect can be evaluated numerically. For SiO_2 and glasses other than BPSG, k_{eff} was found to range from 0.3 to 10 cm/s. For the BPSG processes examined, it varies over a narrower range of 1.1 – 2.5 cm/s. It is seen that the silane and the TEOS process have similar characteristics for silicate glasses of similar compositions.

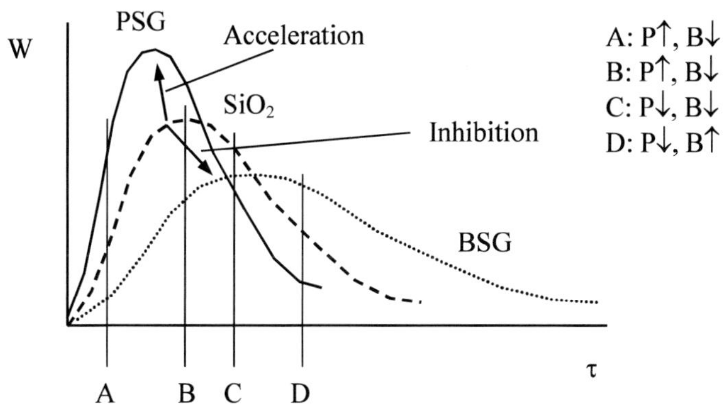

Figure 1.14. Typical changes in kinetic curves for silane-based oxidation (SiO2) with addition of phosphorus (PSG) and boron (BSG) compounds, causes either acceleration or inhibition of oxidation reaction.

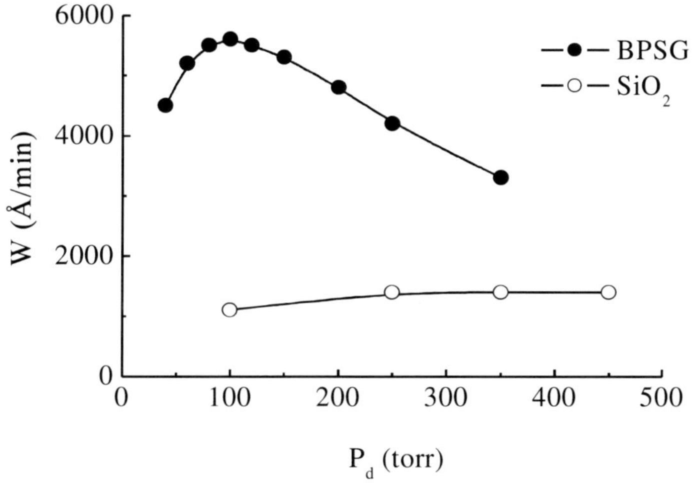

Figure 1.15. Acceleration effect for TEOS-based SA CVD BPSG films vs. total pressure in chamber (re-drawn with simplification from [184]).

This analysis shows that PSG and BPSG thin film deposition kinetics look significantly closer because of the acceleration phenomenon. The kinetic data relating to phosphorus compounds strongly indicate that all CVD processes of BPSG (as well as PSG) are governed by a closely related reaction mechanism. It is believed that it involves free radicals produced by interaction between the phosphorus compound and the oxidizing agent. In explaining these features, it is necessary to note that the oxygen-based oxidation of silane, phosphine, and other phosphorus compounds occurs by the free-radical chain mechanism, see reviewed literature in [3,305]. These findings suggest that the acceleration should be taken to indicate the free-radical chain mechanism of the oxidation. Furthermore, it has been established that the acceleration of oxidation lessens in a sequence: $PH_3 \rightarrow PCl_3 \rightarrow$ organic esters of phosphoric or phosphorus acids. This becomes clear because PH_3 oxidation is a well-known example of a typical free-radical (chain-brain) reaction; oxidation of other phosphorus additives studied was significantly poorer. The decrease of the deposition rate along with the addition of boron compounds implies that the oxidation by the free-radical chain mechanism is being inhibited. The conclusion is supported by a reduced deposition rate when silane oxidation is used, as well as by lower phosphorus content in glasses obtained by adding boron compounds to the source gas. By contrast, the analysis presented in the literature about kinetic data showed that boron compounds accelerate the relatively slow oxidation of TEOS.

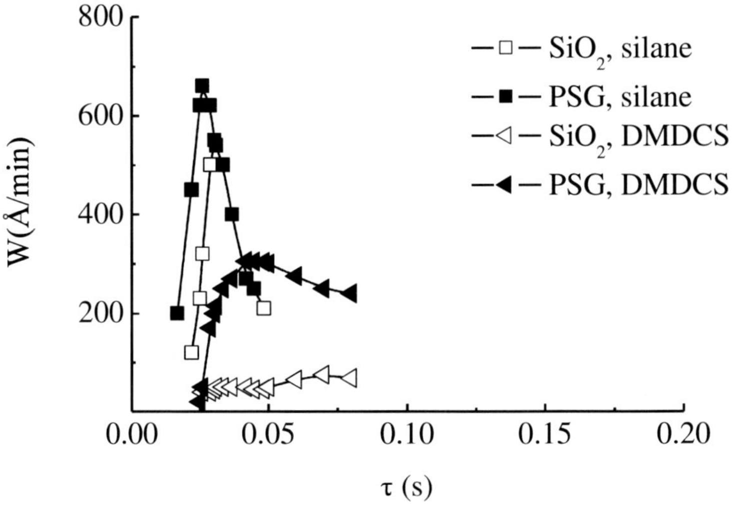

(a)

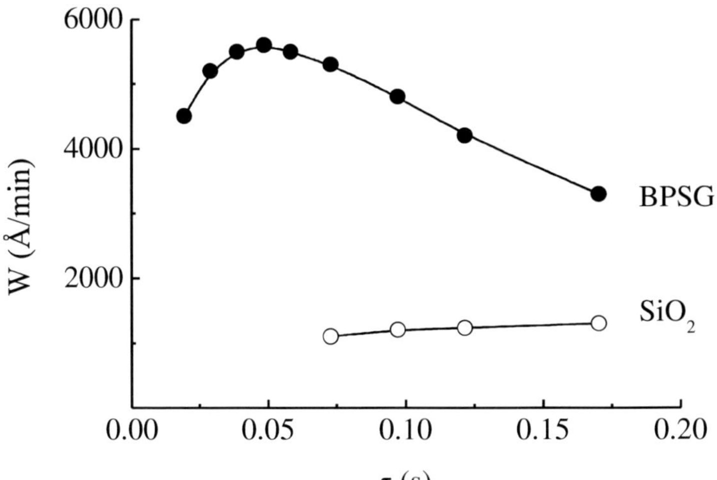

(b)

Figure 1.16. Comparison of acceleration effects for oxide-based silane and DMDCS compounds (a) and TEOS-ozone compounds (b) (re-calculated from data presented in Figure 1.15).

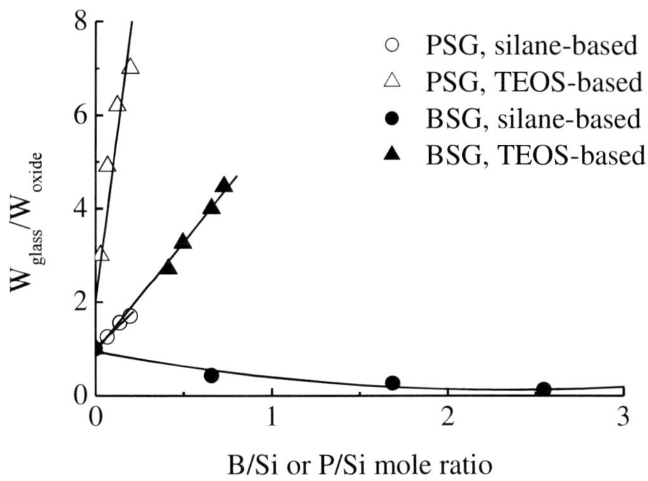

Figure 1.17. Summary of acceleration and inhibition effects of oxidation by boron and phosphorus compounds: (a) the ratio of the glass deposition rate W to the SiO2 deposition rate W0 as a function of B/Si or P/Si mole ratio in the gas phase.

3.4. Basic Aerosol Formation Features at Glass CVD

It is well-known that aerosol (micro-particle) formation is a major problem of the silicon-dioxide and silicate glass thin film CVD processes. Aerosol contamination of thin films is generally unacceptable for microelectronic technology. The technology requirements become harder with the IC technology development. For instance, particles with > 0.1 μm size need to be under systematic control in the IC device manufacturing.

The particle formation problem is very complicated. This is due to a variety of reactors, chemical compounds and process conditions used for the CVD film deposition. Until recently, the reasons and mechanisms of micro-particle formation during the oxidation reactions have not been completely understood. At the same time, micro-particles cannot be considered as something unusual. Aerosol is the solid phase material with a similar composition to the thin films. The difference between them is that thin film is formed on the substrate surface, but particles are formed in the CVD reaction gas-phase. Thus, thin film and micro-particles are both the products of the chemical reaction, but the particles can be considered as undesirable reaction by-products. Therefore, the main goal of the researcher and engineer working in the thin film field is to find the reaction conditions to obtain thin film material without micro-particles.

We would like to stress that some special related research reports were published regarding aerosol formation effects during CVD [218,306-308]. The author's research results about this matter are presented in [1-3,309]. Basic conclusions made from the research results can be summarized as follows.

(1) For any studied silicon-based CVD reaction, the increase of deposition temperatures and deposition pressures were found to enhance the appearance of micro-particles in/on the deposited films. It has been observed for different CVD thin film processes

that there are some deposition rate values which cannot be exceeded. Above such values, the films reveal micro particle contamination. Generally, these limiting rates are about 1.5 to 2 times higher than the optimized defect-free deposition rates.

(2) For any studied CVD reaction, especially those realized in tubular hot-wall reaction chambers, particle formation effects appear in the reaction area closer to exhaust.

(3) For oxidation CVD processes, acceleration of reaction kinetics causes the enhancement of aerosol problems during the glass film deposition.

(4) To avoid particle formation, it is necessary to decrease the gas phase volume, simultaneously increasing the ratio of the reaction square to the reaction volume (S/V). In LPCVD systems, this can be achieved by placing the substrates with a distance of about 5 mm. between them. In single wafer chambers, the lowest possible spacing between the showerhead and the substrates should be used.

(5) For oxidation CVD processes, the particle formation effect arises in the areas of deposition rate decrease in time curves presented in Figure 1.14.

According to the above listed features, particle formation effects can be characterized by two basic statements:

a) *particle formation effects are proportional to the deposition rate values* (i.e., T_d, P_d, $[C_{Si}]$, etc) that basically corresponds to the reaction maximum at low residence time τ values. In this case, fine particles are embedded into the films, changing their appearance (for instance, under the dark field of an optical microscope the surface looks like "polycrystalline"). This is also confirmed by atomic force microscopy measurements of the surface roughness for thin films. In particular, the phosphorus acceleration results in higher surface roughness, and the boron inhibition leads to lower roughness;

b) *particle formation effects are proportional to the residence time τ* that basically corresponds to the reduced deposition rate values. This case was found to correspond to the particles with significantly larger size and particle formation on the substrate surface, on the reactor walls, showerhead, etc. It is necessary to stress that it is very difficult to define the residence time responsible for the large particle formation. Roughly, based on the author experience, it can be designated as the area at high τ when the deposition rate decreases to the half of the maximum value.

These two cases are represented by a generalized graph in Figure 1.18 as areas with "Fine particles" and "Large particles". The large particle area is defined as certain τ corresponded to the ½ W_{max}. A solid line represents certain optimized CVD conditions and both substrates positions designated A and B on the time curve provided thin film without particles. Note that wafer positions in the reactor are normally chosen by the experiment and should be close to the time curve maximum. So, from this point of view, both positions A and B presented in Figure 1.18 are acceptable for the film deposition.

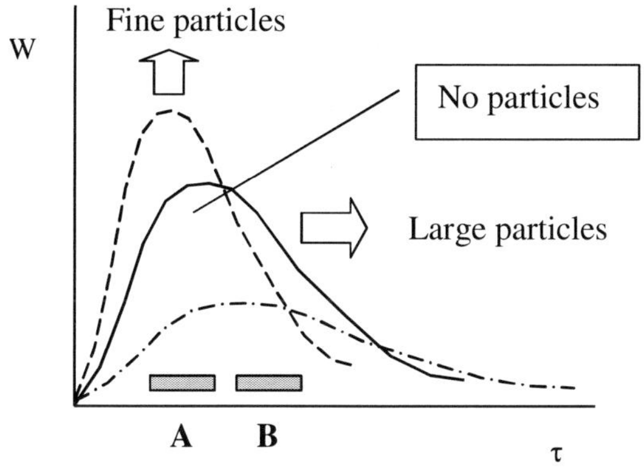

Figure 1.18. Generalized scheme of particle formation effects at thin film CVD (see text for details).

It is important to see what will happen when we change these optimized conditions in order to enhance the process productivity. Accelerating the CVD process, either by process conditions, by adding the phosphorus compound or even by plasma gas-phase activation will change the time curve shape to that presented by the dashed line. In this case, a wafer placed in position A will have a higher deposition rate, but it comes to the fine particle area. Sometimes it is acceptable and these thin films can be used, but generally this is the wrong way to "improve" CVD process. At the same time, the wafer in position B after acceleration can get into the area of large particles that is unacceptable for practical use.

Let us illustrate an example above with the comparison of experimental data for TEOS-ozone based silicon dioxide and BPSG thin film CVD in different single-wafer reaction chambers (APCVD, SACVD, and continuous APCVD reactor). This comparison in part was done by Vassiliev [207,224] after re-calculation of the kinetic data published by world-wide researchers [115,122,184] into the parameter of "residence time". Some necessary numerical data were evaluated based on data taken from the advertisers [310,311]. The final comparison of data is presented in Figure 1.19. It can be seen that with silicon dioxide and BPSG thin films, the CVD processes were realized in a wide range of τ.

The silicon dioxide deposition process was realized in APCVD single-wafer chamber conditions [122] after re-calculation of the drop in the area of the deposition rate decrease with a residence time value of about 0.2 s. At this residence time value, the phosphorus compounds acceleration effect was quite low, not exceeding 30-50 %. This area of chemical reaction corresponds to the "Large particle" area. It is necessary to note that the used single-wafer APCVD chamber had a very specific design with wafer "face-down" position in reactor (see Figure 1.7(a)). We believe that this inconvenient design was chosen to provide the possibility of reducing aerosol formation effects using particle sedimentation phenomenon. At the same time, the APCVD continuous chamber with high gas flow passing through a very narrow single injector [115] allows the provision of very high gas velocity or, vice versa – very low residence time. This corresponds to the effective non-particle area of the time curves (see triangles in Figure 1.19). Kinetic time curves that are presented for silicon dioxide SACVD process conditions in a wide range of τ [184] were found to be quite close to both the above cited APCVD process conditions.

Although they are close for silicon dioxide thin films, these processes differed drastically for BPSG thin films. The value for the maximum in the SACVD BPSG time curve was evaluated to be about 0.05 s. Note that this CVD process was realized in the single wafer chamber presented in Figure 1.7(b) exactly in the area of τ_{max}. There were no data published about aerosol particle problems for SACVD BPSG films. Continuous APCVD BPSG thin films data are presented in Figure 1.19 as fixed data points. The residence time for this case was evaluated to be close enough to the maximum in SACVD BPSG curve. Thus, from chemical point of view the conditions for both cited cases were very close to the maximum yield, i.e. deposition processes were well optimized. Note that in the case of APCVD, this low τ value can be provided by using very large gas volumes. By contrast, vacuum SACVD conditions made the provision of high gas velocity significantly simpler, by reducing the process pressure a few times. Both these examples showed much better CVD conditions for BPSG film deposition as compared to the single wafer APCVD deposition reactor with a residence time value that was about 4 times longer.

This comparison clearly shows the necessity of careful optimization of CVD conditions that can be made on the base of the experimental time curves calculated using the reactor and CVD parameters. In addition, this methodology allows the detection of the CVD process features in order to find proper ways for further process development and optimization.

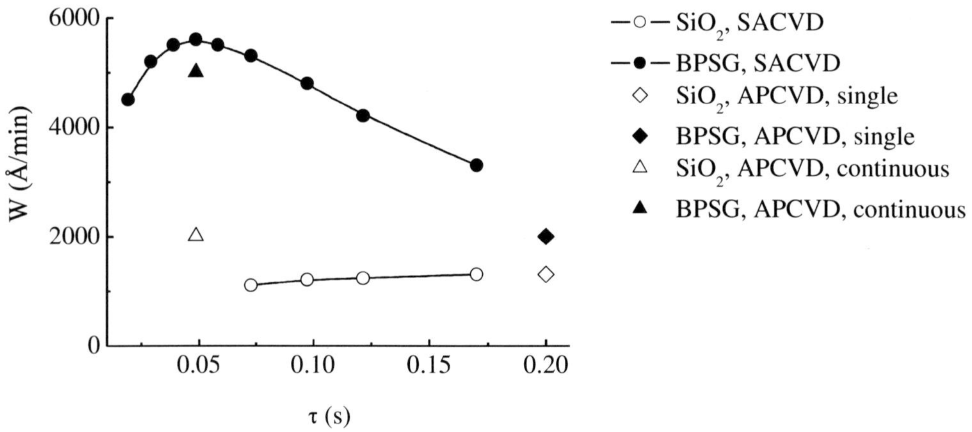

Figure 1.19. Comparison of acceleration effects for TEOS-ozone BPSG thin films obtained using different tools and deposition conditions (see text for details).

Chapter 4

AS-DEPOSITED BPSG FILM STEP COVERAGE AND GAP-FILL

The film deposition on flat substrate surfaces is used for only the CVD process and thin film material characterization. In IC device technology, it is necessary to characterize the films on the real device structures. Real semiconductor substrates in microelectronic technology are non flat because they consist of an IC device element; the latter can be more or less complicated, forming either steps, or gaps, see Figure 1.20(a,b). Thin film growth on IC device structures differs on the device sidewalls depending on the step shape (usually vertical), its geometry and, especially, depending on the geometry and parameters of gaps between neighbor steps.

The main goal of thin film CVD on the device structures is to provide a conformal film covering on the stepped surfaces and to fill the narrow gaps of the device structures with the deposited material without voiding, as shown in Figure 1.21(a,b). Voiding phenomenon shown in Figure 1.21(c,d,e,f) worsens the film properties and uniformity within IC device and, therefore, is undesirable.

In this chapter, the author summarizes the basic results of thin film CVD studies on the stepped IC device structures. The summary that is presented is based on the author's original publications [188,208,231,244,257,261].

4.1. BASIC DEFINITIONS OF STEPPED DEVICE STRUCTURES

The gap between two adjacent device features on a wafer is usually characterized by the so-called "aspect ratio", AR = H/G, where H and G are the depth and width of the gap, μm, see Figure 1.20(b).

The trend toward higher IC device complexity has resulted in gap widths less than 0.1 μm, with gap depths that changed insignificantly. Accordingly, AR may be as large as 10 and even more in modern ICs. A consequence of ongoing device scaling is that the sizes of voids may approach the size of the gap itself. In addition, designers usually plan to have vertical device steps and gap profiles, but actual profiles may deviate from this shape. The shape of gaps was classified using four major types of gap profile: reentrant, tapered, vertical, and tapered-vertical [244]. The film deposition on these types of IC device structures differs drastically. Because of differences in real gap shapes, for numerical characterization of CVD features on stepped surfaces, like gap-fill capability, the author proposed to use the only vertical gap shape. Moreover, it was also proposed that the so-called "structure complexity" parameter, SC = AR/G, or SC = H/G^2, could be effectively used for CVD characterization instead of AR [244,257]. It was believed that the SC parameter serves as a comprehensive parameter.

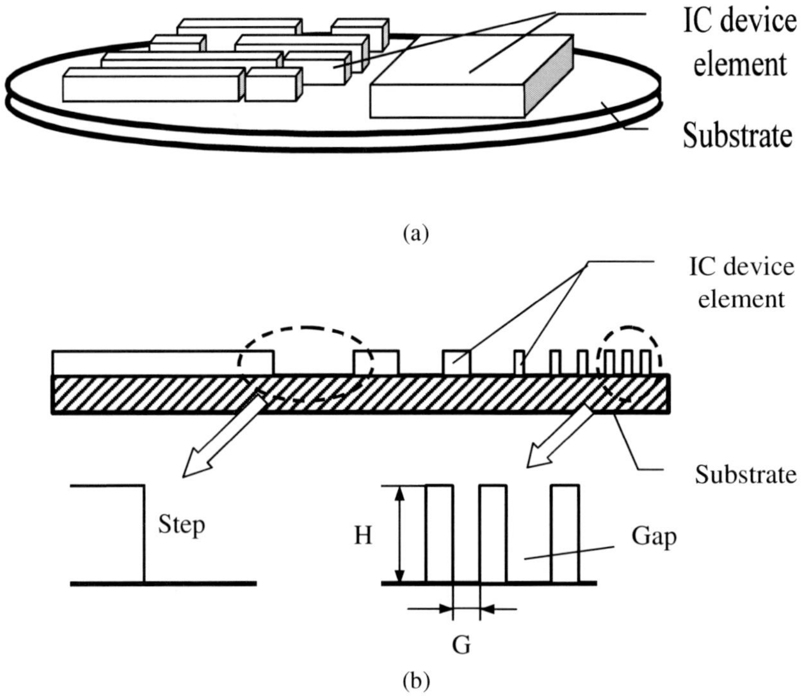

Figure 1.20. Substrate stepped top-view schematic (a) and its cross-section (b) with definitions of step, gap and gap height (H) and width (G).

4.2. Basic Definitions for Thin Film CVD on Stepped Device Structures

To define step coverage and gap-fill, the Scanning Electron Microscopy (SEM) method was used. Samples of the IC device structures with deposited thin film were prepared by cleaving the semiconductor substrate in the perpendicular direction with respect to the structure length followed by cutting the unusable parts of substrate to the sample square sized about 0.5 –1 cm. As-prepared cleavage surfaces are usually subjected to short-term hydrofluoric acid etching: this improves the SEM images making them sharper. A cross-section view of the sample should be similar to the scheme presented in Figure 1.20(b).

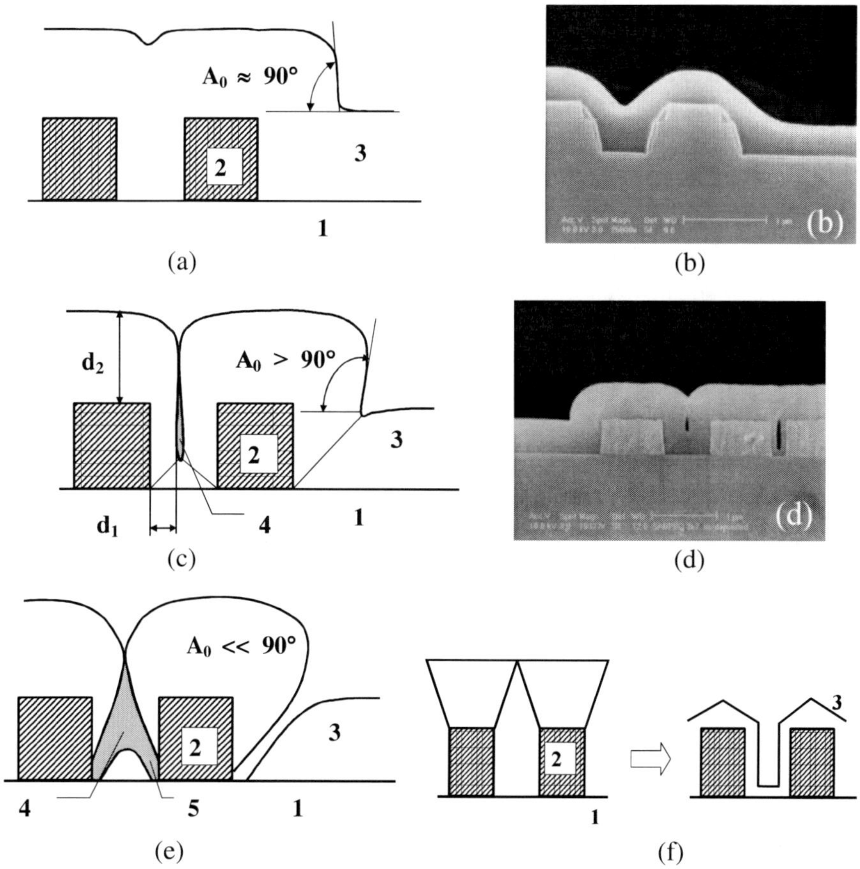

Figure 1.21. Sketches and SEM photos of thin film cross-sections on the IC steps: (a) "conformal", (b) "flow-able", (c,d) non-conformal, (e) very non-conformal, (f) HDP-CVD without bias. Definitions: 1 – substrate; 2 – metal line; 3 – thin film; 4 – voids; 5 – thin film breaking areas; A0 – angle of thin film surface with respect to the substrate surface.

The quality of thin film coverage on device structures can be quantified in different ways. For instance, the angle A_0 can be used that can be more or less 90 degrees (see Figure 1.21). However, the thin film step coverage is commonly quantified in terms of the ratio of film thicknesses measured in relation to the sidewall and the top of a step, respectively, d_2/d_1, % (see Figure 1.21(c)). Analysis of real thin film features on device structures with vertical shape shows that in most CVD cases, the film covering is not conformal. In this case, cross-sectional SEM micrographs of the filled gaps indicate that the film may contain voids, whose shape depends on the CVD technique employed, and process conditions (see Figure 1.21(c,d)). Voids are unacceptable because they seriously impair IC reliability. In the worst case of deposition, especially for the PECVD case, large keyhole-like shaped voids, or even the film discontinuity, can be obtained (see Figure 1.21(e)). However, in contrast, TEOS-ozone silicon dioxide CVD, either APCVD or SACVD, revealed so-called "flow-like" step coverage profiles with excellent gap-fill (see Figure 1.21(b)). However, this profile of the film was found to change to that presented in Figure 1.21(c) for BPSG film deposition at the same process conditions.

The cross-section sketch in Figure 1.21(f) represents HDP-CVD thin film coverage. In the case of no-bias used (the only high-speed plasma deposition is in use), the film growth on the device structures is hardly non-conformal (left side). However, when bias conditions are used (simultaneous deposition-etch processes), the film growth becomes shaped as shown on the right of Figure 1.21(f) providing excellent gap-fill even for very aggressive structures.

Normally, in order to characterize the quality of the film step coverage or gap-fill on device structures researchers state step coverage values or the AR's of gaps that can be filled by the used CVD technique. Our theoretical and experimental investigations into the gap-filling problem have enabled us to define the following framework for evaluating gap-fill capability in the IC technology.

(1) If steps on the surface are at least a few micrometers apart, step coverage can serve as a preliminary characteristic of a CVD technique. However, step coverage is unsuitable for characterizing processes that involve sputter deposition, like HDP-CVD.
(2) The gap profile is a key factor in evaluating gap-fill capability, especially when comparing results reported by different workers. The evaluation must be done for the typical gas shape (vertical).
(3) The best way to evaluate the gap-fill capability of any CVD technique experimentally is as follows. It is necessary to deposit a film on a stepped surface with gaps that have the same profile and H but differ in G (see details below). This allows us to obtain a graph presented in Figure1.22 with closed (well-filled gaps) and opened (gaps with voids) circles. The dashed line in this graph separates good and bad gap-fill and reflects the critical gap that can be successfully filled with a certain CVD technique and film. The slope of this line is fixed "structure complexity" parameter, SC = AR/G (or SC = H/G^2) that was introduced by the author in [244].

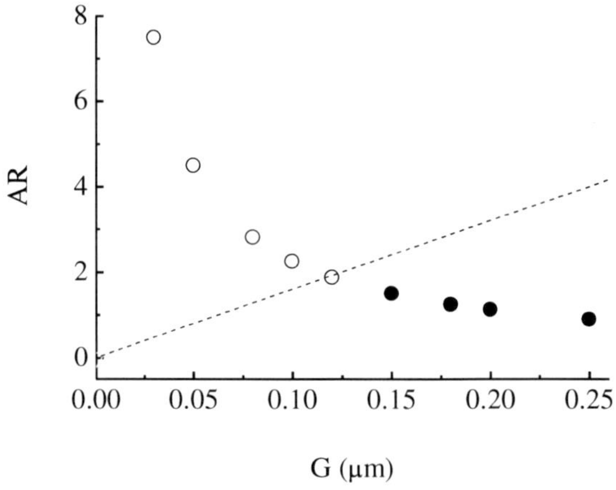

Figure 1.22. Evaluation of gap-fill capability by (a) finding the critical values of gap width and aspect ratio. The case of the vertical gap profile is considered, see Figure 1.21(d).

4.3. Characterization of CVD Processes on Stepped Structures

This film step coverage was found to correlate with the kinetic parameters of the CVD process. In order to obtain an overall measure of deposition kinetics, we defined the effective constant of the deposition rate k_{eff}. It is clear that with the other factors being fixed, the deposition rate grows with k_{eff}. The data in Figure1.23(a,b) indicate that films obtained by rapid deposition (for instance, using RF plasma-activated CVD) exhibit poor step coverage; moreover, voids may occur at the bottoms of gaps. By contrast, step coverage of 100 % or even higher is typical of slow (low deposition rate) CVD processes. These processes are the pyrolysis of TEOS and the oxidation of TEOS with an ozone-oxygen mixture with APCVD or SACVD conditions. The other CVD processes considered were found to yield step coverage well below 100%. In the final analysis, the correlation between step coverage and k_{eff} was explained on the basis of the mechanisms and the classification of silicon-based thin film CVD processes [1-3,224].

It is important to highlight that the above mentioned CVD acceleration effects with phosphorus compounds added to TEOS-ozone mixture caused the acceleration of k_{eff} in a few instances. In parallel, a drastic drop of the film step coverage from flow-like 130 % to slightly non-conformal took place (see Figure 1.21(b) and Figure 1.21(c,d)). The trends of step coverage change with k_{eff} changes are presented in Figure 1.22(b) in general form together with the positions of basic glass thin film CVD processes. Numerical evaluation of step coverage was presented as follows [2,3]: $d_2/d_1 \sim 45.4 \times k_{eff}^{-0.51}$ (where d_2/d_1 is presented in %, k_{eff} is expressed in cm/s).

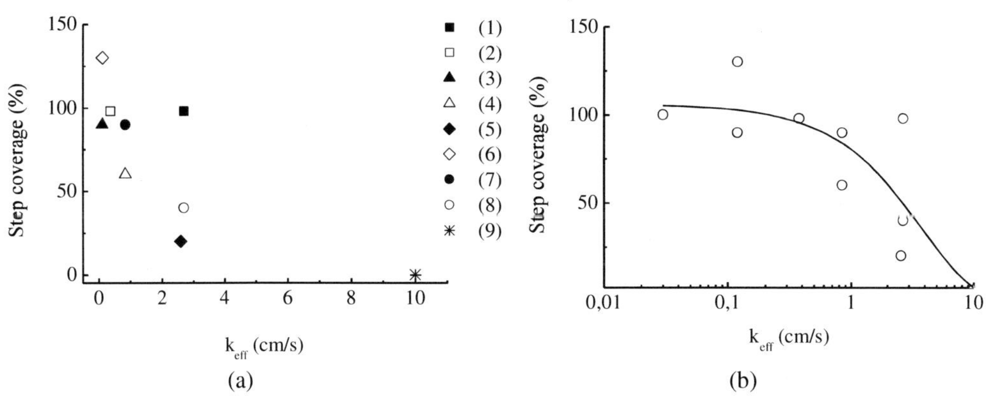

Figure 1.23. Step coverage on a single device step vs. keff for different CVD processes and chemical compound combinations: (1) LPCVD pyrolysis of SiH4 ; (2) LPCVD pyrolysis of TEOS; (3) LPCVD ammonolysis of SiH2Cl2; (4) LPCVD oxidation of SiH4 by N2O; (5) LPCVD oxidation of SiH4 by O2; (6) SACVD oxidation of TEOS by O3/O2 mixture; (7) SACVD oxidation of TEOS-B(OC2H5)3-PO(OC2H5)3 mixture by O3/O2 mixture; (8) PECVD oxidation of TEOS-B(OCH3)3-P(OCH3)3 mixture by O2 ; (9) unbiased HDP-CVD oxidation of SiH4 by O2. The graph in Figure 1.23(b) drawn based on treatment of data presented in Figure 1.23(a).

Step coverage analysis permits an analysis of the differences in CVD processes on the steps. However, in modern IC device structures it is necessary to evaluate and characterize the

CVD process gap-fill. An approach to numerically characterize the gap-fill was developed by Vassilyev and is presented in [244,257]. Using SEM cross-sectional photos, it is possible to determine which of the gaps have voids and visualize the results by plotting a graph as AR=f(G) (see Figure 1.22). The first point representing a void-free gap can be called a critical point. Using the example in Figure 1.22, for a certain value of H, the critical point was found to lie at G = 0.12 μm and AR = 2. For a given deposition technique and gap profile, critical points corresponding to a different H can be regarded as lying on the same straight line through the origin. The slope of the line gives us "structure complexity" $SC = AR/G = H/G^2$ that can serve as a measure of gap-fill capability for the CVD technique considered. It would take much time and money to determine critical points for many values of H. In light of this fact, it was assumed that a single point suffices to evaluate the process. The slope of the evaluation line enables one to quantitatively compare different CVD techniques in terms of gap-fill capability. A deposition process with a larger slope can successfully fill a ULSI structure in which G is smaller and AR is larger. Numerical evaluation of SC was presented in [2,3] as follows: $SC \sim 3.03 \times k_{eff}^{-1.47}$ (where SC is presented in μm^{-1}, k_{eff} is presented in cm/s). More details of SC determination can be found in [244,257].

Another approach to compare CVD processes is to examine the behavior of vertical void size normalized to gap depth (see an example in Figure 1.24). It refers to the deposition of PMD that separates the transistor layer and the lowest metallization layer in ULSIs with ultra-small G and large AR. In most cases, such an insulator is silicon dioxide, PSG, or BPSG, the first two materials requiring no reflow. Figure 1.24 evaluates two CVD processes commonly employed in IC manufacturing: (1) the HDP-CVD of PSG with simultaneous argon-ion sputtering and (2) the TEOS-ozone BPSG. The comparison reveals a marked contrast between the gap-fill capabilities of the two processes. With the SACVD BPSG process, the relative void height increases gradually as G is decreased. With the HDP-CVD process, a rapid increase is observed, implying that there is little or no filler in gaps whose widths are only slightly below critical G.

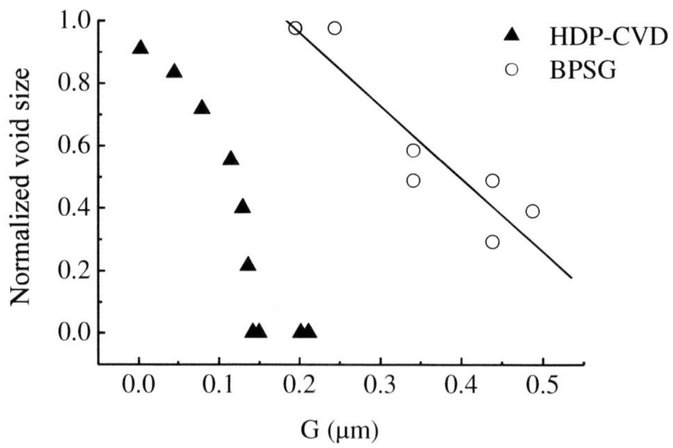

Figure 1.24. Evaluation of the ratio of vertical void size to gap depth against gap width for the deposition of PSG by high-density-plasma SiH4-PH3 oxidation and the deposition of BPSG by SACVD TEOS–TEB-TEPO–O3/O2.

Table 1.2. Step coverage and gap-fill summary for difference CVD processes

Thin film CVD processes and chemistries	Thin film	k_{eff} (cm/s)	d_2/d_1 (%)	SC (μm^{-1})
LPCVD TEOS pyrolysis	SiO_2	<0.5	95-100	<10
APCVD, SACVDTEOS-ozone	SiO_2	<0.5	>100	<10
APCVD, LPCVD silane-based	SiO_2, glasses	1-2	30-40	<4
APCVD, SACVD TEOS-based	SiO_2, glasses	1-2	60-80	<4
PECVD silane-based, TEOS-based	SiO_2, glasses	>2	10-30	<1
HDP-CVD without bias	SiO_2, PSG	>10	0-10	<0.1
HDP-CVD with bias	SiO_2, PSG	NA	NA	>20

Table 1.2 summarizes the major silicate glass CVD process characteristics in terms of k_{eff}, step coverage, and SC; the case of vertical gap shapes is considered. Regarding other types of gaps, none of the CVD processes considered was found to be successful in filling reentrant gaps. With tapered gaps, SC is much larger compared with the case of almost vertical gaps.

4.4. IMPROVEMENT OF GLASS FILM CVD ON STEPPED STRUCTURES

Acceleration of the deposition processes (corresponding to k_{eff} increase) was found to worsen the step coverage and gap-fill. However, the process inhibition showed opposite trends. Thus, boron compounds were found to decrease slightly the k_{eff}, therefore improving the film step coverage a little bit. We would like to focus on two approaches of the high-speed thin film CVD process inhibition:

(1) By special additives like C_2H_4, C_2H_6, NH_3, NO_2, which suppress the free radicals formation during the chain-brain reactions, see reviewed data in [3]. This method, however, has some disadvantages. First, it needs precise control on the very small additive concentrations. Secondly, additives are carbon-containing compounds, which can affect the film composition and properties;

(2) By the use of modified silicon compounds, for instance, partly fluorinated TEOS, $SiF(C_2H_5)_3$, frequently named FTES. This compound in an equal mixture with TEOS showed slightly better BPSG film step coverage [312]. However, in the case of compound modification it is necessary to be sure that the film composition will be acceptable for the use in microelectronic technology. In fact, some species containing fluorine, carbon, water, hydrogen are undesirable in the films used in IC technology.

Chapter 5

GLASS FILM CVD SCHEMES

Experimental data presented above were explained in a number of reports, including the author's publications and reviews [1-3,224,269]. In this chapter, we briefly summarize the simplest and effective explanations of silicate glass CVD commonalities, differences, and schemes of thin film CVD processes.

5.1. GENERAL THIN FILM CVD PROCESS SCHEME

The observed glass thin film CVD features were explained with a simplified multi-route and multi-step kinetic scheme developed for silicon-based thin film CVD processes. Repinski [42] proposed a multi-route approach for CVD processes for the first time in 1985 considering as an example silicon nitride thin film CVD. This scheme was expanded by Vassiliev on polysilicon, silicon dioxide and glass thin film CVD processes [1,2,224,269]. Figure 1.25 represents a proposed general CVD thin film kinetic scheme with three reaction routes: a heterogeneous route (Route 1, HTG 1), a two-step reaction route (Route 2) of thin-film formation, and a route of particles (aerosol) formation in the gas phase (Route 3) Route 2 includes two first order consecutive reactions of formation of an intermediate reaction product (IMP, 1st step) and its further consumption (2nd step).

A simple scheme of consecutive reactions is expressed as follows: A → IMP → P. Here A represents a silicon-based compound, IMP is an intermediate product, and P is a silicon-contained thin film. A homogeneous IMP formation (HMG) is assumed to occur in the gas phase, and a heterogeneous IMP consumption (HTG 2) with the film formation is supposed to occur on the substrate surface. This very general and simplified scheme is capable of explaining all of the basic experimental data. According to the scheme, the reaction route differs depending on the compound combination and the deposition conditions. For instance, the CVD features of the TEOS pyrolysis reaction can be explained using the heterogeneous reaction route HTG 1. Despite a few explanations of a possible reaction mechanism of TEOS-ozone CVD process (see, for instance, [313]), we have also chosen the heterogeneous route HTG 1 to explain this very specific "surface-sensitive" process [122]. In fact, despite some differences, a very good film thickness uniformity as well as excellent step coverage and gap fill are the common features of these TEOS-based deposition processes.

In contrast, silane-based APCVD/LPCVD oxidation and PECVD processes were explained with the same scheme but with a two-step reaction route. Their differences in the observed thin film kinetic features were explained by the difference in the limiting step of the two-step reaction route II. The latter is assumed to be either a step of homogeneous formation HMG of intermediate IMP (1st step), or a step of its heterogeneous consumption HTG 2 (2nd step). For silane-based oxidation processes and PECVD processes, the reactions are supposed to be limited by the 1st step, i.e., step of gas-phase intermediate product formation (HMG). In this case, little pieces of the glass material forming in the gas phase diffuse to the surface and build-up into the growing thin film. Therefore, the basic reason for bad film thickness uniformity and non-conformal film step coverage for these processes is the following. The ratio between the IMP diffusion rate and the rate of IMP consumption on the surface causes a non-uniform IMP distribution in the CVD reactors (as well as in the narrow gaps). We believed that this model is more suitable for the explanation of the gas-phase deposition reactions as compared to the traditional explanation based on a mean-free path and surface migration considerations. Therefore, the acceleration of the oxidation reactions with phosphorus compounds could be explained in most cases as a change of reaction route. For instance, the TEOS pyrolysis process was explained by the heterogeneous route I (HTG 1) after acceleration switched to the two-step route II with the 1st IMP-formation-limiting step of the accelerated TEOS oxidation (HMG). Furthermore, the strong acceleration of the oxidation reactions by the phosphorus compounds initiated route III of an aerosol formation in the gas phase.

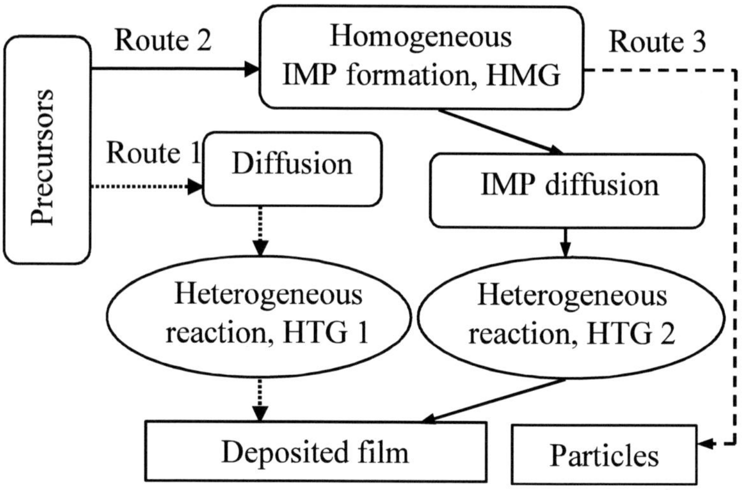

Figure 1.25. Proposed general multi-route and multi-step kinetic scheme of the silicon-based thin-film gas-phase deposition processes (see text for details).

5.2. CVD Thin Film Deposition Processes Classification

To find links and interrelations between observed gas-phase deposition process features, the author classified those features. A classification of the silicon-based CVD thin-film gas-phase deposition processes was discussed in [1-3,257] and presented in simplified form in Table 1.3.

Classification of the studied reactions is done using a few groups, named A-F. Each of these groups summarizes the studied gas-phase deposition processes with close enough kinetic trends, k_{eff} values, the basic LPCVD process features, typical step-coverage numbers, etc. A certain fixed reaction route or a limiting step of the kinetic scheme is also attributed to every selected group. Changes of the deposition kinetic features with deposition conditions or gas-phase additives, like phosphorus- or boron compounds, mean that a certain deposition process "switches" from one group to another group of processes, as shown by the arrows. This is explained by the change of the reaction route or the limiting step of the deposition reaction. Our assumption regarding the possibility to extend the proposed classification for the conventional PECVD and unbiased HDP-CVD processes were found to be not in contrast. These processes are highly activated and their estimated k_{eff} values significantly higher, by about ten times, as compared to those values of conventional CVD processes. If they are not carefully optimized, they show the worst numbers of film step coverage and gap-fill in IC technology. These non-optimized processes, therefore, should be placed in a group F together with the non-optimized APCVD silane-based processes, which show similar deposition process features. At the same time, an optimization of APCVD and PECVD processes, mainly due to a reduction of k_{eff}, allows improvement of the process characteristics and the placement of these optimized processes in group E.

5.3. Thin Film SiO_2 and Silicate Glass Deposition Processes Schemes

As for silane-based thin film CVD processes, their schemes can be simply explained by chain-brain chemical reaction (see reviewed material in [3]). In terms of the generalized scheme in Figure 1.25, these CVD reactions can be described by the reaction route II with homogeneous limiting step. TEOS-based deposition kinetics is more complicated. Figure 1.26 illustrates how the above scheme works in the TEOS deposition of SiO_2 and glass films [224]. This process is notable for the largest number of compounds and the highest acceleration.

Route I represents the formation of SiO_2 and is implemented as either the high-temperature pyrolysis of TEOS at the surface or the low-temperature reaction between adsorbed TEOS and adsorbed ozone. The latter method yields SiO_2 films of less homogeneous structure (with higher roughness and porosity) compared with ones produced with oxygen serving as the oxidizing agent. If boron and phosphorus compounds are added to the source gas, CVD follows route II. It involves phosphorus radicals (labeled R*), the homogeneous formation of the intermediate products IP_X and IP_Y, and their heterogeneous conversion to the glass. The homogeneous formation of $IP_X$is believed to be the rate-determining step of CVD for PSG and BPSG, whereas the heterogeneous conversion of IP_Y is believed to play the same role for BSG. With route III, the aerosol is thought to be produced as clusters, some of which should contain boron and phosphorus oxides.

Table 1.3. Classification of studied silicon-based thin-film gas-phase processes

CVD groups	A	B	C	D	E	F
Chemical systems	TEOS-O_3 →	SiH_4* → TEOS** →	TEOS-O_3-TEB SiH_2Cl_2-NH_3 $SiCl_4$-NH_3 SiH_2Cl_2-N_2O	DMDCS-O_2 → SiH_4-N_2O →	SiH_4-PH_3*** TEOS-O_2-PH_3 SiH_4-NH_3 TEOS-O_3/O_2-TEPO DMDCS-O_2-PCl_3 SiH_4-N_2O (high T_d) SiH_4-O_2-PH_3 Optimized PECVD	Non-optimized APCVD PECVD unbiased HDP-CVD
k_{eff} (cm/s) (estimated range)	~ 0.1-0.2	<2.7* 0.38**	0.05-0.28	0.12-1.0	~0.37*** 1-2,6	>2
Aerosol formation	No	No	No	No	No	Yes
Step coverage (%)	~130	100	90-100	70-90	40-70	0-10
Reaction route	I	I	II	II	II	II, III
Limiting step	HTG 1	HTG 1	HTG 2	HTG 2, HMG	HMG	HMG

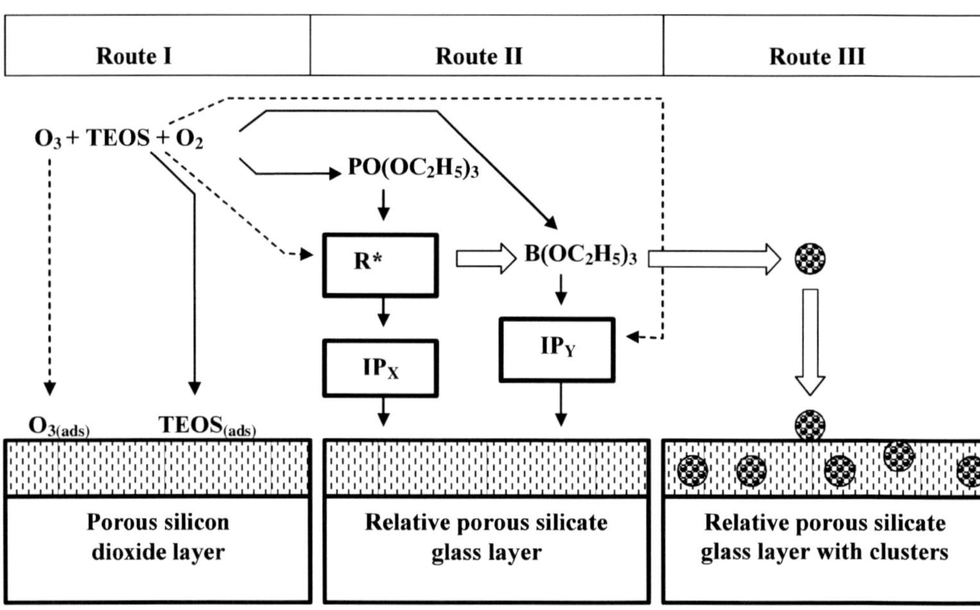

Figure 1.26. Application of general CVD process scheme to the TEOS deposition of SiO_2 and glass films (simplified schematic based on Figure 11 presented in [224]).

In the scheme in Figure 1.27, only low-temperature deposition processes (~ 400 – 500 °C) are considered. The TEOS-ozone CVD process can be placed on the left side of the scheme, reflecting an example of a very slow (low deposition rate) CVD process. It produces the glass film with the disordered structure. On the other pole of the scheme in Figure 1.27 we can place conventional high-speed CVD, PECVD and HDP-CVD processes. The kinetic scheme of these processes assumes IMP formation in the gas phase followed by diffusion to the wafer surface. This type of deposition process provides high film density; therefore, these films are significantly less disordered as compared to the films deposited by CVD processes with a heterogeneous reaction mechanism. Here it is necessary to mention that the studied phosphorus compounds enhance the rate of CVD oxidation processes and according to the diagram in Figure 1.27, and additionally improve the structure ordering.

SiO_2	**Increase of k_{eff}** →	**Glass films**
CVD process with heterogeneous reaction mechanism (HTG 1) or heterogeneous limiting step of 2-step model (HTG 2)	**CVD process changes** • gas-phase particles ↑ • step coverage ↓ • gap-fill ↓	Gas-phase reactions with homogeneous limiting step (HMG)

Figure 1.27.Simplified diagram of relationship between an effective constant of the deposition rate. (↓ and ↑) reflect decrease or increase of process characteristic, respectively.

Chapter 6

BPSG FILM INDUSTRIAL CVD TECHNOLOGY

In Chapter 2, some data were presented about BPSG film deposition tools and deposition methodology. We focused primarily on the brief characterization of CVD reaction chambers in order to demonstrate some important differences between their designs. Note that not all of them were adopted by the semiconductor industry. In this chapter, we designate industrial CVD methods as Method 2 to Method 9 depending on the tool used for deposition. The other mentioned CVD tools were issued as a pilot or small series tools and will not be consider below. Surprisingly, many types of the CVD tools were used for glass film deposition. This can be explained by the changes in technology requirements along with IC development that stimulated new requirements for the tools and CVD processes. For instance, thin film gap-fill issues at the end of 1980th gave rise to new TEOS-ozone APCVD, SACVD, and HDP-CVD methods and production tools appearance.

A summary of BPSG CVD production methods is shown in Table 1.4.The numbers in the round brackets represent a method number. Here we selected the deposition chemistry as a key factor of the method classification. All these CVD methods yielded BPSG films in which boron and phosphorus content was for about 6 – 8 wt % each. These methods were also suitable for the deposition of SiO_2, PSG, and BSG films.

Table 1.4. Summary of industrial BPSG thin film deposition methods

Hydride-based SiH_4, PH_3, B_2H_6		TEOS-based Silicic, boric, and phosphoric acid ethers	
O_2	N_2O	O_2	O_3/O_2
APCVD (2)	PECVD (6,7)	LPCVD (3)	APCVD (4)
LPCVD (3)	-	PECVD (8)	SACVD (5)
HDP-CVD (9)	-	-	-

Briefly characterizing data in Table 1.4, we would like to note the following. Both silane-based and TEOS-based processes can be implemented in a few versions. Some of them used SiH_4 in combination with the other hydrides, chlorides, or esters at 350 – 450 °C at atmospheric or at low pressures. The others employed TEOS (silicic acid ether) in combination with ethers of boric, orthophosphorus, or orthophosphoric acid. Silane-based methods were realized at APCVD, LPCVD, PECVD conditions. The last HDP-CVD method combines hydride-based LPCVD with argon sputter deposition using high density plasma (HDP). Although this approach was designed to make SiO_2 or PSG films from hydrides, it is suitable for BPSG deposition. As for TEOS-based chemistry, it could not be realized at HDP-CVD conditions because of the very low deposition pressure used. The TEOS-based processes were implemented in three versions: (1) LPCVD at 650 – 725 °C, (2) PECVD at about 400°C, and (iii) APCVD or SACVD at 400 – 500 °C using an oxygen-ozone mixture with 4 – 12 wt % of ozone.

More data regarding industrial BPSG thin film deposition methods are combined in Table 1.5 and Table 1.6 for CVD and PECVD processes, accordingly. The author would like to note that he does not compare numerous characteristics of the tools, like substrate size, tool productivity, deposition uniformity, added particle density, boron- and phosphorus concentrations, and so on. In fact, as long as all these tools were dedicated for semiconductor production, their characteristics were optimized well.

Table 1.5. Basic information about the most important industrial CVD processes

Process parameter	**BPSG thin film deposition method**			
CVD method No.	Method 2	Method 3	Method 4	Method 5
Condition	APCVD	LPCVD	APCVD	SACVD
Reactor type	Continuous	Multi-wafer	Single-wafer	Single-wafer
Chamber sketch *Company name* "Tool name" Chemistry Oxidizer T_d, °C P_0, torr W, Å/min [Example reference]	Figure 1.4(c) *WJ* "WJ-1000" Hydride-based O_2 400 – 430 760 ~ 5000 [26,69,115]	Figure 1.4(d) *Thermco* "VLTO" Hydride-based O_2 ~ 350 – 450 ~ 1 ~ 100 [58]	Figure 1.7(a). *Quester* "APT-4800" TEOS-ozone O_3/O_2 400 760 2000 [101,122]	Figure 1.7(b) *AMAT* "P 5000" TEOS-ozone O_3/O_2 480 – 550 100 – 200 Up to5000 [114,161,289]
Chamber sketch *Company name* "Tool name" Chemistry Oxidizer T_d,°C P_0, torr W, Å/min [Example reference]	Figure 1.4(c) *WJ* "WJ-1000" TEOS-ozone O_3/O_2 500 760 ~ 5000 [195]	Figure 1.4(e) *Anicon* "FS CVD" Hydride-based O_2 350 – 450 ~ 0.1 150 – 170 [59 – 61]	-	-

Notes: AMAT – Applied Materials; WJ – Watkins-Johnson.

In these tables, we only summarize the deposition temperature and pressures. Typical deposition rate values were presented either by tool manufacturers or evaluated using reference data. Because there were many methods of BPSG deposition, these data allow us to find and to properly explain the differences between BPSG thin film structure, composition and properties.

Kinetic data presented in Chapters 3 – 5 clearly showed that hydride-based and TEOS-based chemistries revealed opposite behavior when silicon dioxide films are deposited. Therefore, we can reasonably expect that the properties of the films, deposited using these chemistries, will significantly differ. However, we have analyzed the CVD acceleration phenomena caused by even small additives of phosphorus-contained substances used for PSG and BPSG film deposition. This chemical reaction feature decreases the differences between hydride-based and TEOS-based chemistries. In particular, the deposition rates of PSG and BPSG films deposited using different CVD methods and chemistries are very close to each other. Nevertheless, the properties of BPSG thin films deposited in different conditions and tools differ and will be discussed in detail in Part 2 of the monograph.

Table 1.6. Basic information about the most important industrial PECVD processes

Process parameter	BPSG thin film plasma-enhanced deposition method			
PECVD method No.	Method 6	Method 7	Method 8	Method 9
Condition	PECVD	PECVD	PECVD	HDP-CVD
Reactor type	Multi-wafer	Rotating plate	Single-wafer	Single-wafer
Chamber sketch *Company name* "Tool name" Chemistry Oxidizer T_d,°C P_0, torr W, Å/min [Example reference]	Figure 1.5(a) *ASM America* "P-CVD" Hydride-based N_2O 350 – 415 0.8 – 1.5 250 – 500 [24,25,51,52]	Figure 1.5(b) *Novellus* "Concept One" Hydride-based N_2O 400 2.4 3500 – 4500 [9,288,290]	Figure 1.6(a) *AMAT* "P-5000" "Centura" TEOS-based O_2 380 – 420 5 – 13 Up to 10000 [95,129]	Figure 1.6(b) Novellus "Concept-2 SPEED HDP" Hydride-based O_2 ~ 625 ~ 5×10^{-3} ~ 3500 [297]
Chamber sketch *Company name* "Tool name" Chemistry Oxidizer T_d, °C P_0, torr W, Å/min [Example reference]	-	-	-	Figure 1.6(b) AMAT "Ultima HDP-CVD Centura" Hydride-based O_2 ~ 520 ~ 5×10^{-3} ~ 3500 [298]

Note: AMAT – Applied Materials

FINAL REMARKS ON PART 1

Concluding Part 1, it is necessary to highlight that BPSG film growth can be performed by many CVD methods using a variety of tools, chemical compounds, and process conditions. Phosphorus compounds accelerate the thin film CVD process features, making the oxidation process faster. However, this worsens the film step coverage and gap-fill and enhances the aerosol processes. The latter can be inhibited by special additives. These additives decrease the deposition rate and slightly improve the step coverage decreasing the aerosol formation processes.

Obtaining the kinetic time curves for thin film CVD allows the acquisition of basic kinetic information about the process, such as optimized deposition rate, reaction time, and effective constant of the depsoition rate. Considering the need for stepped IC device structures, PECVD and silane-based CVD processes were found to be the worst case. However, TEOS-ozone based CVD processes, especially SACVD processes in single-wafer chambers, were recognized to be the best one. All accumulated CVD BPSG knowledge is nesessary for understanding of BPSG thin film composition, structure, and propertries in Part 2 of the monograph.

PART 2. BPSG FILM COMPOSITION, STRUCTURE, AND PROPERTIES

Part 2 (Chapters 7 to 14) is devoted to a description of BPSG thin film composition, structure, and basic properties. As long as the CVD thin film technique is not the traditional glass forming technique, special attention is taken to find the links between BPSG film structures and properties with definitions of the classical glass chemistry. This allows us to see the relationships between glass film composition, structure, and properties with BPSG thin film CVD process features.

Analyzing BPSG film composition, structure and properties, the author refers to the main BPSG film production processes, designated in Part 1 of this book as Method 1, Method 2, etc. This allows focusing on the most common CVD techniques used in the electronics industry.

One more introductory remark is about terminology. As we have already mentioned above, being added to silicon dioxide to form the glass structure using any known method, the boron and phosphorus oxides form bonds like Si – O – B, Si – O – P. Therefore, they cannot be named "dopants" like many authors do. In order to simplify the text, we use the term "additives" below that include both boron and phosphorus, existing together in BPSG films.

Chapter 7

BRIEF REVIEW ON BPSG THIN FILM ANALYSIS TECHNIQUES

7.1. ANALYTICAL METHODS OVERVIEW

As we have shown in Part 1, BPSG thin films with different boron and phosphorus content are deposited in flow-type CVD reactors. The film is a result of oxidation chemical reactions. A variety of silicon-, boron- and phosphorus-containing compounds can be used for the film synthesis. The techniques and the basic kinetic aspects of the glass film deposition are presented in Part 1 of the monograph. Silicon dioxide, borosilicate, phosphosilicate, and borophosphosilicate glass thin films can be deposited in the range of about 200 °C to 800 °C. When necessary in the device technology, the glass films can be thermally treated at anneal temperatures (T_a) of 700 – 950 °C for 10 – 60 min in diffusion furnaces (named below as furnace anneal, FA) or subjected to rapid thermal annealing (below named as RTA) for 0.3 –2 min.

Chemical compounds used for silicate glass thin film deposition are inorganic (such as hydrides and, sometimes, halides) and organic (such as ethers, alkiles, etc.). This means that the real glass film composition can differ from the obviously expected composition with the only glass formation elements such as Si, B, P, O. As long as the chemicals used in the electronics industry are well-purified, the author excludes the possibility of any impurity to be included in the film due to its presence in the chemical compounds. We have to take into account the presence of hydrogen (H), hydroxyl (OH), water (H_2O), as well as carbon-contained groups in the glass films as a reaction by-products trapped by the growing glass film during the oxidation reactions. In order to analyze the real CVD glass composition, we need to analyze Si, B, P, O, H, OH, H_2O, carbon groups, etc.

A number of thin film analytical methods have been implemented to study glass thin films [64,77,81,83,91,94]. To analyze the film composition and properties, non-destructive and destructive sets of analytical techniques can be used. The difference between them is that the first set does not require the breaking of the semiconductor wafer or damaging the thin film. For instance, these are optical methods. This set of non-destructive methods is used in experimental development and semiconductor manufacturing. Most of these methods are included in the semiconductor manufacturing process flow and placed directly inside semiconductor Fab clean-rooms. It is very important to highlight that modern analytical methods used in semiconductor manufacturing and development allow measuring the film parameters at many points within the semiconductor wafers. This statistic approach provides mapping of the parameter within the wafer, or 3D image of the parameter such as the thin film thickness uniformity within the wafer. Mapping gives an excellent capability to improve the CVD processes by decreasing the process non-uniformity down to a few percent within wafers. For instance, in a series of our work [186,187,190,211-214,224,230,232] the film thickness and its uniformity within the wafer were calculated from a total of 49-points using the Prometrix UV-1050 tool by Tencor Instruments, Inc. (USA). The typical BPSG film thickness standard deviation value did not exceed± 2 % within the 200mm silicon wafer and between wafers.

By contrast, the second set of destructive techniques (for instance, secondary ion mass spectrometry, SIMS, scanning electron microscopy, SEM) requires either breaking the wafers in order to prepare small-sized samples, thin film damaging, or thin film removal (such as wet chemical etch and ion chromatography). These techniques, mostly quantitative, are often used for some special purposes. For instance, ion chromatography was used for the calibration of manufacturing clean-room analytical tools. Calibration curves for additive analysis in Fab clean-rooms can also be created by inductively coupled plasma (ICP) analysis [224]. As a result, some special and expensive destructive analytical methods are mostly performed in special analytical centers, equipped with modern tools, like the well-known Balazs Laboratory (US) [314], or Charles Evans and Associates (US) [315].

To study the glass film composition and properties, the film thickness on the silicon wafer is normally chosen in the range of 0.5 – 1.0 μm. The thickness of silicon-based thin films is measured using optical methods (ellipsometry or refractometry). In addition to the film thickness, these methods allow the determination of the refraction index (RI) value, which is normally used as a characteristic of thin film stoichiometry (i.e. film composition). For instance, silicon dioxide films revealed RI values in the range of 1.44 to 1.46. On thermal processing above 750° C, the refractive index of BPSG films becomes sensitive to phosphorus concentration, causing a small increase of RI up to 1.52 [52,82]. This property was at one time used by some researchers for analytical purposes. Rough evaluation of RI with the phosphorus content changes in BPSG films was evaluated to be about 0.003 per one wt % of phosphorus [93]. At the same time, the boron concentration increase did not affect the RI values.

In addition to the film thickness allowing the calculation of the film deposition rate, boron [B] and phosphorus [P] concentrations are the essential parameters of the boron- and phosphorus-contained glass thin films. The studied range of BPSG film compositions is shown by data points in Figure 2.1. The additives content is usually expressed in terms of elemental weight percent or oxide weight percent (wt %). In order to obtain the weight percent of oxides having data for elements, the given values should be multiplied by ~ 2.3 for phosphorus and ~ 3.2 for boron. There were a few methods proposed and used for the additive concentration analysis (see description below). These methods provide their own accuracy. This is necessary to be taken into account, especially while comparing the film composition data obtained using different analytical techniques. With external quantitative calibration of the analytical tools at the optimized film thickness, the accuracy of analytical techniques used in Fab clean-rooms for boron and phosphorus concentration measurements was evaluated to be about ± 0.2 wt. %.

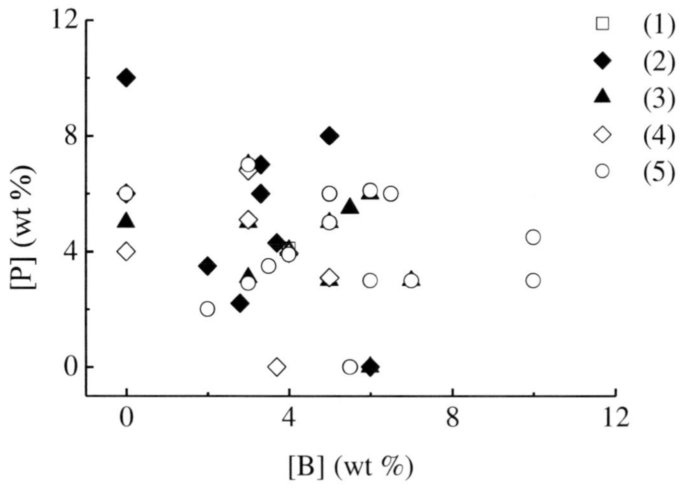

Figure 2.1. Composition range of phosphorus- and boron-containing silicate glass thin films (PSG, BSG, BPSG) prepared by different CVD techniques with the use of different chemical compounds (the author accumulated data for a long term studies). Designations of CVD reactors and chemical compositions:
(1) APCVD, 420 °C, $SiH_4 - O_2 - (CH_3O)_3PO - (C_3H_7O)_3B$;
(2) LPCVD, 420 °C, SiH4 – $O_2 - (CH_3O)_3PO - (C_3H_7O)_3B$;
(3) PECVD, 400 °C, $(C_2H_5O)_4Si - O_2 - (CH_3O)_3PO - (CH_3O)_3B$;
(4) APCVD, 400 °C, $(C_2H_5O)_4Si - O_3/O_2 - (C_2H_5O)_3B–(C_2H_5O)_3PO$, and LPCVD, 800 °C, $(CH_3)_2SiCl_2 - O_2 - (C_2H_5O)_3B–PCl_3$;
(5) LPCVD, 440 °C, $(C_2H_5O)_4Si - O_3/O_2 - (C_2H_5O)_3B - (C_2H_5O)_3PO$.

Below the author has summarized information regarding the non-destructive (Table 2.1) and destructive (Table 2.2) analytical methods used in his BPSG thin film research for a long period of time (detail method descriptions can be found in available related technical literature).

Table 2.1. A summary of Fab clean-room analytical methods used for non-destructive APCVD and SACVD BPSG thin film analysis [224,232,245,317]

BPSG film property	Analytical method	Example of tool type (manufacturer)	Data points within wafer with 200 mm diameter
Film thickness, RI	Ellipsometry	Prometrix UV 1050 (Tencor Instruments)	Wafer mapping
Additive content	IR spectrometry	QS 500 Fourier Transform IR Spectrometer (BIO-RAD)	9
	XR fluorescence (XRF)	X-ray PW 2800 Analyzer (Philips)	5
Mechanical stress in the film	Wafer curvature measurement	FSM 128 (Frontier Semiconductor Measurements)	NA

Table 2.1. (Continued)

BPSG film property	Analytical method	Example of tool type (manufacturer)	Data points within wafer with 200 mm diameter
Film surface dark/bright fields	Optical microscopy	Axiotron (Zeiss/Sony)	NA
Surface defect monitoring	Laser scattering	TPC 8500 Model (Inspex) Surfscan 7700 (Tencor)	Wafer mapping Wafer mapping
Surface roughness	Atomic Force Microscopy (AFM)	D 5000 Scanning Probe Microscope (Digital Instruments)	NA
Water absorption by the film	IR spectrometry	QS 500 Fourier Transform IR Spectrometer (BIO-RAD)	NA
Total charge in the films	Surface Photo Voltage (SPV) or Surface Charge Analysis (SCA)	Surface Photo Voltage CMS 4000 (Semiconductor Diagnostics) Quantox tool	Wafer mapping

One more set of analytical techniques includes some special methods for electrical measurements. Most of them require manufacturing of special test structures on the wafers (i.e., capacitor structures with sputtered and patterned metal layers, for instance, aluminum dots) for measurements of Capacitance-Voltage (C-V) curves, dielectric breakdown and leakage current (I-V) curves, etc.

Some details and examples of analytical results for BPSG thin films are presented below.

Table 2.2. An example of summary of destructive analytical methods used for APCVD and SACVD BPSG thin film analysis [224,232,245,251,316]

BPSG film property	Analytical method	Example of tool type (manufacturer)
Etch rate	Chemical etch in HF diluted HF buffered	NA
Solid BPSG film defects on the surface and inside the film	Scanning Electron Microscopy (SEM)	XL 30 FEG (Philips)
Film conformality on device steps, voiding phenomenon	Scanning Electron Microscopy (SEM)	XL 30 FEG (Philips)
Quantitative additive (B, P) content in BPSG film	Inductively-Coupled Plasma (ICP)	Balazs Laboratory tool
Quantitative silicon (Si), additive (B, P), and residues (C,H) distribution within BPSG film thickness	Secondary-Ion Mass Spectrometry (SIMS)	PHI-6600 Quadrupole Mass Spectrometer (PHI Electronics) Charles Evance and Associates

7.2. FTIR Method Implementation BPSG Film Composition Analysis

FTIR spectrometry is a powerful, relatively simple, and widely used method for thin film analysis. An example of as-deposited SACVD BPSG FTIR spectra is presented in Figure 2.2.

The following main common peaks can be clearly seen in Figure 2.2: the Si – O bond at 775 – 835 and ~ 1100 cm^{-1}, the P = O bond at 1300 – 1350 cm^{-1}, the B – O bond at 1300 – 1450 cm^{-1}, etc. The ratios of the latter peaks to the Si – O peak are used for quantitative characterization of additive contents in the films. As long as the glass film structure is formed directly during the film synthesis at relatively high temperatures, the peaks of mixed bonds, such as Si – O – B, Si – O – P, and P – O – B, are expected to be in the spectra too. This is true for the Si – O – B bond, which occurs at around ~ 920 cm^{-1}. However, according to Kolesova [316] the peaks for Si – O – P (~ 1145 cm^{-1}) and P – O – B (~ 1100 cm^{-1}) bonds are over-lapped by the strong Si – O peak at ~ 1100 cm^{-1} and therefore, they cannot be seen. A more detailed description of peaks observed using FTIR techniques, can be found in Table 2.3 cited from reference [271]. This table accumulates the author's experimental data and data from many other references. Note that interaction of the film with the ambient moisture can cause a number of changes in the FTIR spectra and will be a matter of special discussion below.

BPSG film thermal treatment at elevated temperatures causes some changes in the FTIR spectra, mostly reflecting a certain film densification. Basically, FTIR spectra analysis gives the conclusion that the glass film structures can be considered to be similar for all CVD methods. Therefore, for further analysis it can be stated that the film structure forms completely during the film deposition (need to note that this statement is not valid for very low temperature CVD film growth methods, like low temperature PECVD methods).

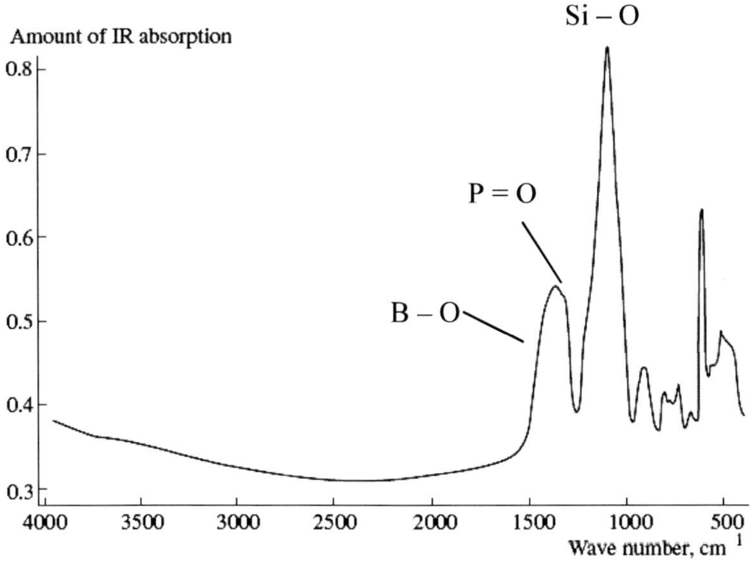

Figure 2.2. Typical as-deposited on silicon BPSG thin film IR spectrum.

Table 2.3. Transmission FTIR vibration frequencies and bonding assignments for B and P-contained silicon oxide films [271]. Reproduced by permission of ECS – The Electrochemical Society

Active Bond	Compound	Wavenumber (cm^{-1})
O – Si – O	BPSG	809
		810
B – O – Si	BPSG, BSG	925
	BPSG	918
	BSG	930
Si – OH ... HOH	BPSG	955
Hydrogen bonded		930 – 940
Si –O	BPSG	1072 – 1090
	BPSG	1089
P – O	Phosphoric acid	910 – 1040
	P – O – P	950 – 1100
Si – O – P	BPSG bulk	~ 1145
P – O – B	BPSG, BPO_4 bulk	~ 1100
Si – O – Si	BPSG	1190 (shoulder)
	SiO_2	1100 – 1200
P = O	BPSG, PSG	1330
	BPSG	1325
B – O	BPSG	1395
	BPSG	1394
B – OH	$B(OH)_3$ gas	1429
		1450
	$B(OH)_3$ etch residue	1433
	$B(OH)_3$ film	1429
H – O – H scissors	BSG	1630
	HOH/Si(100)	1630
PO – H	BPSG	2785
		2525 – 2725
H – O – H stretch	BPSG	2900 – 3500
	BSG	3400
	HOH/Si(100)	2900 – 3600
PO – H	PSG, BPSG	3270
		3227
BO – H	BPSG	3330
	BPSG etch residue	3334
	$B(OH)_3$ aqueous	3194
	$B(OH)_3$ solid	3200, 3213
SiO – H...HOH	BPSG	3445
Hydrogen bonded	BPSG	3450
SiO – H	BPSG	3535
	SiO_2	3550
SiO – H	BPSG	3625
	SiO_2	3635
	BPSG	3650
	HOH/Si(100)	3660

7.3. Laser Scanning Method

This non-destructive analytical method allows the definition of the number of surface micro-particles (defects), their positions on the wafer, and the relative size of defects. Typical results are presented in Figure 2.3(a,b) for the SACVD BPSG film sample after its storage in the Fab clean-room air. One can see the area of surface defects that appeared on the wafer surface (on the left) after 6.5 hours (see Figure 2.3(a), on the left). A histogram with the relative size of defects with their total concentration 6.31 defects per square centimeter can be seen on the right. These data, being monitored for a long period of time for SACVD BPSG films with different additive contents, allowed the study of the kinetics of the BPSG defect appearance and, eventually to propose the defect formation mechanism [2,187].

7.4. Microscopy Methods

Optical microscopy is an in-line clean-room method. It is used for quality control, particularly, for defect observation on the surface of thin films. Examples of optical microscopy data for BPSG films are shown in Figure 2.4(a,b).

Atomic Force Microscopy (AFM) allows the observation of some small details regarding surface of the films, such as: surface roughness (RMS) and surface imperfection [225,226]. The latter can be surface defects (see the author data in Figure 2.5(a,b,c)).

The roughness measurements of BPSG samples revealed some challenges as these films contain surface imperfections, or surface defects, in the form of liquid droplets (see Chapter 13 for more details). These liquid-state defects, whose images can be seen in Figure 2.4(a) and Figure 2.5(a,b,c) start to form immediately after BPSG film deposition and, especially, after the film anneal. They subsequently grew as a result of continual film moisture absorption from the ambient air. As the size of these defects become larger, they pose a source of error in the measurement of surface roughness, as reported for the first time in [131]. This source of error is especially difficult to detect in as-deposited films as these defects would appear to look like the surface of the film. However, under the AFM phase mode, these liquid state defects are visible and they can be identified quite easily as areas with different contrast. In order to reduce this surface error, measurements should be made immediately after deposition or furnace anneal since these defects grow rapidly in the first few hours after furnace anneal. Another method to eliminate/reduce this source of error is by subjecting the wafer to a cleaning step to completely remove these defects. As it was discovered, these defects were actually liquid in nature, and thus, rinsing the sample with deionized water would easily remove them. A comparison between the same sample before and after the cleaning treatment is shown in Figure 2.6(a,b) and Figure 2.6(c,d), respectively.

It was found that the roughness of the film (rms) with defects (before cleaning) was 3.16 nm as compared to the same sample after the water rinsing step which was 2.74 nm. This clearly demonstrates the degree of error incurred due to defects and obviously, the roughness of smoother films would be greatly affected by this.

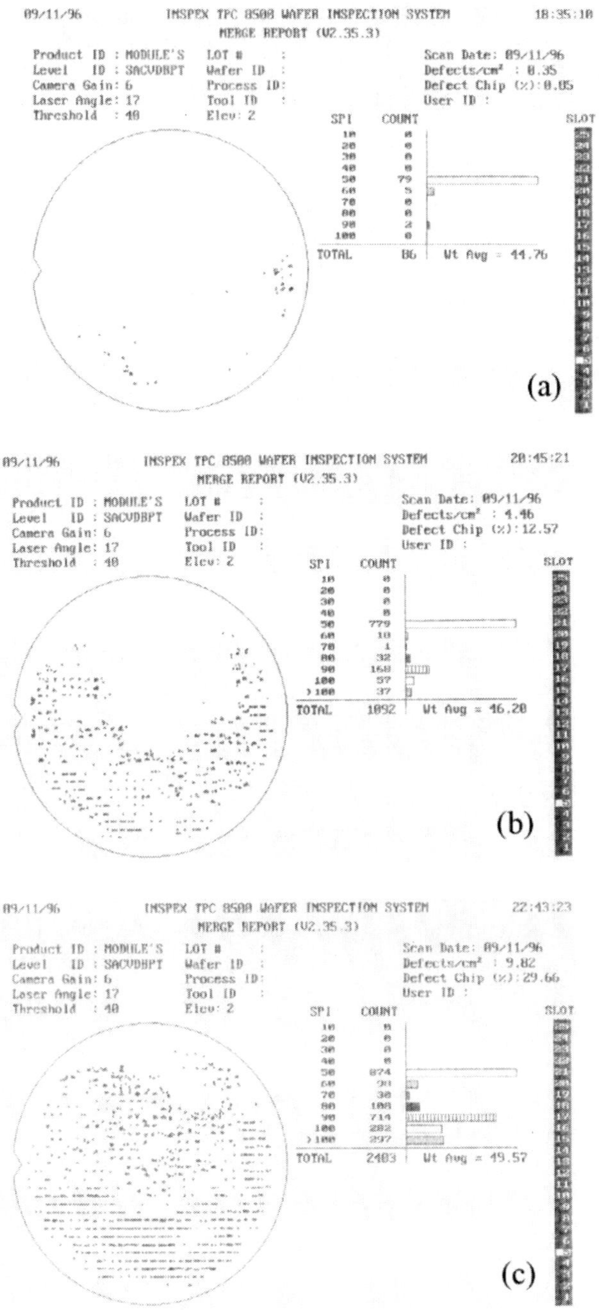

Figure 2.3. Three examples of BPSG film changes in the surface defect density and defect size on the 200 mm wafer registered with the use of INSPEX TPS 8500 wafer inspection tool after 2 (a), 4.5(b) and 6.5 (c) hours since the film thermal anneal. Changes in the defect size histogram can be seen on the right parts of the datasheets.

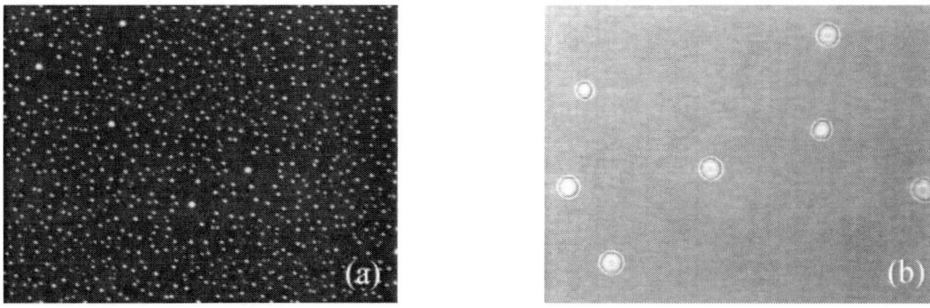

Figure 2.4. Examples of BPSG surface defects in dark (a) and bright (b) fields of an optical microscope. Magnification is 50$^{\times}$.

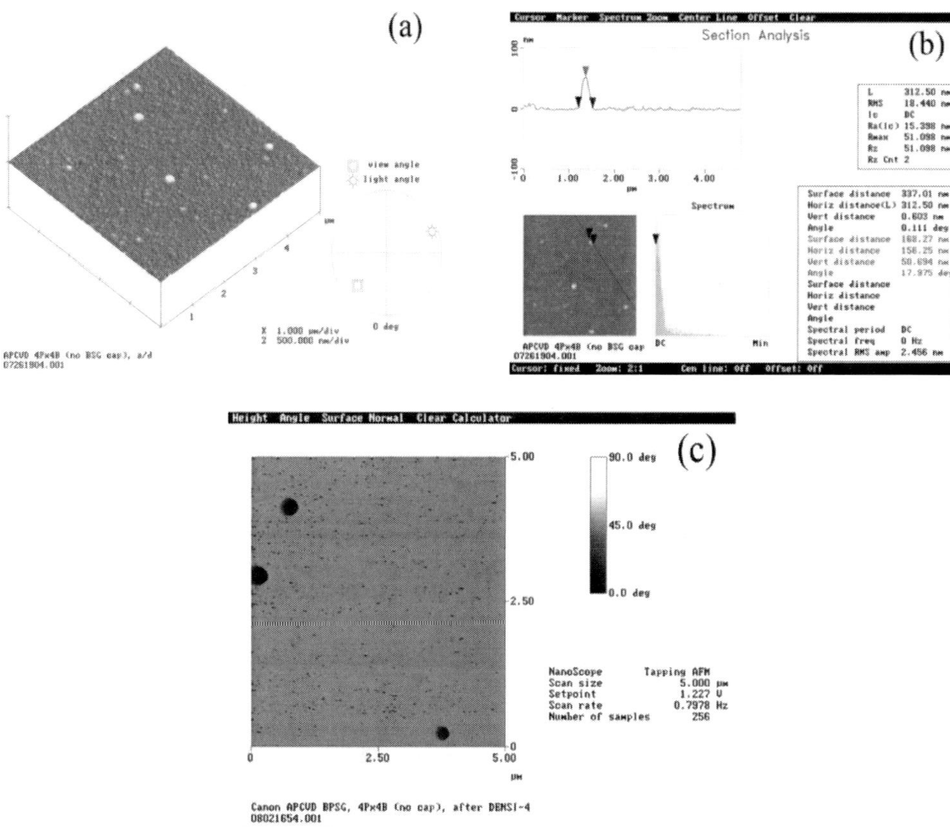

Figure 2.5. AFM data for 4 wt % [B] – 4 wt % [P] APCVD BPSG film with surface defects obtained with D 5000 Scanning Probe Microscope: 3-d view (a), cross-sectional view (b), and phase mode view (c).

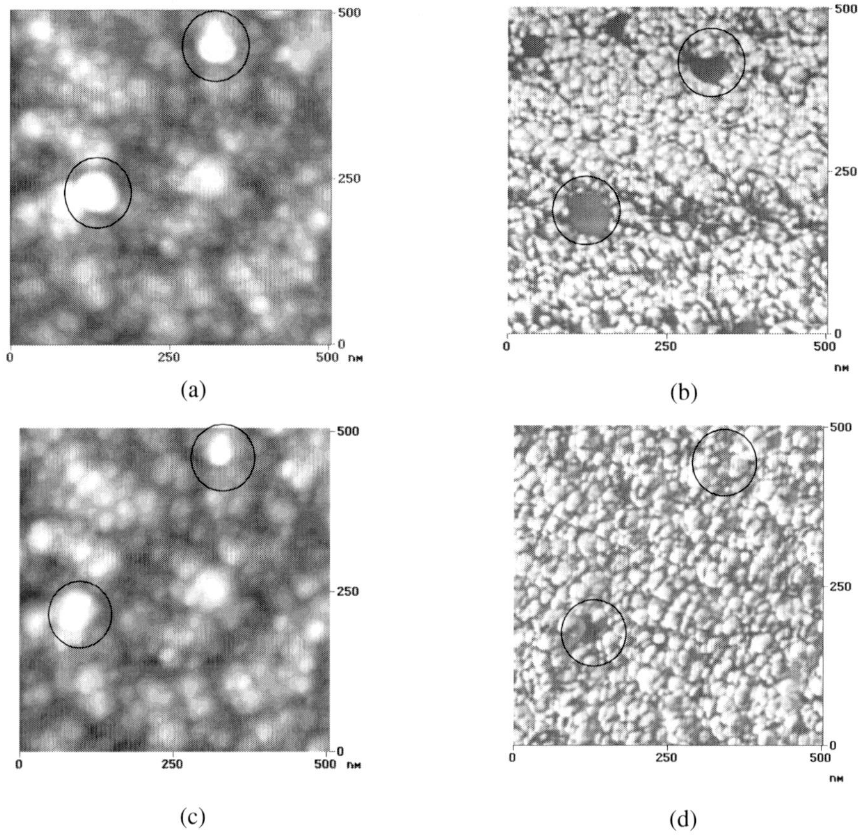

Figure 2.6. AFM images of silane-based BPSG film before water rinsing in scan (a) and phase (b) AFM modes and after water rinsing in scan (c) and phase (d) AFM modes. Circuits indicate liquid state defects. Phase mode of (a) clearly indicate contrast difference between liquid defects and BPSG film.

Scanning Electron Microscopy (SEM) allows the viewing of the surface of the film and, using a cross-sectional technique, some imperfection can be seen inside the film, see Figure 2.7(a,b).

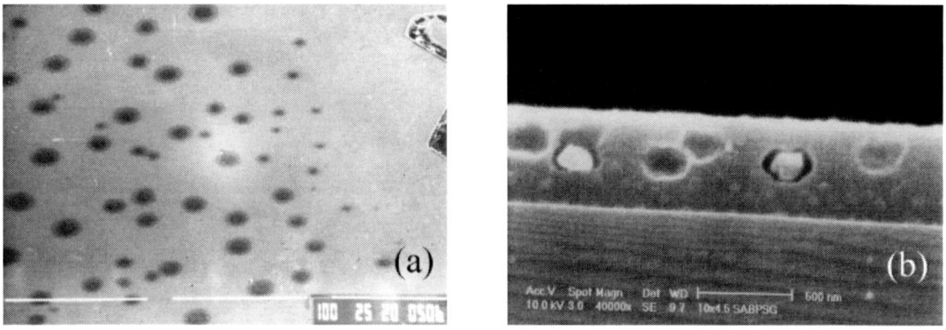

Figure 2.7. SEM photos of the film surface with defects (a) and micro-particle defects inside the film (b).

7.5. Total Charge Mapping Methods

Surface Photo Voltage (SPV), also named Surface Charge Analysis (SCA) is a method for fast and non-destructive observation of the total charge in dielectrics. This method is used in ULSI manufacturing as a rapid characterization of chemical cleaning quality and thermal oxide quality [318-321]. First time results for SPV technique implementation for BPSG films deposited on silicon substrate were published by Vassiliev et al. [214]. An example of 200mm wafer mapping is presented in Figure 2.8(a,b). One can see a large difference of total charge value within wafers that indicate high standard deviation (StD) values that obviously should be minimized.

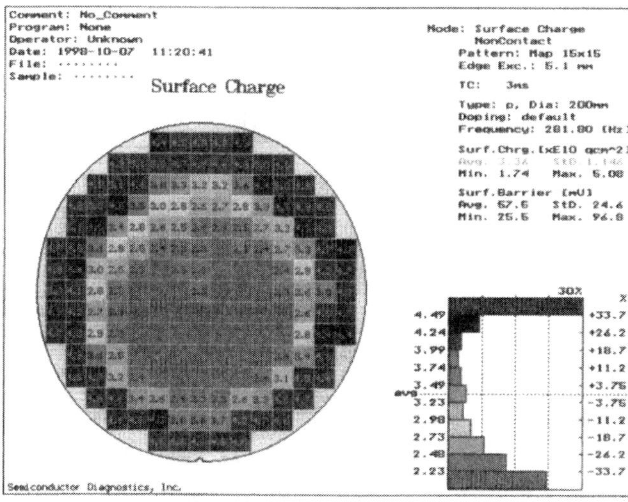

(a)

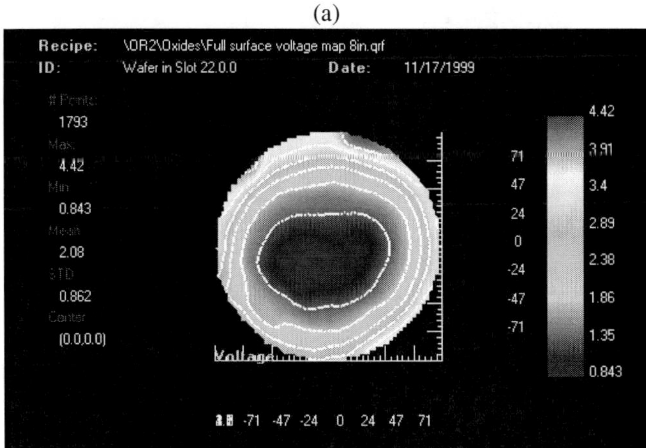

(b)

Figure 2.8. Data for total charge measurements in the CVD films on silicon substrate using SPV CMS 4000 tool by Semiconductor Diagnostic, Inc. (a) and SCA Quantox tool (b). Presented at 4th Intern. Dielectric for ULSI Multilevel Interconnection Conf. (DUMIC) [214] and 17th Int. VLSI Multilevel Interconnection Conf. (VMIC) [317], accordingly.

Comparison of data showed that in some cases, the CVD process (especially, PECVD) was not optimized properly with respect to the total charge in the films. Therefore, some effort should be made in order to understand the reasons of this charge behavior and to optimize its uniformity within the wafer.

7.6. Secondary Ion Mass Spectrometry Method

This destructive analytical method permits the calibration of the composition of the films and the detection of the additives distribution within the film thickness (i.e. profiling). This method does not have any reasonable limitations. Light elements, such as hydrogen, can be analyzed quantitatively which is not possible by the other techniques. In Figure 2.9(a), one can see a typical example of additive analysis in BPSG films. This case demonstrates that boron and phosphorus in the BPSG films revealed uniform additive concentration within the film depth. However, in some cases the additive distribution can differ greatly. For instance, an abrupt boron concentration peak is seen in Figure 2.9(b) showing non-optimized boron injection start into the deposition chamber.

Similar profiling data can also be obtained for carbon and hydrogen elements.

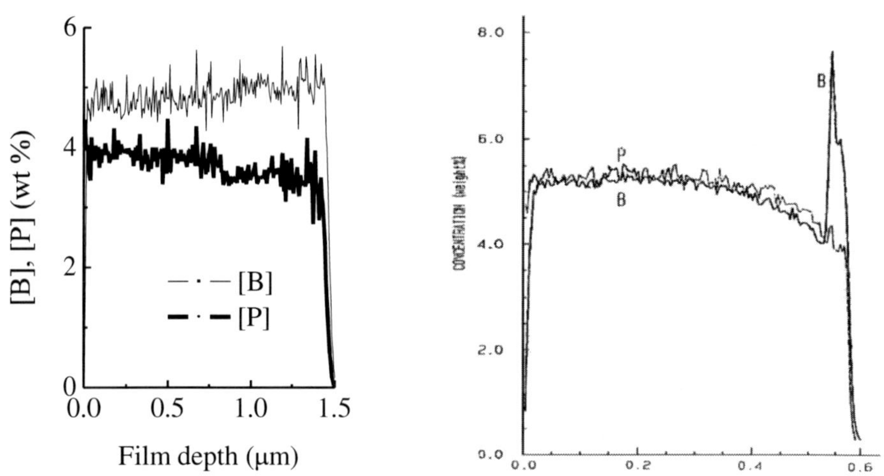

Figure 2.9. Examples of SIMS additive concentration profiles within BPSG films: left side - uniform [2]; right side – extremely non-uniform. Presented at 3rd Intern. Dielectric for ULSI Multilevel Interconnection Conf. (DUMIC), 1997 [186].

Chapter 8

BPSG FILM COMPOSITION

Silicon content. The referenced CVD methods produced the stoichiometric films without any excess of silicon. This primarily means that the film deposited the excess of the strong oxidizer (O_2, O_3/O_2 mixture) with respect to the silicon-, boron-, and phosphorus-contained reactant mixture). The silicon content could be easy analyzed using optical methods, which showed refractive index values between 1.44 and 1.48 depending on the additive content.

The only method which can cause a slight excess of silicon with respect to the stoichiometry is the PECVD method with the use of a relatively soft oxidizer, such as N_2O [52,53,85,256,288,290]. To overcome this negative issue, the authors of cited papers proposed to add some oxygen additive to the basic N_2O flow in order to suppress the appearance of silicon excess in the film [52].

Boron and phosphorus content. Additives (B and P) concentration in BPSG film was a matter of particular technology requirements. In BPSG thin films, the range of [B] and [P] varied between 2 – 8 wt % each (see Figure 2.1). The total additive concentrations in the films and their distribution within the wafer sample are normally measured in a few points on the silicon wafer with the use of the in-line FTIR method(see Figure 2.2). Unfortunately, FTIR data are not that informative with respect to the additive, because they represent total additive values in the films taken from a relatively large sample area. Using this technique, it is possible to see the total additive content as well as its increase or decrease with respect to the standard (preliminary calibrated) sample.

The SIMS profiling technique is more informative for BPSG films. This method allows the discovery of some additive deviations from the expected uniform distribution (see Figure 2.9(b)). The typical BPSG thin film CVD process can be divided into three phases: process start, film growth and process finish. In terms of additive and oxidizer flows coming to the reaction chamber, the middle period is normally stationary. This indicates a uniform additive distribution within which the film thickness could be expected. However, the additives distribution within the film is uniform if the CVD process has been optimized properly. Some peaks or depletions of the additive concentration can be observed in the film near the film surface, orthe substrate surface(see Figure 2.9(b)). The first case corresponds to the process finish phase, when additive compounds are switched-off.The second case is the consequence of the process start phase when additive compounds are switched-on either separately, or together. The latter is normally the consequence of badly optimized additive injection in the beginning. This can happen, for instance, in the case of unexpected additive flow pulse due to the overpressure in the gas manifolds. The second case of intentional additive compounds switching-off was proposed in [109] as a simple solution of undesirable BPSG film moisture absorption phenomenon. Switching-off the additives in the last few minutes of the CVD process allowed the growing of a very thin protective silicon dioxide layer (frequently called "cap-layer") with significantly reduced additive content. This layer suppressed moisture interaction with the film and enhanced its stability (see more detailed discussion below).

It was found [224] that, in some cases, the behavior of additives distribution in the cap-layer can differ. For example, in SACVD BPSG films (Method 5), the boron concentration decreased much faster than the phosphorus concentration after additives compounds switch-off while depositing the thin silicon dioxide cap layer. In several cases, it was observed that the phosphorus concentration remained stable or increased slightly through that thin cap layer after compound switch-off. This is clearly shown in Figure 2.10(a). These data are in agreement with previous data for APCVD TEOS-ozone BPSG films deposited using Method 4 at atmospheric pressure [171]. Therefore, in accordance with the SIMS analysis, near the film surface of both atmospheric and subatmospheric pressure BPSG films, the phosphorus content is much higher than boron content.

In contrast, PECVD BPSG film grown according to Method 8 did not show a similar behavior (see Figure 2.10(b)). It can be seen that both additive concentrations in the film decreased. This was explained by the lower process pressure during PECVD film deposition. The lower pressure provided a considerably lower "time constant of the deposition chamber" of PECVD vacuum system as compared to SACVD and, especially, APCVD systems (this is analyzed in Part 1 of the monograph). Due to the low time constant, after compound switch-off, the residual concentration of additives in the gas phase is reduced much faster than that of SACVD conditions.

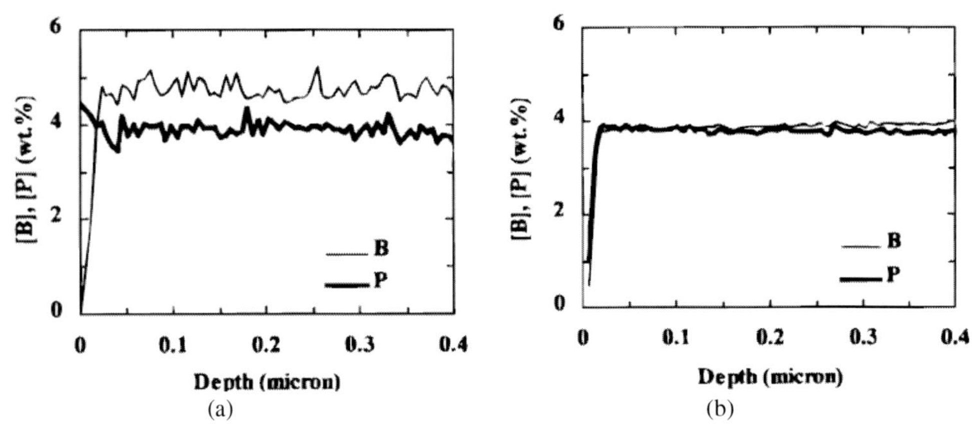

Figure 2.10. SIMS profiles of boron and phosphorus in as-deposited SACVD BPSG (a) and PECVD TEOS-oxygen BPSG film (b) [224]. Reproduced by permission of ECS – The Electrochemical Society.

Another observed specific feature of SIMS profiles in BPSG films deposited using Method 5, is the higher deviation of additives concentration in SIMS profiles as compared to PECVD TEOS-oxygen BPSG films (Method 8) (as can be seen in Figure 2.10(a,b)). A similar result was shown in [171] for APCVD BPSG film deposited using Method 4. Analysis of these differences in terms of standard deviations revealed about three to five times higher deviation values for SABPSG films than those values in PECVD BPSG films.

Film anneal at enhanced temperatures was found to cause some additive decrease (1.5 times) in a 200 – 300 Å region near the surface; the highest decrease of additives was found to be after steam anneal of BPSG films. In both cases, this depletion is considered to not be that important.

Additive state in the CVD glass films. According to classical chemistry, the complete oxidation of any boron and phosphorus compounds results in boron oxide B_2O_3 and phosphorus oxide P_4O_{10} (or usually simplified to P_2O_5). Normally, for practical CVD, an oxidizer is used in significant excess with respect to the reactant mixture. There has been no published data regarding the different status of boron in the glass films instead of B_2O_3. Thus, the CVD oxidation product of any boron compounds can be assumed to be the only B_2O_3.

The situation with phosphorus could be different because of many variations with the compounds used. The author would like to highlight one very important aspect of phosphorus analysis in the films, obtained using CVD Methods 1 – 3 [56,83,61] and PECVD Methods 6,7 with the use of N_2O as an oxidizer [52,53,85,256,288,290]. Analyzing the phosphorus state in the films deposited with the use of a strong oxidizer, such an oxygen and ozone-oxygen mixture, most of researchers showed that phosphorus was incorporated in the films in a form of phosphorus oxide P_2O_5 with phosphorus in P^{+5} state. However, depending on the PECVD conditions, phosphorus was found to exist in different forms: such as P_4O_6 (P^{3+}), P_4O_{10} (P^{5+}), PH_3 (P^{3-}), and phosphorus directly linked to a silicon (P^0) state. This observation was confirmed by the comparative measurements between the total amount of phosphorus in the films and less than that phosphorus oxide quantity in the films [63]. Using the ion chromatography method, the P_2O_3 additive in the films was detected with the content below 1 wt % [52]. The free phosphorus state (P^0) detected with the ESCA method as Si – P bond (130 eV) together with P – O bond (134 eV) was also reported in [322].

These different phosphorus states were due to the incomplete oxidation of compounds with different initial phosphorus states. For instance, Robles et al. [161] used $(C_2H_5O)_3PO$ as a compound in reaction with TEOS-O_3/O_2 mixture and, checking the content of P_4O_{10}, free phosphorus and P_4O_6 in the films, concluded that the content of the latter two was almost nonexistent. It could be predicted, as long as the possibility of P^0 or P^{3+} existence in SABPSG films seems to be impossible due to the P^{5+} state of phosphorus already in the TEPO compound molecule along with a strong oxidizing (ozone-oxygen mixture) ambient in the reaction chamber.

In summary, it is possible to conclude that careful optimization of the CVD process allows obtaining the stoichiometric BPSG film composition. It is important to note that post-deposition film anneals help to increase the concentration of the P^{+5} state in the film. The case of stoichiometric BPSG film compositions will be considered below, because PECVD Method 7 was used relatively rarely as compared to the others.

Carbon content. The presence of carbon impurities in the deposited films is an additional characteristic of vitreous materials prepared using the CVD technique and organic compounds. The carbon content was measured by SIMS. Its concentration was uniform within the film depth with a little increase near the film surface that can be explained by the film contact with the ambient.

The average concentration of carbon in SACVD BPSG films measured by SIMS was found to be less in films with a higher phosphorus content (see Table 2.4). For instance, the carbon concentration in the phosphorus-rich glass films (7 wt % [P] and 3 wt % [B]) produced by Method 5 was uniform over the film thickness. It was approximately three times lower than that in the films containing approximately equal concentrations of phosphorus about 4 wt. % (~ 1.05×10^{19} atoms/cm^3 and ~ 3.19×10^{19} atoms/cm^3, respectively).

It is necessary to note that thermally activated CVD processes with a relatively large excess of oxidizer with respect to compounds always produced a relatively low content of carbon.

In contrast, PECVD methods were found to cause a problem of IC device damage during its manufacturing known as "Plasma-Induced Damage", PID [323-326]. High radial charge non-uniformity was found for PECVD systems with diode types of reactors, particularly using SPV/SCA analysis. It was found to depend on the oxygen to silicon compound ratio. Using the SIMS method, it was found that the increase of O_2/TEOS mole ratio from 2 to 40 caused the exponential decrease of carbon content from ~ 9.5×10^{20} atoms/cm^3 до ~ 1×10^{20} atoms/cm^3 [327]. At the same time, the authors found an almost linear correlation of charge in the range of 0.1×10^{10} до 2.5×10^{10} ions/cm^2 in the silicon dioxide film with the increase of carbon content from 0.3×10^{20} до 5.5×10^{20} atoms/cm^3. As a result of the study, the authors have chosen a O_2/TEOS ratio to be at least 30 that was about 3 times higher than that required according to a scheme of proposed TEOS oxidation reaction.

Hydrogen content. Hydrogen was found in the glass films prepared by the referred CVD methods. Generally, the hydrogen concentration is uniform within film thickness. It increases noticeably when the deposition temperature becomes lower than 400 °C. Comparative study of hydrogen concentration in sequentially deposited CVD films used for ULSI technology was performed intentionally in [317]. For one-time SIMS hydrogen profiling, the following films were sequentially deposited in stacked structure: thermal oxide, LPCVD polysilicon (with following boron implant and anneal), CoSi, PECVD silicon nitride layer (PECVD - SiN), SACVD BPSG, PECVD silicon dioxide (PECVD SiO_2), HDP-CVD SiO_2 and PECVD - SiN. Experimental data are presented in Figure 2.11.

It can be seen that hydrogen concentration in as-deposited SACVD BPSG film (Method 5) containing 4.5 wt. % [B] and 4 wt. % [P] was found to be uniformly distributed having an average value as low as 3.0×10^{20} atoms/cm^3. This was found to be not higher than that of PECVD silicon dioxide film deposited using Method 8 without additives (about 6×10^{20}atoms/cm^3), or HDP-CVD SiO_2 film using Method 9 without additives (about 4×10^{20} atoms/cm^3).

After the film RTA at 850 °C for 30 seconds, hydrogen concentration was found to be about 1.3×10^{20} atoms/cm^3.

The hydrogen concentration in as-deposited and non-capped SACVD BPSG films stored in clean-room conditions for 3 weeks was found to be increased up to about 3×10^{21} atoms/cm^3, reflecting the moisture absorption effect. This observation corresponded to the adequate changes in FTIR spectrum moisture-related areas (see detailed discussion in Chapter 12). Thus, hydrogen can exist in the film either as free hydrogen, or in a hydroxyl group, or in the trapped water.

Table 2.4. Average carbon concentration in SACVD BPSG films (× 10^{19} atoms/cm^3) [224]. Reproduced by permission of ECS – The Electrochemical Society

Additives content	As-deposited	FA, N_2, 45 min	FA, O_2, 30 min
4.5 wt % [B] – 4 wt % [P]	3.19	0.92	0.72
3 wt % [B] – 7 wt % [P]	1.05	0.36	0.17

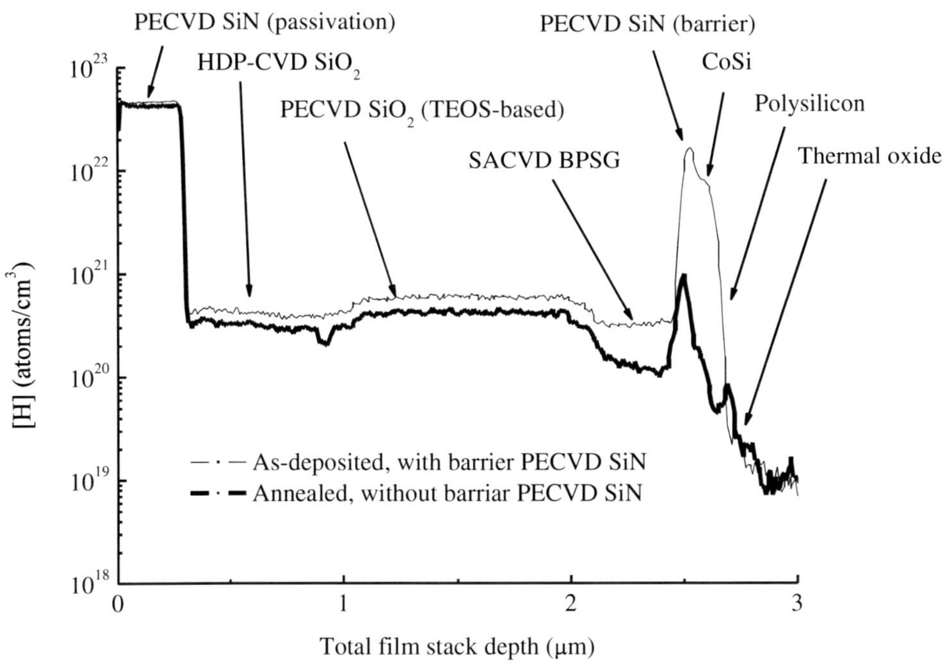

Figure 2.11. Hydrogen concentration profile in studied stacked films. Presented at 17th Intertn. VLSI Multilevel Interconnection Conf. (VMIC), 2000 [317]).

Chapter 9

BPSG FILM STRUCTURE

In Part 1 of the monograph, we analyzed some aspects of BPSG film growth under a variety of CVD methods and process conditions. It is important to emphasize that the CVD technique is not a usual glass forming technique. Its major feature is the formation of a complete glass structure during film deposition in the additive ranges presented in Figure 2.1. On one side, the CVD temperature is high enough to form the film structure (> 400 °C). However, on the other side the post-deposition anneal in device technology is performed at a lower temperature than that of glass crystallization temperature and there is a trend to soften the anneal conditions. Thus, the as-deposited BPSG structure can be considered as a relatively stable glass structure consisting of Si, B, P, O atoms.

Note that the summary presented in Chapter 8 showed that the concentrations of carbon and hydrogen in CVD glass film are low enough to affect the glass structure. Therefore, these impurities can be simply ignored for glass structure considerations. However, BPSG film properties differ from one glass type to another one due to the difference in CVD features, such as deposition conditions and the chemical compounds used. BPSG as a ternary glass is a complicated glass material.

Therefore, basic glass composition (quantity of Si, B, P, O) is a very important factor strongly affecting film properties. In fact, it was shown that film properties depend on phosphorus and boron oxide concentrations and their ratio in the film.

Despite the active and long application of BPSG films in IC technology, BPSG film structure-related data have been published very rarely. To date, there are only a few summaries about this matter written by the author [2,269,270,276,281]. In Chapter 9, spread related information found in a wide range of literature sources is consolidated. Some data have also been taken from classical glass books and encyclopedias [328-331], as well as from chemical literature [332-338].

The latter helps to understand the basics of the glass ingredient structures. Rare information about boron-phosphorus contained glass compositions was collected basically for bulk glass compositions. This summary helps to explain the peculiarities of the observed BPSG film properties and their dependence on the CVD methods from the structural point of view.

9.1. Brief Review of Silica Structure: Basics and Terms

There are two main classes of solids that are named crystalline and amorphous. In addition to the terms amorphous, solid and glass, other terms are non-crystalline solids and vitreous solids. Amorphous solids and non-crystalline solids are more general terms, while glass and vitreous solids have been historically reserved for an amorphous solid prepared by the rapid cooling of a melt. In fact, according to classical chemistry, glass is considered as a kind of supercooled liquid formed from mixed oxides by the melt cooling method. The difference between crystalline and amorphous solids is the nature of their atomic-scale structure.

The tendency of various substances to form glass was explained by Zachariasen in 1932 [332] in terms of the possibility of specific linking of the elementary structural polyhedra. A continuous random-network glass model that he developed assumed that glasses have a disordered spatial structure. According to Zachariasen, in order for a given oxide A_mO_n to form a glassy solid, it must meet the following criteria: a) the oxygen should be linked to no more than two atoms of A; b) the coordination number of the oxygen about A should be small, on the order of 3 or 4; c) the cation polyhedra must share corners only and d) at least three corners must be shared.

In vitreous silica, the elementary structural polyhedron is a SiO_4 tetrahedron with a silicon atom in its center. Thus, in amorphous SiO_2, each silicon atom is bonded to four oxygen atoms, and each oxygen atom is bonded to two silicon atoms. The nature of bonding in this structure is covalent, bond energy is estimated to be about 434.7 kJ/mol, and the distance between Si and O atoms in tetrahedron is 1.61 Å [338]. All SiO_4 tetrahedra are bonded by their vertices by means of Si – O – Si bonds resulting in the formation of a three-dimensional spatial network. Using Myuller's structural group symbols approach [336], this elementary glass structure is represented by group symbol $SiO_{4/2}$. Therefore, the silicon atom has a fourfold coordination with respect to oxygen and the nature of bonding in this structure is covalent. The network structure is useful to introduce very important terms: short-range order and medium-range order. Atomic positions in a crystal exhibit a property called long-range order or translational periodicity; positions repeat in space in a regular array. In an amorphous solid, translational periodicity is absent, i.e. there is no long-range order. In the glasses each atom has three nearest-neighbor atoms at the same distance (called the chemical bond length) from it, just as in the corresponding crystal. Both crystalline and amorphous solids exhibit short-range order, which is a consequence of the chemical bonding between atoms and which is responsible for holding the solid together. The length of short order (in terms of its radius) is about 1.6 – 2.8 Å. As for medium-range order, its radius is estimated to be 2.8 – 5.5 Å [338]. This term characterizes the structural disordering, particularly, showing the reason of a porosity of a material.

9.2. Brief review of Boron Oxide and Phosphorus Oxide Structures

This has been illustrated by the use of a simple silica vitreous state structure. Chemical species such as phosphorus, boron and germanium (which like silicon enter into the structure of the network by forming strong chemical covalent bonds with oxygen atoms) are called network formers. Another chemical species such as sodium and calcium, which do not bond directly to the glass network, are called network modifiers. This case is out of our consideration. It has been mentioned above that in vitreous silica the elementary structural polyhedron is a SiO_4 tetrahedron, where the silicon atom has fourfold coordination with respect to oxygen. However, P_2O_5 and, especially, B_2O_3 structures are different. Information about boron- and phosphorus silicate glass structures is related to bulk glass compositions. Until recently, there has been no data about thin film glass structures. We believe that the borophosphosilicate glass structure can be understood based on the separate analysis of SiO_2, B_2O_3 and P_2O_5 structures described for bulk glasses. This allowed us to collect information about these oxides followed by the creation of the ternary glass structure. The latter is believed to be dependent on glass composition. The brief review below is a consolidation of data from [328-338].

The phosphorus oxide network consists of PO_4 tetrahedra, where every PO_4 tetrahedron is bounded to three other tetrahedra. PO_4 tetrahedron has three P – O (1.62 Å) and one P = O (1.39 Å) bonds, and the bond energy is about 339.4 kJ/mol. The elementary structure of vitreous phosphorus oxide, usually represented by chemical formula P_4O_{10}, is expressed by group symbol $4(O^-P^+O_{3/2})$ [336].

Boron oxide is a significantly more interesting ingredient of BPSG. Considering the physical and chemical properties of B_2O_3, it is necessary to highlight the following. Many physical properties of B_2O_3 are sensitive to a small amount of residual water and thermal history. Boric oxide exists as vitreous glass and as two crystalline forms. The vitreous form is a colorless, hard, tough, glass like solid, commonly prepared by the dehydration of boric acid. B_2O_3 is an excellent Lewis acid: it coordinates even weak basis to form fourfold coordinate borate species. Vitreous B_2O_3 does not crystallize in the absence of seed crystals or increased pressure. In vitreous boric anhydride, the network is believed to consist of two structural units forming an almost planar or layer-type structure of continuous B – O – B covalent bonds. The distance between layers is about 4.3 Å. The first structural units are random trigonal $BO_{3/2}$ units with a boron atom in their centers, which are linked with neighboring ones by three oxygen atoms. The second one is so-called boroxol $(- B - O -)_3$ rings. The distance B – O in boroxol groups is about 1.37 Å and the bonding energy is about 469.3 kJ/mol. The elementary structure of vitreous boron oxide B_2O_3 is represented by group symbol $2(BO_{3/2})$.

There is a certain disagreement over the relative amounts of these two structural units that are due to the presence of residual moisture. It is believed that approximately 2/3 of the boron atoms form the boroxol rings. This network breaks down as the glass melts. It was proposed that above 800 °C, the liquid consists of discrete, but strongly associated, small molecules, conceivably the same monomeric form of B_2O_3 units that are observed in the vapor phase. The most common crystalline B_2O_3 form has a hexagonal or α-B_2O_3 and is crystallized at 200 – 250 °C. A less common monoclinic β-form can be crystallized at high pressure 4000 MPa and 600 °C. The beta-form consists of a highly compact network of BO_4 tetrahedra. The acidic character associated with trigonal BO_3 groups is thus masked in β-form B_2O_3. For tetrahedral-type boron-contained glasses with alkali modifiers, the elementary structure unit is designated by group $B^-O_{4/2}$. The distance between boron and oxygen atoms in tetrahedron is about 1.48 Å. According to Nemilov [338], combining $BO_3/_2$ and $B^-O_4/_2$ groups results in the formation of complex groups, such as pentaborate $[B_5O_8]^-$, tribotrate $[B_3O_5]^-$, etc. Thus, based on a brief review of the literature, the preferable effect of the boron oxide additive and its structure on the BPSG glass structure and properties is undoubtedly expected.

9.3. BPSG Composition and Structure

Considering structures of boron- and phosphorus-contained silica is needed to highlight some important features presented in the literature [333-335]. The studied range of BPSG film compositions (expressed in elemental wt % [B] and [P]) is shown by the data points in Figure2.1. Normally, the total boron and phosphorus content in the glass films did not exceed 12-14 wt %. The reason of this in case of BPSG films is the following. Englert and Hummel studied bulk $SiO_2 - P_2O_5 - B_2O_3$ compositions at high temperatures and found the crystallization areas in the diagram presented in Figure 2.12(a). Thus, boron phosphate BPO_4 compound can form at high temperatures in the mixture of oxides.

According to the reference books, boron phosphate is a crystalline material with a tetragonal structure and a sublimation temperature of 1600 °C. Under normal conditions, boron phosphate exhibits a cristobalite structure with BO_4 and PO_4 alternating tetrahedral united by the sharing of oxygen atoms to form a three dimensional structure. BPO_4 can be prepared by different synthesis procedures, in particular, by heating a boric and phosphoric acid mixture at temperatures below 1000 °C [328,331]. According to [337], in the glasses of the $BPO_4 - SiO_2$ system, crystalline boron phosphate is formed at 950 – 960 °C and silicon oxide does not crystallize in this case. Therefore, a high annealing temperature is a necessary condition for the formation of BPO_4. It should be noted that post-deposition high-temperature BPSG film thermal anneal is a technology requirement to provide glass softening and flow. Therefore, taking into account the BPO_4 formation possibility and BPSG film composition should be carefully optimized in order to avoid BPO_4 crystallization area.

Data presented in Figure 2.12(a) were re-calculated by the author in a convenient form (see Figure 2.12(b)). This graph clearly defines the main area of BPO_4 crystal formation. Comparison of this graph with data presented in Figure 2.1 allows the drawing of Figure 2.13, where all studied BPSG film compositions are plotted together with BPO_4 crystallization area. Most of the studied BPSG film compositions lie quite close to the BPO_4 boundary. This is mainly due to the necessity to provide the best film softening and, therefore, the film planarization on IC device stepped surface, especially at low anneal time, like RTA. However, these compositions were prone to form BPSG defects, either on the surface, or inside the film. This matter will be discussed below in detail in Chapter 13.

Considering data for bulk borophosphosilicate glasses, it is necessary to take into account the results of research performed by Kolesova [316,337]. Using IR spectroscopy, she investigated borophospate and highly concentrated borophosphosilicate bulk glasses with different compositions. It was shown that for boron-rich compositions, the glass structure consisted of triangles and boroxol groups. In both cases, boron had threefold coordination. It is very important that graduated adding of P_2O_5 to the B_2O_3 or $SiO_2 - B_2O_3$ mixture resulted in adequate changes to the glass structure. The boron oxide network become broken and the boron changed coordination from threefold to fourfold with respect to oxygen atoms. Eventually, the boron phosphate BPO_4 glass structure (equal concentration of B_2O_3 and P_2O_5) was characterized as a tetrahedral-based PO_4 and BO_4 network.

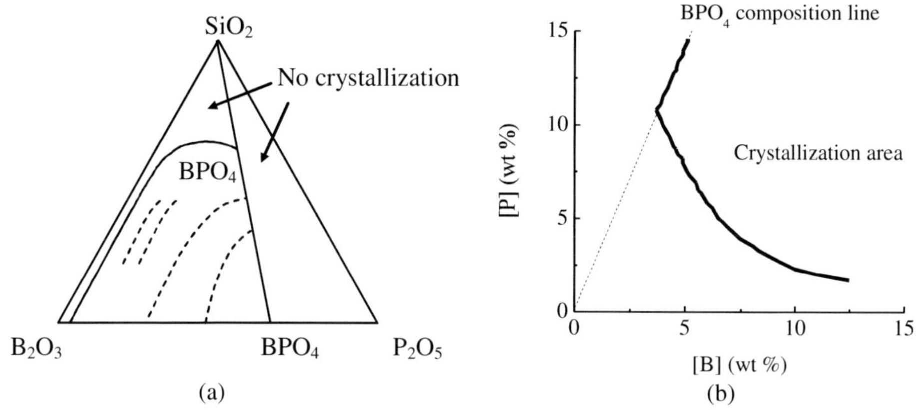

Figure 2.12. Diagram for bulk SiO2 – P2O5 – B2O3 system (re-drawn with simplification from Ref. [334]) (a) and its re-calculation into a convenient format made by the author (b). Dashed lines in Figure (a) from left to right represent temperatures 1000 °C, 1100 °C, 1200 °C, and 1300 °C, accordingly.

Similar data were obtained by Aitken et al [284], who studied ternary bulk BPSG samples prepared from mixtures of previously prepared 30 % P_2O_5 – 70 % SiO_2 and 30 % B_2O_3 – 70 % SiO_2 glasses. Using the Nuclear Magnetic Resonance (NMR) technique, the authors showed the presence of a very small percentage (≤ 3 %) of the total Si in the form of both fifthfold and sixfold coordinated species in the most P-rich glasses. The same glasses revealed an axially symmetric $O = PO_{3/2}$ species, whereas the spectra of B-rich glasses showed only symmetric $PO_{4/2}$ tetrahedra. As boron was progressively replaced by phosphorus, B coordination changed monotonically from fully trigonal to dominantly tetrahedral, with the tetrahedral B fraction ranging from 80 to 95 % for the most P-rich glasses. Overall, the results are consistent with a structural scenario of associated $BO_{4/2}$ and $PO_{4/2}$ tetrahedra comprising a fully polymerized silica-like network for the glasses of intermediate composition. The greater degree of network connectivity was caused by the preferential association of B and P to form BPO_4-like groups.

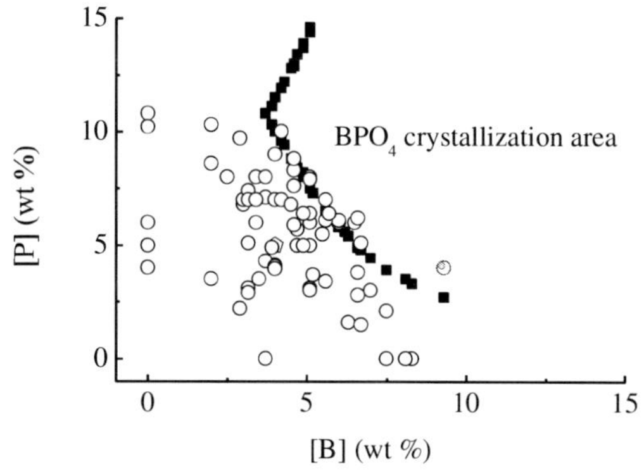

Figure 2.13. Correlation of studied BPSG thin film compositions (open circles) with the BPO_4 crystallization boundary taken from Figure 2.12(b) [269]. Reproduced by permission of ECS – The Electrochemical Society.

9.4. Differences in BPSG Composition and Structure

Since studies concerning structures for glass thin film materials are absent in the literature, the hypothetical structures of borophosphosilicate glasses are discussed below with allowance made for the data for bulk borophosphate and borophosphosilicate glasses enriched in phosphorus and boron oxides. In this context, the experimentally found that composition dependence of the reactivity for the borophosphosilicate glass films (for example, with respect to the air moisture) is explained by the corresponding dependence of the fraction of boron atoms threefold coordinated with respect to oxygen in the glass network, as well as the specific film structures presented in Figure 1.26.

We select and analyze three important BPSG glass compositions, shown in Figure 2.14, defined as follows: [B] >> [P], [B] $\approx$ 0.4 [P], [B] << [P].

The first type of glass composition involves borophosphosilicate glass films strongly enriched in boron, i.e. [B] >> [P]. In the structure of monolithic glasses with high boron content, boron atoms are in a threefold-coordinated state with respect to oxygen. The glass structure can contain different groups, such as B_3O_6 boroxol rings and BO_3 triangles. Note that according to the classical literature, pure boron oxide has a layered structure with an interlayer distance of approximately 4.3 Å. For monolithic borophosphate glasses, it was experimentally demonstrated that a gradual increase in the concentration of phosphorus oxide in a batch (and, hence, an increase in the phosphorus concentration in a glass) leads to the destruction of boroxol rings and triangles. In this case, threefold-coordinated boron with respect to oxygen transforms into fourfold-coordinated boron. Therefore, it is reasonable to assume that thin borophosphosilicate glass films enriched with boron should have a different structure due to the formation of the glass structure containing boron in the threefold-coordinated state with respect to oxygen.

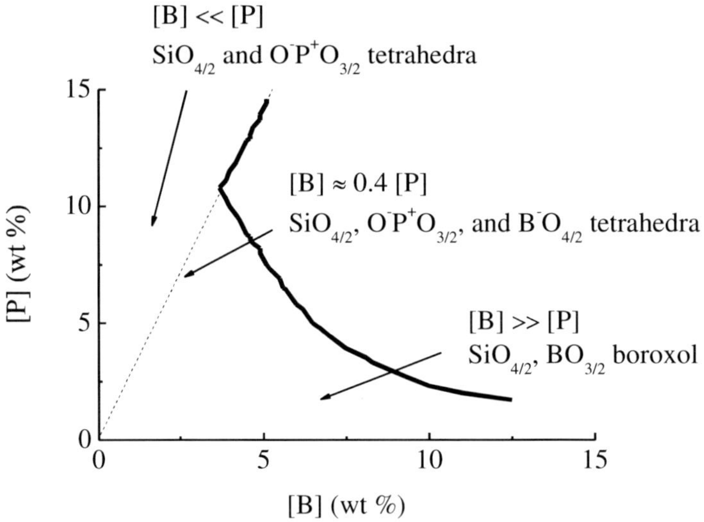

Figure 2.14. Indication of proposed BPSG structures for three selected BPSG compositions (see text for details) [269]. Reproduced by permission of ECS – The Electrochemical Society.

Based on these properties, the largest structural differences are the feature of CVD Methods 4,5. These methods were characterized by having the smallest effective constant k_{eff}. The formation of these films is described by the two stage scheme of sequential reactions with the limiting heterogeneous stage of consumption of intermediate products of the reaction between tetraethoxysilane, triethyl borate, and ozone. Apparently, these are structures with threefold-coordinated boron which provide a way of penetrating water into the bulk of boron-rich glass films. In particular, this reasonably explains the experimental data regarding moisture penetration in the glass films obtained in [178]. A simple comparison of the effective size of a water molecule (the distance between hydrogen atoms is approximately equal to 1.5 Å) with the hypothetically maximum interlayer distance in the structure of boron oxide (4.3 Å) confirms the consistency of the above assumption.

It is known that according to the Lewis theory, boron oxide is a hard acid and coordinates even soft bases with the formation of structures with a tetrahedral coordination. On this basis, a high reactivity of the boron-rich borophosphosilicate glass films with respect to water vapor could be explained by the fact that threefold-coordinated boron atoms $(\equiv Si-O-)_3B$ in the glass structure serve as acceptors of electron pairs. The water molecules, which are bases according to the Lewis theory, play the role of donors of electrons pairs. However, it is known that water oxygen is not a good donor of electron pairs and cannot form a donor-acceptor bond with boron and transform it into the tetrahedral coordination. Specifically, metaboric and orthoboric acids are the sole products of the addition of water to boron oxide. Therefore, the existence of stable $(\equiv Si-O-)_3BOH_2$ structural groups, in which boron would have a tetrahedral coordination with equivalent bonds between the boron atom and all four oxygen atoms, seems to be unrealistic. Moreover, according to [178], water is weakly bound and tends to migrate from the surface of the glass film into the bulk, i.e., from one boron atom to another boron atom. In this respect, we can assume the formation of a $[(\equiv Si-O-)_3B] \bullet H_2O$ group, in which water exists in a way similar to crystal hydrates.

The introduction of phosphorus compounds into the mixtures of alkoxy compounds and oxygen leads to a multifold increase in the rate of oxidation reactions. This can be characterized by an increase in the effective constant k_{eff} and to an increase in the effective density of borophosphosilicate glass films. This increase is proportional to that in the phosphorus concentration in the films. A decrease in the absorption of water by borophosphosilicate glasses with an increase in the phosphorus concentration is explained by the simple decrease in the number of threefold coordinated boron atoms with respect to oxygen.

The second type of glass compositionis the most interesting. It includes glasses characterized by a low reactivity with respect to the moisture at the concentration ratio [B]/[P] $\approx$ 0.4. This ratio corresponds to equivalent contents of P_2O_5 and B_2O_3 oxides in borophosphosilicate glass. Note that this is close to the BPO_4 composition line in Figure 2.14. It is assumed that all boron atoms in the glass are fourfold coordinated with respect to oxygen. The structure of this glass is similar to those of silicon oxide and boron phosphate. It can be represented as a three dimensional random network composed of SiO_4, BO_4, and PO_4 tetrahedra. In PO_4 tetrahedra, three bonds are bridging with Si atoms and the fourth bond is the double bond. The structural group of this glass in which boron and phosphorus atoms are linked by oxygen bridges can be written in the following form $(\equiv Si-O-)_2OP-O-B(-O-Si\equiv)_3$. It is obvious that a glass of this type has a much better ordered structure compared to glasses enriched with boron. This implies that the glass of the composition under consideration insignificantly absorbs the air moisture and weakly interacts with it.

The third type of glass composition involves glasses enriched with phosphorus, i.e. [B] << [P]. Phosphate glasses have a structure in the form of a three-dimensional random network comprised of SiO_4 and PO_4 tetrahedra. The interaction of water vapors with glasses can be described in the same manner as in the above case.

Therefore, in light of proposed BPSG film structure-composition correlations, the properties of the borophosphosilicate glass films studied can be determined by several factors. Among them are:

- kinetic features of the chemical vapor deposition methods, which are responsible for the porosity of the synthesized glass films, as shown in Figure 1.26;
- occurrence of a certain number of boron atoms in the threefold coordination with respect to oxygen in the glass structure;
- total oxide concentration in the glass;
- concentration and sizes of phosphorus oxide clusters embedded into the glass structure.

The heat treatment of glasses at temperatures above their softening point leads to a decrease and disappearance of pores and the incorporation of clusters into the glass structure and their evaporation from the films. As a consequence, after heat treatment at high temperatures, BPSG films prepared by different CVD methods that have identical compositions become similar in their film properties.

Chapter 10

BASIC BPSG FILM PROPERTIES

The properties of BPSG films were reported in many publications. There were a few examples of comprehensive characterization of BPSG film properties [50,52,78,93,110,129,161,207,224,245,269,270,276,281,289]. In this chapter, we briefly describe the most common results of studies of BPSG film properties. It is done for the main CVD processes used over the years of BPSG film implementation in IC device technology. Some details regarding CVD processes are presented in Table 1.5 for CVD processes and Table 1.6 for PECVD processes. BPSG films used in IC technology usually require additional post-deposition thermal anneal either for the film planarization, or for the film densification. Therefore, the film properties after the film anneal are also important for consideration and anneal conditions are also indicated below.

10.1. SURFACE MORPHOLOGY OF THIN GLASS FILMS

The film morphology is evaluated quickly by eye-control characterizing the film surface as glance, mate, dusty, etc. An opportunity to qualify the film surface morphology numerically appeared at the beginning of 1990 due to the development of the Atomic Force Microscopy (AFM) technique (see Chapter 7).

To describe the BPSG film surface morphology, the author used a Digital Instruments NanoScope III Large Stage Tapping Mode Atomic Force Microscope system. The surface roughness of different types of BPSG films was characterized [224,225]. The surface roughness, defined as the root mean square (RMS), was measured by taking the average roughness of several scans measured under various length scales, from 0.5 to 2 μm. The film surface roughness was found to be the basic physical difference of as-deposited thin film morphology. To obtain this conclusion, films with various additive concentrations and thickness of about 500 – 600 nm were deposited using APCVD, LPCVD and PECVD methods. The roughness was also investigated for the films annealed in a vertical furnace at temperatures of 850 °C and 900 °C for 30 minutes and 12 minutes, respectively.

Table 2.5. Surface roughness (RMS) and deposition rate of $SiO_2$and glass films deposited by various CVD мethods

CVD Method	Additive content [B] - [P] (wt %)	Thickness (nm)	W_d (nm/min)	Roughness RMS (nm) As-deposited	Roughness RMS (nm) Annealed 850 °C
2	[4] – [4]	600	N.A.	12.39	1.13
3	SiO_2	500	10 ± 2	2.51	-
	6 % BSG	500		1.10	-
	4 % PSG	500		2.34	-
	[2] – [3.5]	580		1.98	1.42
	[2.8] – [2.2]	600		2.50	-
	[3.7] – [4.3]	1050		2.83	1.34
	[3.3] – [6]	560		1.86	-
	[3.3] – [7]	440		1.84	0.40
8	SiO_2	600	300	0.53	-
	7.2 % BSG		396	0.39	-
	6 % PSG		685	0.69	-
	[2] – [7]		775	0.77	-
	[4.4] – [6.3]		980	0.85	-
	[5.9] – [5.5]		920	0.84	-
	[6.7] – [3.2]		790	0.84	-
4	5 % BSG	500	84	0.65	-
	5 % PSG		102.5	1.20	-
	[4] – [4]		140	1.43	0.24
	[3] – [5]		130	2.38	0.32
	[5] – [3]		120	1.80	0.23
	[3] – [7]		145	3.22	0.24
5	SiO_2	600	Figure 2.15(a)	Figure 2.15(b)	-
	6 % BSG		200	0.45	-
	6 % PSG		245	0.73	-
	[3] – [5]		370	0.72	0.35
	[3] – [6]		380	1.22	0.21
	[4] – [5]		400	0.55	0.18
	[5] – [3]		340	0.70	0.19
	[5] – [4]		420	0.69	0.18
	[5] – [5]		513	0.89	0.17
	[6] – [3]		410	1.35	0.27
	[6] – [6]		470	1.35	0.42
	[10] – [3]		340	0.72	0.81

Correlation of film surface morphology with film deposition kinetics. The measured surface roughness values of the films produced at different deposition conditions are summarized in Table 2.5. Analysis of these values allows the highlighting of the following features for as-deposited films that can be used for additional confirmation of the proposed CVD reaction schemes.

In general, it was confirmed that the silane-based chemistry (Method 2) produced films with higher surface roughness than that of TEOS-based chemistry (Methods 3,4,5,8). The silane-based films deposited at a low pressure revealed RMS values six to seven times lower than the films produced at atmospheric pressure conditions. Films deposited at atmospheric pressure had the highest RMS values and these values were comparable to those obtained in [131]. The high RMS value was explained by the difference in silane concentration used for deposition. In order to obtain a higher deposition rate, the concentration is about ten times higher at atmospheric pressure CVD conditions than under low pressure conditions. It is known that for both atmospheric and low pressure conditions the increase of the silane concentration causes the silica micro-particle formation due to undesirable gas-phase processes [3].

It is also evident from the results in Table 2.5 that the BSG film surface roughness is 1.5 – 2 times lower than that of PSG, BPSG and SiO_2. This is consistent with the conclusion made in Part 1 that boron compounds were found to suppress the deposition rate of silane-based films (see Chapter 3). It is therefore likely that gas phase reactions were also reduced and, as a result, boron-contained films appear smoother.

The TEOS-based glass film revealed lower RMS values than the silane-based oxide, despite the similarity of the deposition rates at atmospheric pressures. However, at a reduced pressure, the deposition rate of TEOS-based glass films was significantly higher than the silane-based glass films. In the case of PECVD deposition, the RMS values of the BSG film were lower than all other types of films. Comparing data obtained for TEOS-ozone deposition chemistry in a single-wafer chamber for SACVD and APCVD conditions, BSG surface roughness was also found to be lower than other glass films. In this case, it is also obvious that boron suppressed gas-phase reactions, which led to smoother films. In contrast, there was a significant difference between the roughness of SiO_2 and glass films.

The roughness of SiO_2films were also reported to be strongly dependent on the CVD process conditions. In case of silane-based chemistry, an increase of the process temperature, pressure, and silane concentration, either separately, or together, tends to increase the film deposition rate [1-3]. This always causes an increase in the micro-particle formation phenomenon in the gas-phase [2]. As a consequence, the film surface becomes rougher. This silane-based film feature looks obvious. However, in the case of TEOS-based chemistry, the roughness of SiO_2films was also reported to be strongly dependent on the process conditions [225]. Figure 2.15(a) shows that the increase of TEOS flow caused an increase of the deposition rates. At the same time, the increase in the TEOS flow rate also increased the surface roughness to a maximum at a flow rate of 600 mg/min and then decreased gradually at higher flow rates (see Figure 2.15(b)). The maximum rms value corresponded to the highest surface sensitivity effect (the difference between the film deposition rate on silicon and silicon oxide substrate), as shown by the dotted line in Figure 2.15(a). In contrast, under similar process conditions, glass films revealed significantly lower rms values, but at higher deposition rates (see Figure 2.15(a,b)). Moreover, both additives in any combinations were found to suppress the effect of surface sensitivity. The reason behind this was explained in

Part 1 of the monograph (see Figure 1.25 and Figure 1.26). The reaction mechanism quickly changes due to the presence of additives in the gas-phase. These changes mainly explain the appearance of new gas-phase limited reaction routes for glass film growth. These reaction routes are the result of phosphorus compound oxidation in the gas phase in accordance with the chain-brain mechanism resulting in the formation of highly active intermediate products.

Results in Table 2.5 show that for all oxidation reactions studied, boron compounds caused slight or no improvement to the deposited film surface morphology even with respect to SiO_2 films. Thus, there is no evidence that boron compounds altered the reaction mechanism from heterogeneous to that of gas-phase limited even when the whole deposition process was accelerated. In contrast, all phosphorus containing films showed higher surface roughness values. This confirms a change of reaction mechanism from surface-limited to gas phase with material species formation in the reaction volume.

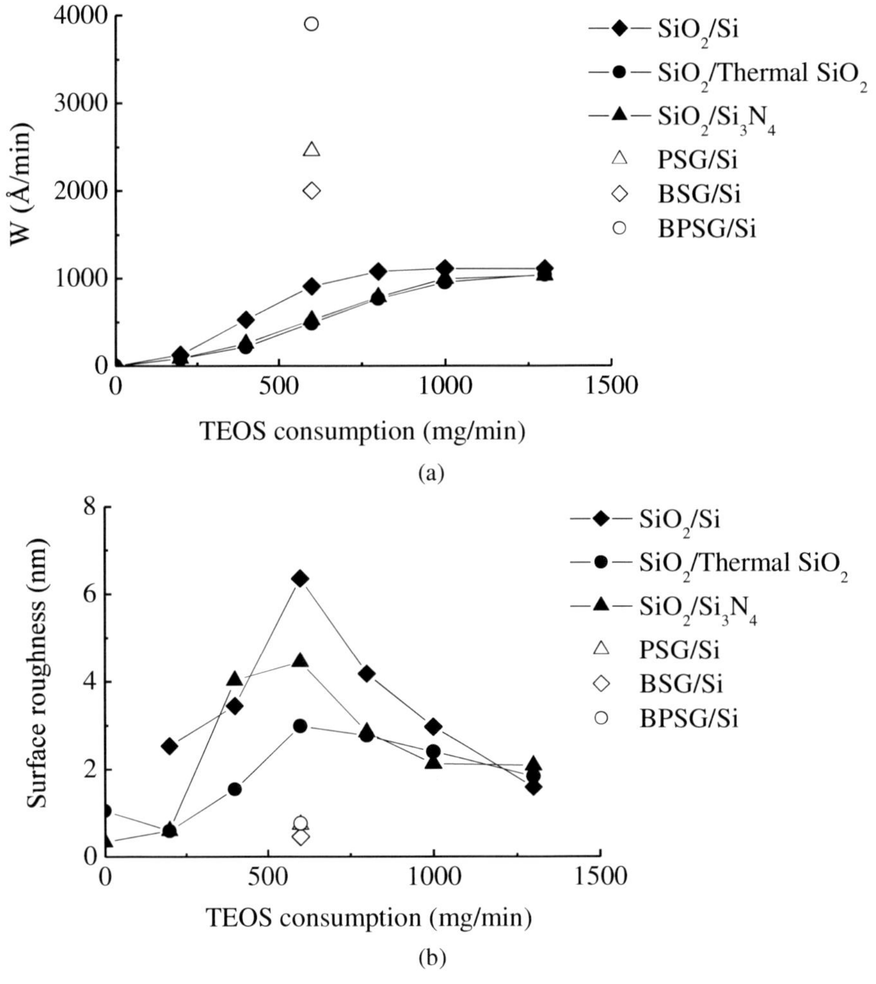

Figure 2.15. Deposition rate (a) and surface roughness (b) of SACVD TEOS-ozone SiO_2 and glass films vs. TEOS flow (re-drawn based on Ref.[225]).

In addition, the comparison of RMS data in Table 2.5 shows that the surface roughness of SACVD TEOS-ozone films is about 2.5 – 3 times less than that deposited at APCVD. It is necessary to note that the deposition rates of APCVD BPSG films were about 3 times less than those of SACVD BPSG films. This observation is in contrast with that observed in silane-based chemistry. This difference was explained by the implementation of different carrier gases (He at SACVD deposition and N_2 at APCVD deposition). According to [184], this produced differences in the diffusivity of intermediate gas-phase clusters to the film surface, which subsequently combined in the gas-phase to form large species. Another reasonable explanation lies in the longer residence time of the APCVD process that increased the impact of the gas-phase reaction in the total deposition process (see comparison in Figure 1.19). The residence time value of the APCVD condition is reported to be about 5 times higher than the SACVD condition. In this case, it would seem that SACVD deposited films are better in terms of a smoother surface.

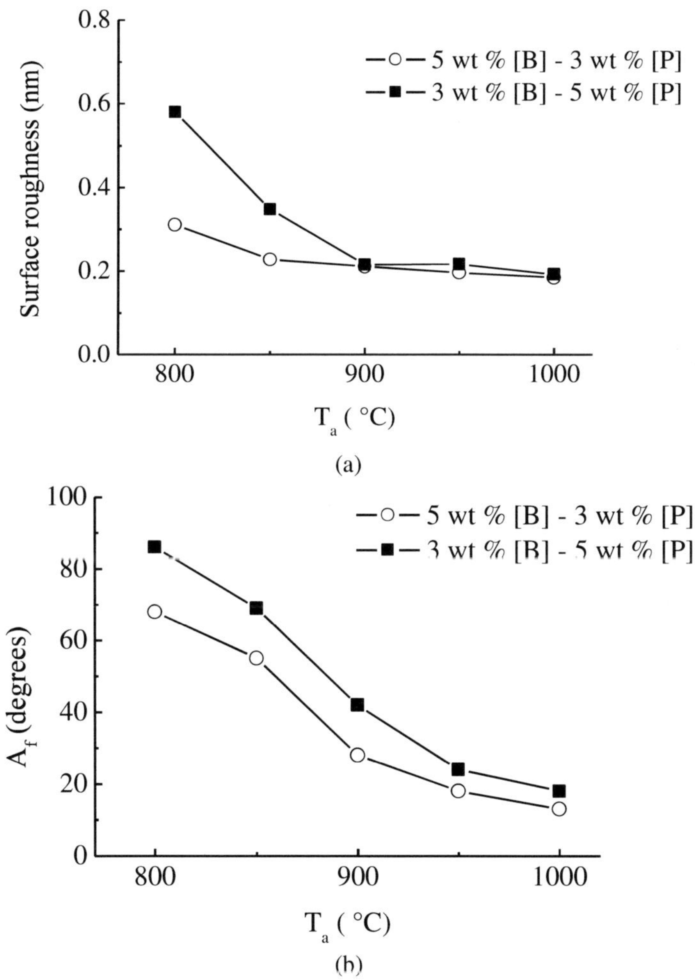

Figure 2.16. Surface roughness (a) and flow angle (b) of atmospheric pressure TEOS-ozone BPSG films after 30 minutes anneal in argon ambient. (re-drawn based on Ref.[225]).

Changes of BPSG film morphology with thermal annealing. BPSG film anneal at temperatures higher than the softening temperature causes the film flow and, therefore, subsequently smoothens the edges of device structures covered with flow-able glass film (see Figure 1.1). Flowing of the BPSG film was found to also change the film surface morphology and to produce a smoother film. The correlation between the BPSG film flow angle on the device step and the surface roughness of annealed BPSG films observed in this investigation are presented in Table 2.5.

The changes of BPSG film morphology after the film annealing can be clearly seen in Figure 2.16(a,b). This figure shows the correlation of the surface roughness and the flow angle values A_f with the increase of the anneal temperature. Using the AFM technique, the annealed BPSG film surface roughness was estimated to be about 3 – 4 times lower than that the as-deposited films. These data clearly demonstrate the planarization and smoothening of film surfaces by annealing at elevated temperatures.

10.2. Film Mechanical Stress

Film stress measurements can be carried out using different tools. For instance, the FSM 128 tool by Frontier Semiconductor Measurements, Inc. was used in the author's studies. Low-temperature SiO_2-based CVD films tend to be in tensile stress as deposited and the addition of phosphorus oxide or boron oxide to CVD SiO_2 films is known to decrease tensile stress [50]. Similarly, as deposited at 430 °C CVD, BPSG films were in very low tensile stress in the range of 9×10^8 dyne×cm^{-2}. However, as-deposited PECVD films are known to reveal slightly compressive stress values 8×10^8 dyne×cm^{-2} [129] and $< 0.5\times10^9$ dyne×cm^{-2} [95]. We also found that, for instance, for two popular production processes designated above as Method 8 and Method 9, typical stress values for as-deposited BPSG films were in the range of $(4.3 \div 6.5)\times10^8$ dyne×cm^{-2} (tensile), but $(4.0 \div 6.0)\times10^8$ dyne×cm^{-2} (compressive), respectively. Note that the compressive stress values in these PECVD films were adjusted by adjusting RF power density [2].

However, considering BPSG film PMD applications, one can realize that as-deposited film stress value is not that important which is due to the film high temperature annealing processing used in the main IC device technologies. For instance, post-deposition RTA at 850 °C for a couple of minutes of mentioned above BPSG film deposited using Method 8 and Method 9 allowed eliminating stress differences due to the film flow; stress in both cases become compressive at $< 1\times10^9$ dyne×cm^{-2}. Similar low compressive stress values $< 8\times10^8$ dyne×cm^{-2} were found for furnace annealed APCVD BPSG film deposited using Method 1 in a wide range of process parameters [57].

10.3. Film Density – Film Porosity – Film Shrinkage

Until recently, there was no direct published data regarding any correlation of the BPSG film density with the change of CVD process parameters. This is possibly due to insufficient film thickness for precise density evaluation. According to some references, silicon dioxide thin film density was estimated to be around 2.25 g/cm^3 that is typical quartz value. It is necessary to note that the film density value itself does not really play an important practical role in IC technology. On the other hand, the reciprocal parameter of this quantity - the film porosity - plays a very important role in some applications, for instance, for low dielectric constant films (the latter is very important in modern technology, but this is far from the BPSG film consideration in this monograph).

For practical IC technology, it was found that the heat treatment makes it possible to differentiate thin glass films in their effective density (or the film porosity). This can be indirectly judged from the shrinkage of silicate glasses and their dissolution rates in aqueous solutions of hydrofluoric acid. The shrinkage (difference between post annealed and initial film thickness divided by the latter, %) was found to be in the range of a few percentage points. The highest shrinkage (up to 5 – 7 %) was found for TEOS-ozone silicon dioxide and borosilicate glass films prepared by Method 4 and Method 5 that goes along with so-called "surface sensitivity effect" [99,101,106,120,121]. The lowest shrinkage value < 0.1 % was obtained for the films synthesized by plasma-activated methods: Method 8 (PECVD) and, especially, Method 9 (HDP-CVD) [298-301]. It was shown [281] that the shrinkage of the phosphorus-rich films with a ratio [B]/[P] < 0.5 is approximately two times less than the shrinkage of the films with [B] ≥ [P]. This indicated a somewhat higher density of the former films. A higher porosity of the films synthesized by Methods 4 and 5 is indirectly evidenced by higher dissolution rates of films prior to and after heat treatment.

10.4. Film Wet Etch Rate

Since the first BPSG-related paper appeared [50], the BPSG thin film wet etch rate was the topic of many studies. The specific values obtained were found to depend on the method of film deposition, film density (as-deposited or annealed film), film composition, type of chemical solution and degree of dilution by water. In conjunction with the IR data and the XRF measurements for phosphorus, the etch-rate values assist in rapid composition control. Chemical etch-rate measurements in liquid etchants, such as buffered HF (BHF) and unbuffered HF (so-called "P-etchant") provided an additional convenient test for rapid composition control. This test could be performed in minutes by ellipsometric determination of the film thickness followed by measuring the film dissolution time. The etch rate in both cited etchants increased with the increase of phosphorus concentration in BPSG film as for binary PSG films. In contrast, the etch rate in BHF decreases with the increase of boron content as in the case of BSG and BPSG films [50,93]. The etch rate data for standard BPSG films in both BHF and in the more composition-sensitive P-etchant are summarized in Table 2.6 for both as-deposited and densified films. These trends, used for the express-analysis of additives concentrations earlier, are also presented in Figure 2.17 [67]. If the boron or phosphorus content in the BPSG film is kept constant, the other variable component can be easily determined.

Etch rate data should be taken at a controlled temperature preferably under fixed isothermal conditions, such as 25.0 °C. The effects of temperature on the etch rate in BHF of BPSG films and thermally grown SiO_2 for comparison can be found in [50]. Therefore, a correlation can be made between the etch rates and different temperatures. The differences in the slope of the temperature dependencies indicated different activation energies for the etching mechanism of SiO_2 and BPSG films.

In VLSI and ULSI technology, diluted etchants are used. Dilution allows a decrease in the etch rate making etch process more controllable with thinner dielectric films. For instance, in annealed 4.5 [B] – 4 [P] the SACVD BPSG films etch rate in buffered 20:1 HF was about 5 Å/s, while in 100:1 diluted HF it was about 4 Å/sec.

In addition, instead of wet etch rate (WER) analysis, which gives the absolute values of a film solubility at certain conditions, the so-called "wet etch rate ratio parameter, WERR" is in use. This dimensionless parameter, often presented in material characterization publications, was introduced to characterize the difference in studied thin film etch with respect to the thermal oxide etch value in the same etchant and with the same conditions. These values can indirectly reach 1.5 – 2.0 and, therefore, indicate the differences in the film density.

Differences in wet etch rates of silicon dioxide-based materials were also found using low pressure HF vapor etching [157,158] and named "selective etching". These differences were clearly identified using stacked structures with different silicon dioxide-based thin films separated by silicon layers.

In fact, comparison of SEM cross-section images for non-etched and etched structures clearly shows that PSG and BPSG films etched quickly in a lateral direction while silicon dioxide was almost non etched by HF vapor. Obviously, this severe etch rate difference can be effectively used in some IC device technology applications.

Table 2.6. Chemical tech rate ranges of SiO_2, PSG, BSG and BPSG. Treated data from Ref. [50]

Film type, additive content	Film condition	Etch rate (Å/s) at 25 °C	
		BHF*	P-etchant**
SiO_2	As-deposited	92	15–16
SiO_2	Thermally-grown	16	2
PSG, 6.5 wt % [P]	As-deposited	130–140	120–130
	Densified	90–100	65–75
BSG, 6.5 wt % [B]	As-deposited	6–7	50–70
	Densified	4–5	15–20
BPSG, 4 wt % [B]–4 wt % [P]	As-deposited	25–30	50–60
	Densified	12–15	18–23
BPSG, 5 wt % [B]–4 wt % [P]	As-deposited	12–16	100–120
	Densified	8–11	30–35
BPSG, 6 wt % [B]–3 wt % [P]	As-deposited	11–14	150–170
	Densified	7–9	55–64

*1 vol. HF 49 % + 6 vol. NH_4F 40 %.

** 3 vol. HF 49 % + 2 vol. HNO_3 70 % + 60 vol. H_2O.

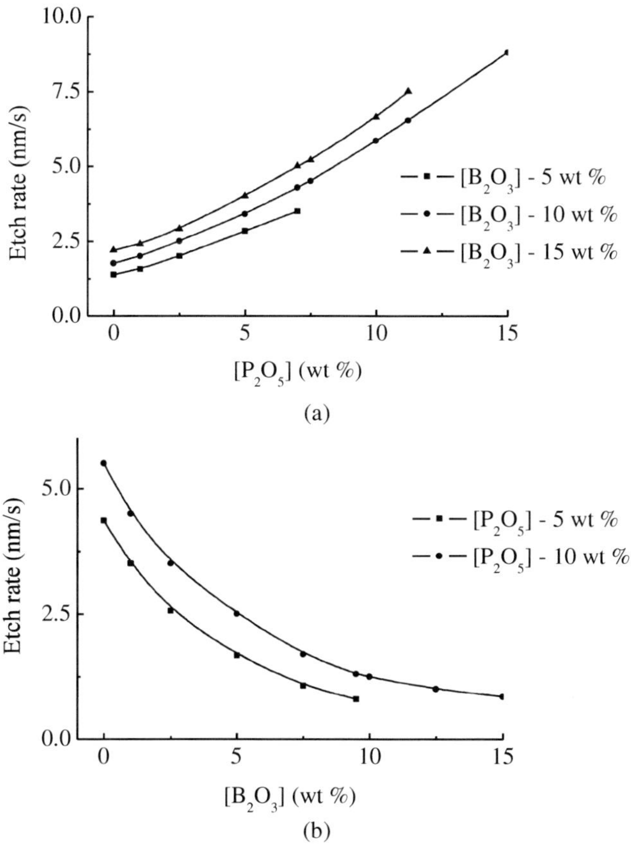

Figure 2.17. Trends of as-deposited silane-based BPSG film etch in P-etchant (a) and buffered hydrofluoric acid (b) (re-drawn from [67]).

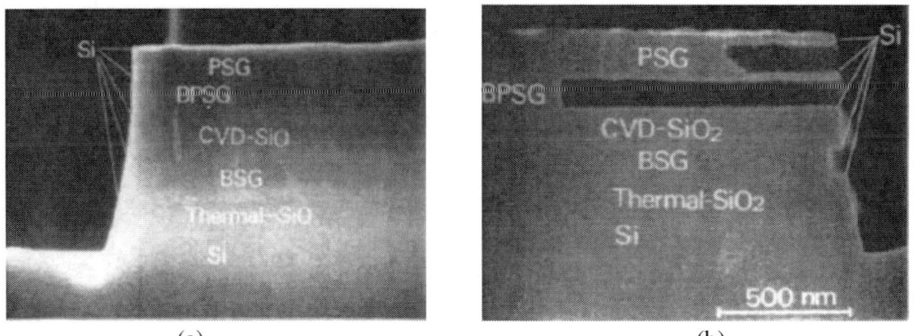

Figure 2.18. Cross-sectional SEM photographs showing the effect of the low pressure vapor HF etching; (a) before HF etching and (b) after HF (600 Pa) etching [157]. Reproduced by permission of ECS – The Electrochemical Society.

10.5. Gettering Properties of BPSG Films

Phosphosilicate glass films are known as getter against sodium ions [8,9,15-17]. The only paper with a demonstration of gettering properties was the paper published by Kern et al. [50] for atmospheric pressure BPSG films deposited using Method 1. To check BPSG film effectiveness against sodium ions, the authors performed two types of tests: (1) radiotracer measurements of Na^{22} penetration through the films on heating using SIMS profiling, and (2) capacitance-voltage bias-temperature (CVBT) stress measurements on metal-gate MOS transistors fabricated using BPSG or PSG films as the passivation layer. In the first series, the sodium was applied to the film surface in the form of a $Na^{22}Cl$ solution. Radioactive tracer analyses with Na^{22} applied on the surface of glass followed by a drive-in diffusion step in N_2at 500 °C for 16 hours demonstrated the superior protective capability of as-deposited or annealed 4 wt % [B] – 4 wt % [P] BPSG films over 3 wt % [P] as-deposited and 6 wt % [P] annealed PSG control samples.

In the second series, the CVBT electrical stress measurements were carried out with aluminum-metallized metal-gate MOS transistors. The control samples were also 3 wt % [P] PSG passivation overcoats; 4 wt % [B] – 4 wt % [P] BPSG films were used as the test passivation layer. The surfaces of the glass were uniformly contaminated with sodium ions from 0.1 molar $NaHCO_3$ solution. Electrical measurements were made and graphically recorded initially and after successive heat treatments at 300, 350, 400, and 450 °C. Devices with several types of 1.8 μm thick BPSG overcoats survived heat treatments up to 400 °C, whereas controls with standard 1.2 μm thick 3 wt % [P] PSG film failed at temperatures as low as 350 °C (large shifts in flatband voltage).

Thus, the experimental results from the low-concentration carrier-free Na^{22} radioactivity measurements and those from electrical measurements with samples contaminated with sodium at many orders of magnitude higher concentrations agree in demonstrating the protective capability of BPSG films. Explaining the mechanism of such protection, Kern and Smeltzer [57] performed experiments with low-energy surface ion-implantation of 1×10^{14} Na^+/cm^2 into the BPSG film followed by 900 °C anneal and SIMS profiling of sodium. Analysis showed that after film anneal, sodium was found to be almost entirely in the BPSG film demonstrating gettering effectiveness compatible to that of 3 wt % [P] and 6 wt % [P] PSG films.

10.6. Electrical Properties of BPSG Films

Dielectric and electrical properties of BPSG films have rarely been published. This seems to be due to the specific application of the film, which is preferably implemented in VLSI and ULSI device technology as a passive insulator between a level of gates and a first level of metal. Nevertheless, introducing new BPSG film CVD techniques, these properties were briefly characterized with simple capacitor structures which consisted of metal (aluminum alloys), dielectric (BPSG film), and silicon (substrate).

The author would like to highlight that an analysis of accumulated references regarding BPSG thin films showed that the electrical properties of BPSG films deposited by PECVD Method 7 were studied in comparatively more in-depth [256]. We believe this was due to some specific features of this thin film, mostly because of soft oxidizer (N_2O) use in PECVD. In fact, this film is known to consist of fully oxidized phosphorus species P^{+5}, and non-completely oxides ones, such as P^{+3} [52,288,290]. In addition, elemental phosphorus P^0 was also considered in the film [322]. However, because of this Method 7 was used a long time ago, and was relatively rare as compared to the others. As a result, we will not analyze specific data obtained by its researchers. Details can be found in the cited publications.

Dielectric properties. The dielectric constant of BPSG films was presented in only a few papers. Any studied BPSG film composition revealed an undoubtedly good dielectric property that is probably the reason for such limited research interest in this matter. Typical values (1 MHz) in as-deposited and densified films were varied in the range of 3.8–4.5 [57], 3.88–4.11 [79] and 5.0–6.0 [110], depending on the CVD method. Thus, the values obtained are quite typical for SiO_2-based dielectrics.

Electrical properties. The dielectric breakdown voltage values, being measured with thick BPSG films in MOS capacitors, were found to be in the range of 4.0 – 9.2 MV/cm. Generally, they reflected a trend of higher values 7.0–8.0 MV/cm for high temperature film deposition [78] and a trend of lower values 4.7–6.6 for low-temperature deposition.

Electrical conductivity (at E = 1 MV/cm) of BPSG films deposited using Method 1 was reported to be ~ $(0.5–1)\times10^{-8}$ A/cm^2 after deposition and ~ 1×10^{-8} A/cm^2 after film anneal [110]. The lowest values were found to correspond to the lowest deposition temperature. Films deposited using Method 5 revealed values of $(4–8)\times10^{-9}$ и $(4–13)\times10^{-9}$ A/cm^2 before and after film anneal, respectively at E = 0.5 MV/cm [224,245]. Electrical defect density for 1 μm thick BPSG films was found to be 0.15 cm^{-2} (V = 130 V, capacitor square 0.114 cm^2) [52] and 0.3 – 0.6 cm^{-2} (V = 25 V, capacitor square 0.092 cm^2, leakage current > 500 nA) [110].

BPSG film net charge measurement features. The well-accepted method of film evaluation using the MOS-capacitor technique is time consuming. As a result, only a few experimental variables can be verified using specially fabricated wafers. The studied dielectric film can also be affected by treatments such as metal sputtering, etch etc. The monitoring of film properties during wafer processing using the MOS technique is not possible. Reported values of charge in BPSG films with 4 wt % of each element measured using the capacitor-voltage method were found to be $(2–4)\times10^{11}$ cm^{-2} for Method 1 [110] and $(5.3–5.7)\times10^{11}$ cm^{-2} for Method 5 [224,245]. The necessity of using a non-destructive method for the film charge monitoring during IC technology flow is therefore clear.

The author described the use of fast and non-destructive Surface Photo Voltage (SPV), also known as Surface Charge Analysis (SCA) technique to characterize BPSG films deposited by a few different methods [214]. Under the exposure of an electrically neutral wafer–film system to laser radiation, the potential correlating with the net charge in a dielectric film is induced on the film surface.

It was believed that this technique could be considered as an additional method for monitoring and characterization of the film properties, where the total charge in CVD films, as well as charge deviation due to heat treatment, could be obtained using the same sample. In fact, this analytical method allowed the discovery of a new way to characterize BPSG film features as well as their differences between the films deposited by different methods. The author first revealed the differences in the net charge values and charge changes during technology treatment in BPSG films prepared on single-crystal silicon wafers by different CVD methods. The SPV CMS 4000 tool by Semiconductor Diagnostic was used for measurements of the total charge values at nine pre-defined points on a wafer (namely, center, 4 randomly placed points with a radius of 64.1 mm and 4 randomly placed points with a radius of 86.9 mm). BPSG films with a thickness of 0.3–1 μm deposited on the silicon wafers using Method 4, Method 5, and Method 8 were studied.

The SCA values for all types of as-deposited films were found to be in the range of (6.0 – 7.0)$\times10^{10}$ q/cm^2 with good uniformity (standard deviation StD was in the range of 0.06 – 0.08). No correlation of SCA with the film thickness or additive concentrations was found. The trends of very weak growth of mean SCA values with the approximate rate of 4×10^6 q/cm^2 per hour or less was calculated irrespective of the film type.

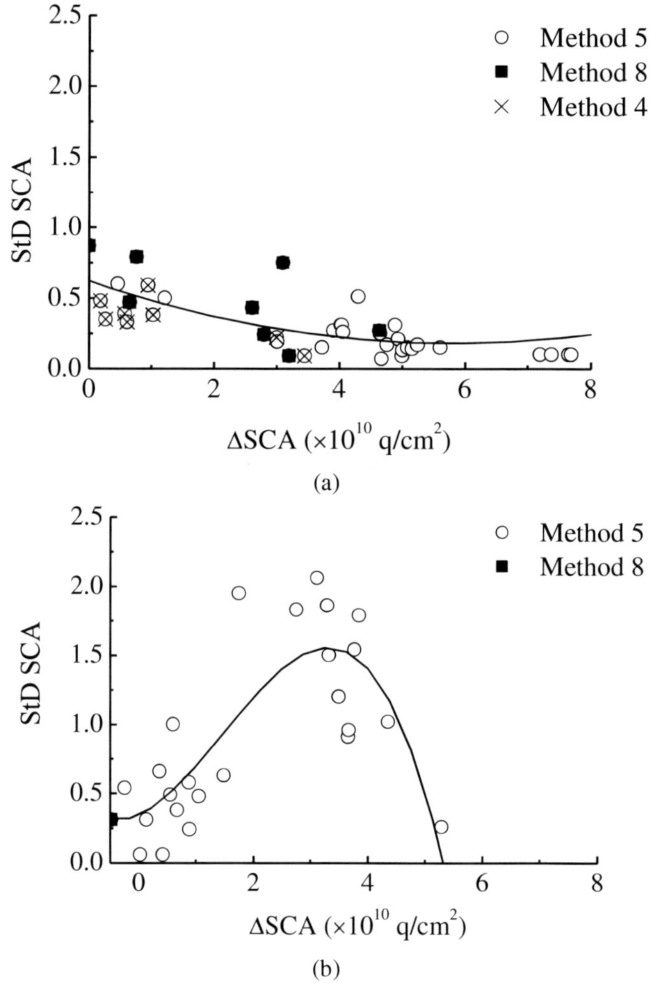

Figure 2.19. StD vs. ΔSCA for BPSG films annealed in a furnace in N2 at 850 – 900 °C (a) and in O2 at 900 °C (b). Presented at 4th Intern. Dielectric for ULSI Multilevel Interconnection Conf. (DUMIC), 1998 [214].

After heat treatment of the same samples performed either in a furnace for 10–60 minutes and nitrogen or oxygen ambient, or under rapid thermal conditions, the net change substantially decreases, mostly in films produced by Method 5. The changes of SCA in BPSG films were found to be dependent on the total additive concentration. The higher charge decrease corresponded to the high additive content in the films. Along with net charge change expressed as ΔSCA (difference in charge values measured for annealed and as-deposited film), the strong standard deviation StD up to 1.5 of the net charge within films was observed, as shown in Figure 2.19(a,b).

The greatest differences of net charge and charge deviation were observed in films produced by Method 5 at RTA anneal and furnace anneal in oxygen, while films deposited using Method 8 revealed very small changes in these net charge-related parameters. These data obviously confirmed the differences in the film structure and properties that was explained by differences in the film porosity including that of within the wafer. More data can be found in our paper [214].

10.7. Summaries of BPSG Film Properties for Production Processes

Summary of the properties of the films deposited using some production methods are shown in Tables 2.7, 2.8, 2.9 for Method 1, Methods 4 and 5, and Method 9, respectively. Note that data in Table 2.9 for HDP-CVD PSG films include the original data and some references to the similar publications. This allows the conclusion that HDP-CVD films deposited using different tools revealed similar properties.

Table 2.7. Properties of silane-based BPSG films deposited by Method 1 (LPCVD) [110]

Film property	Values obtained
Refractive index	1.44–1.48
WER, as-deposited(Å/sec): • P-etchant • BHF	 15–500 2–150
Film porosity (electrographic method, cm^{-2}) for 0.85-1.0 μm thick BPSG film on silicon substrate: • as-deposited • thermally annealed	 2–4 ≤ 1
Electrical defectivity on capacitor structure, def/cm^2, S = 0.092 cm^2, V = 25 V, at I > 500 nA, after film thermal anneal	0.3–0.6
Dielectric strength, MV/cm	4–6
Dielectric constant: • as-deposited • thermally annealed	 5.7–6.0 5.0–5.3
Electrical conductivity, as-deposited, A/cm^2 (E = 1 MV/cm)	$1\times10^{-8} - 5\times10^{-7}$
Charge in film (with thermal oxide underlayer) measured with C-V technique ($\times10^{11}$, cm^{-2})	2–4

Table 2.8. Properties of TEOS-ozone BPSG films deposited by Method 4 and Method 5 [224]. Reproduced by permission of ECS – The Electrochemical Society

BPSG film property	**Obtained BPSG film properties**			
BPSG film type and thickness	APCVD, 5000 ± 250 Å		SACVD, 6000±300 Å	
BPSG thermal anneal (dry gas)	**No**	Yes	**No**	Yes
Refractive Index	1.467–1.477	1.465–1.466	1.466–1.467	1.465–1.467
BPSG film angle on the step of vertical structure with respect to the substrate surface: a) as-deposited b) anneal 850°C, N_2, 45 minutes (below is defined as [a]) c) anneal 900°C, O_2, 12 minutes (below is defined as [b])	≈ 90° – –	– ≤ 70° –	≈ 90° – –	– ≤ 70° ≤ 20°
Voids in the structure gaps	Yes	No	Yes	No
Surface roughness (nm)	1.4–2.4	0.2–0.3	0.6–0.7	0.15–0.2
Film shrinkage (%)	–	~ 4–5.5[a]	–	~ 4–5[a]
Etch rate for 4.5 wt % [B] – 4 wt % [P] BPSG: a) diluted 100:1 HF (μm/min) b) diluted buffered 20:1 HF (μm/min)	– –	– –	– –	~ 0.024 [b] ~ 0.030 [b]
Charge in film measured with: a) C-V technique ($\times 10^{11}$ cm^{-2}) b) SCA technique ($\times 10^{10}$ cm^{-2})	– 6.10–6.80	– 5.00–5.90[a]	5.53–5.70 6.30–6.90	5.31–5.39[b] 3.90–6.20[a]
Leakage current at E = 0.5 MV/cm^2 ($\times 10^{-9}$ A/cm^2)	–	–	4 – 8	4-13
Defect density in the film immediately after deposition with defect size ≥ 0.2 μm (cm^{-2})	≤ 0.1	–	≤ 0.1	–
The time of defect appearance in as-deposited 5wt % [B] – 4 wt % [P] BPSG film under optical mycroscope with the magnification about 500[x](hours)	≤ ~ 15–20	–	≤ ~ 15–20	–
Tensile stress ($\times 10^8$ dyne/cm^2)	–	–	4.3–6.5	–

Table 2.9. Properties of HSP-CVD PSG films deposited by Method 9 [301]. Reproduced by permission of ECS – The Electrochemical Society

Film property	**Author data [297,298] (T=625°C)**	**Summarized published data(T<550°C) see referencesin [301]**	**Summarized HDP-PSG film property**
Thickness non-uniformity (1σ)	1.25	~1.0–1.5	Very uniform
Studied additive range (wt %)	0–7	0–7	Wide additive range
Dopant non-uniformity (wt %)	0.13	0.1~0.2	Very uniform
Stress ($\times 10^9$ dyne/cm^2) *↑	-1.33	<-1.5	Compressive stress
Refractive index *↑	1.4637	1.455–1.463	No silicon excess
P_2O_3 content (%)	Not found	Not found	Full PH_3 oxidation
Shrinkage (%)	<0.1%	-	Very low
Moisture content (%) *↑	<1%	-	Very low
WERR *↑	~2.5–3	2.5–3.5	Dense film
Dry etch rate (Å/min)	~3600	-	Similar to BPSG

Film property	Author data [297,298] (T=625°C)	Summarized published data(T<550°C) see referencesin [301]	Summarized HDP-PSG film property
Surface roughness (RMS, nm)	~0.53	-	Smooth surface
Average surface charge value and its uniformity ($\times10^{10}$ q/cm^2)	~ 6.0±0.1	-	Uniform charge distribution
Gap-fill capability (AR, gap size, structure type)	AR=3.1, G=0.16 μm, rectangular; AR>4, G >0.1 μm, rounded	AR ≈ 3.6–4, G=0.18–0.2 μm, rectangular; AR ~ 5, G~0.07 μm,rounded	Good gap-fillcapability. Similar capability achieved usingdifferent tools (see text for detail)

* Increase (↑) of value with the increase of the phosphorus concentration in HDP-PSG film.

Chapter 11

BPSG Film Flow Capability

Historically, in multilevel VLSI and ULSI circuit technology, BPSG films are used as a planarized Pre-Metal Dielectric (PMD). The major BPSG film property, employed in integrated circuit technology, is a capability to soften (to flow) when being deposited on stepped device surfaces followed by being heated at enhanced temperatures. This allows smoothening (planarization) of high as-deposited film steps and, therefore, improvement of the quality of a metal film, sputtered over the smoothed BPSG film, see Figure 1.1. Sufficient BPSG film planarization can be achieved at significantly lower post-deposition anneal temperatures, as compared to that of PSG films, either at isothermal furnace anneal or rapid thermal anneal conditions. In this chapter, we briefly describe the basics of the glass flow capability, which are necessary for understanding of BPSG film softening features on device structures. The latter is described quantitatively in Part 3 of the monograph.

11.1. Brief Description of Glass Flow Basics

According to classical chemistry, bulk glass is considered as a kind of supercooled liquid formed from mixed oxides by the melt cooling method. The so-called "glass transition temperature" (T_g) (it is also called "glass transformation temperature") is one of the major characteristics of bulk glasses. This temperature defines a temperature region where many of the glass properties change significantly. This T_g temperature is associated with the slowing down of molecular rearrangements in the glass structure during the cooling of glass melt. As an example of T_g experimental measurement for a bulk glass sample, a thermal expansion curve with T_g indication is shown in Figure 2.20. The rate of molecular rearrangements is closely associated with viscous flow and T_g occurs in all oxide glasses at a viscosity of about 10^{13} Poise. The softening point of a glass is the temperature at which it has a viscosity of $10^{7.6}$ Poise.

It is known that glass viscosity varies exponentially with the temperature:

$$\mu = \mu_0 \times \exp(E_a / RT), \qquad (2.1)$$

where μ and μ_0 are glass viscosities, E_a is an activation energy, R - gas constant and T is temperature. In many cases, the viscosity in the region between T_g and temperatures corresponding to a viscosity of $10^8 - 10^{10}$ Pouse is better described by the empirical so-called FVT relation [54]:

$$\mu = const \times \exp[B/T - T_0], \tag{2.2}$$

whose validity, as well as B, T_0 and the values of the constant, depends on the length of the approximation interval.

Nassau et al [54] studied a number of glass compositions and showed, that a flow temperature T_f in thin film glass material would be expected to be close to the glass softening point and can be linked to T_g using an empirical equation:

$$T_f = T_g + 288 \pm 16^{\circ}C \tag{2.3}$$

11.2. Glass Structure Features and Glass Flow Capability

When added to silicon dioxide, the phosphorus, boron, arsenic and germanium oxides decrease the glass transition temperature. Examples of T_g decrease in the boron- and phosphorus-contained bulk silicon dioxide are shown in Figure 2.21(a,b). Boron oxide was found to be the most effective additive, which decreased T_g significantly [54,284].

Analyzing the reasons for additive oxides influence on silica T_g, need to mention the following. The strength of the oxygen-central atom bonds in SiO_2, GeO_2, B_2O_3, $P_2O_5$is about 418 kJ/mol for all oxides. These values, therefore, do not correlate with the experimental viscosity values for these oxides, shown in Figure 2.22. At the same time, the melting points of the crystalline oxides goes along with the variation in viscosity (1710 °C for SiO_2 cristobalite, 1116 °C for GeO_2hexagonal, 580 °C for P_2O_5, and 450 °C for B_2O_3). This is strong indication that the additive oxide structure is a very important factor affecting glass viscosity.

A brief review of the information regarding vitreous oxide structures allows us to make the following conclusions. In vitreous silica the elementary structural polyhedron is a SiO_4 tetrahedron with a silicon atom in its center. These SiO_4 tetrahedra are bonded by their vertices by means of $Si-O-Si$ bonds resulting in the formation of a three-dimensional spatial network. The same structure is in GeO_2. However, structures of P_2O_5 and, especially, B_2O_3 are different. In the phosphorus oxide network, every PO_4 tetrahedron is bounded to three others, because the fourth tetrahedron vertices is $P=O$ bond. Therefore, the lower viscosity of P_2O_5 is due to its structure. In vitreous boric anhydride, BO_3 triangles with a boron atom in their centers are linked with neighboring boron atoms by three oxygen atoms. As a result, an almost planar structure of continuous $B-O-B$ bonds is formed. In this layer-type structure, approximately two-thirds of the boron atoms form the so-called boroxol groups. Thus, the linked-ring structure of boron oxide is believed to be the reason for the low viscosity of this oxide.

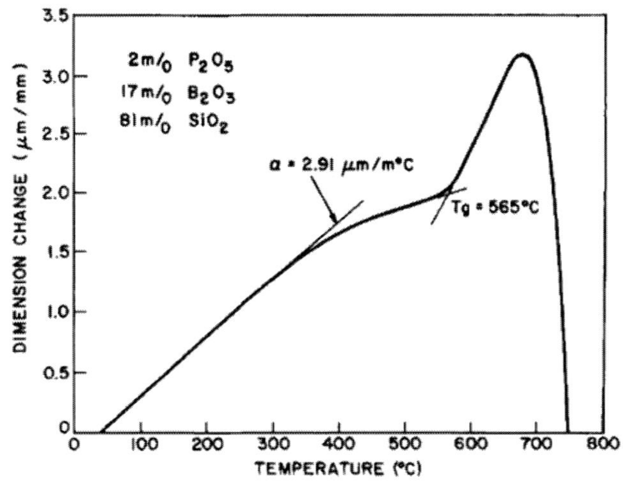

Figure 2.20. Experimental observation of Tg. for BPSG bulk sample [54]. Reproduced by permission of ECS – The Electrochemical Society.

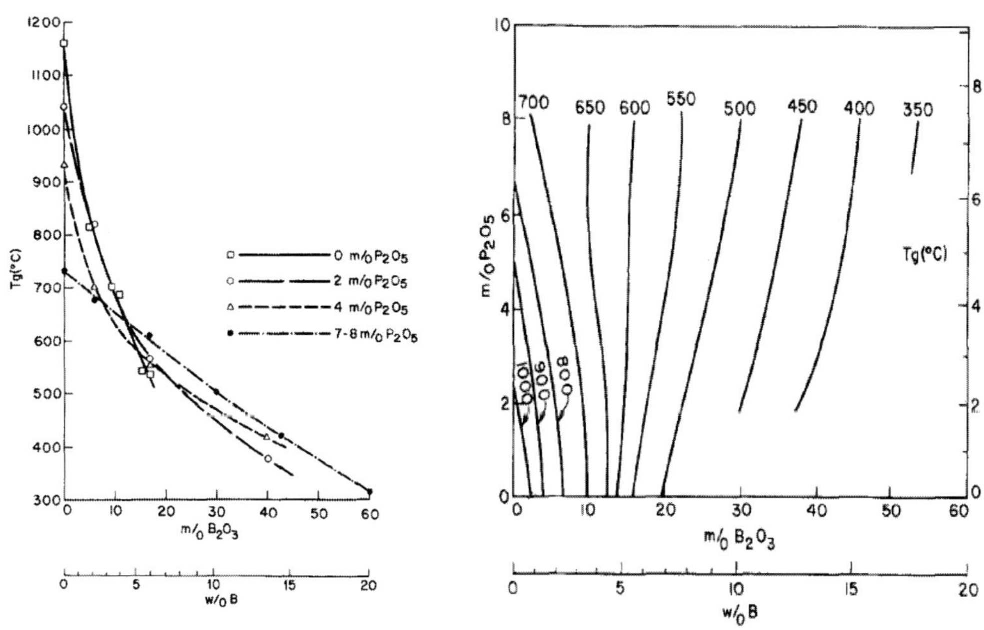

(a) (b)

Figure 2.21. Effect of B2O3 on the Tg for different glass compositions (a) and Tg isotherms for BPSG films (b) [54]. Reproduced by permission of ECS – The Electrochemical Society.

As a general sequence of the above consideration, glass viscosity is linked to glass structure. During the addition of the oxides B_2O_3 and P_2O_5 to SiO_2, the silica network becomes broken resulting in local distortions of the structure and a lowering of the viscous flow of glass. Undoubtedly, oxides with strongly different structures from SiO_2, such as B_2O_3, are supposed to be the most effective in distorting the SiO_2 structure and, therefore, in lowering the glass flow temperature.

11.3. Application of Bulk Glass Parameters to Glass Thin Film

At the first glance, borosilicate glass can be considered as the best flow-able glass film to solve planarization issues in integrated circuit devices. However, the P_2O_5 additive is necessary to keep gettering properties of a silicate glass with respect to alkaline metals. Thus, both boron- and phosphorus-containing glasses were chosen to serve as the best option for flow-able thin film glass material applicable for IC technology [49,50,54,55].

In this section, we discuss the validity of the term T_g for CVD glass films [2]. FTIR data showed that a more or less perfect glass structure of thin film is forming at the deposition temperature being used.

Following the discussion of BPSG film deposition scheme in Figure 1.25, a formation of glass structure can be assumed to occur in the gas-phase through the formation of reaction intermediates followed by their diffusion to the substrate surface and film built-up. In fact, an analysis of the FTIR spectra of as-deposited and annealed films does not show any significant structural differences, which would confirm serious changes of glass structure and, therefore, glass properties at post-deposition heating.

Using this approach, T_g and T_f values can be evaluated for practically used BPSG film compositions, as shown in Figure 2.23. Glass composition is expressed as the total additive concentrations in elemental wt % assuming three simplified possible additive ratios in glass film, namely [B] > [P], [B] = [P] and [B] < [P]. Calculated data points reflect bulk glass film composition and for these glass compositions T_g represents a certain temperature, where many bulk glass properties change significantly.

Considering thin film glass materials, it can be seen that most of the known thin film CVD methods are performed at deposition temperatures T_d of about 450 °C. This is significantly less than that of T_g. After deposition, BPSG films require additional heating (anneal).

High temperature post-deposition BPSG film thermal anneal corresponds to T_f conditions, as shown schematically by the dashed arrow in Figure 2.23. In contrast, low thermal budget anneal, especially at RTA conditions, is expected to be closer to the calculated T_g, as shown schematically by the solid arrow in Figure 2.23. The low thermal budget BPSG anneal conditions used in device technology in recent years in most cases fall into the gap in between glass transition temperature T_g and glass flow temperature T_f. In this case, post-deposition film anneal at temperatures higher than T_g causes a densification of film, but does not change the film composition and bonding significantly. Therefore, implementation of bulk glass characteristic T_g to CVD glass films characterization is done for an artificial reason.

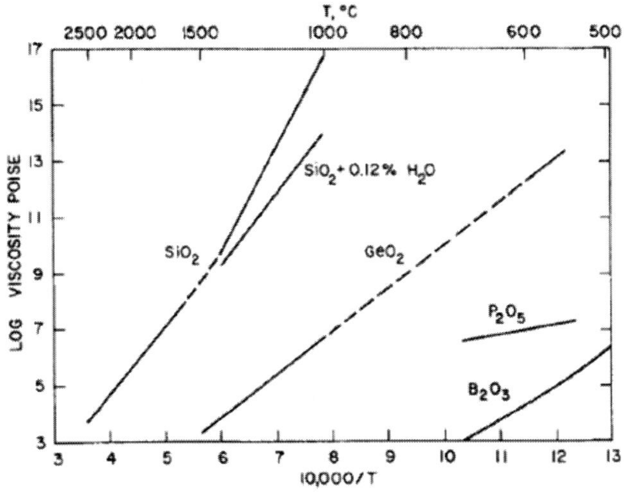

Figure 2.22. Viscosity vs. reciprocal temperature for different glass-forming oxides [54]. Reproduced by permission of ECS – The Electrochemical Society.

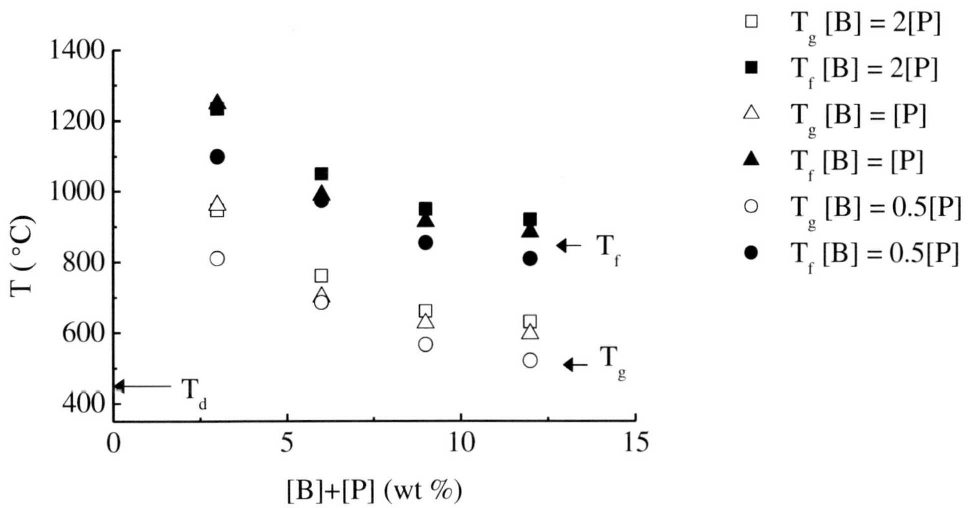

Figure 2.23. Temperature diagram for BPSG films: deposition temperature T_d, calculated glass transition temperatures T_g according to [54], and calculated flow temperature T_f for the same glass film composition. Arrows indicate assumed high (dashed) and low (solid) thermal budget anneal conditions after film deposition.

Chapter 12

BPSG Film – Ambient Moisture Interaction

Ambient air contains more or less water vapor (moisture). For instance, standard clean-room ambient moisture is 40 % that is sufficient to observe moisture interaction during the boron- and phosphorus-contained thin film exposure to the ambient air.

Glass materials are known to interact with moisture which is sometimes called the "moisture absorption" phenomenon. Depending on the glass composition and glass treatment conditions (moisture concentration, temperature, exposure time etc.), the results of such interactions can differ significantly. Base glass material under consideration is silica (SiO_2) with less than 30 wt % totally of boron and phosphorus additives in the form of oxides. Silica is a relatively inert material with respect to moisture while boron and phosphorus oxides are chemically active substances. Therefore, the main idea of moisture interaction with boron- and phosphorus-contained films is that these additives, being included into the glass matrix as network formers, can react with ambient moisture changing the glass material composition and properties. For instance, the ability of additives to leach from the BPSG film is a known phenomenon [50].

The phenomenon of moisture absorption by a bulk glass material is also valid for a CVD glass thin film material. However, there are a few features, which indicate that the CVD film materials should include many more factors for analysis. One feature is that CVD glass films are formed using non-traditional glass-forming techniques. The formation of the glass structure occurs directly during thin film synthesis, strongly depending on the compounds and deposition conditions. The concentration of residual moisture in the film, which is a side product of CVD reactions, can differ from one CVD method to another one. Secondly, the film deposition temperatures for silica-based thin films can differ from room temperature to about 900 °C, significantly affecting the film structure and properties. Third, practically used glass materials can differ from binary to ternary film compositions, therefore overlapping a wide range of additive content from a few wt % up to about 30 wt % of oxides in silica that can additionally affect the film structure. All of these features make CVD glass film materials different with respect to bulk materials and, therefore, very interesting for studies.

The results of experimental studies of ambient moisture-CVD thin film interaction were rarely published rarely. There are only a few excellent papers devoted to the deep studies of moisture absorption performed for the particular samples of BPSG films [178,217,271]. There were no summaries or publications until recently which link moisture absorption effects to the other film properties and, especially, to the film structures. However, this matter is very important. It has been reported that PSG and BSG thin films with high phosphorus and boron content react easily with water, causing VLSI reliability problems [7-9]. Device-related issues caused by moisture in the CVD glass films used as PMD were also reported. PSG with high phosphorus content causes aluminum corrosion in a moist ambient. Water penetrating through the PSG film causes a threshold voltage shift in a metal oxide semiconductor field effect transistor during high-temperature stress tests [178]. Moisture directly affects the electrical behavior of an oxide by enhancing its electrical permittivity and conductivity. Moisture and moisture related species (e.g., H, OH) can diffuse from the PMD oxide to the gate oxide of a metal oxide semiconductor device, thereby degrading its reliability by enhancing its susceptibility to hot electrons [217]. Water was also found to catalyze the etching of oxide films in both gas- and liquid-phase mixtures containing HF [157,158], therefore affecting the film etching process.

In this chapter we summarize and review published data about moisture absorption effects in boron-, phosphorus-contained films deposited by different methods and treated under different thermal conditions. The CVD methods for film deposition were realized at relatively high deposition temperatures above ~ 400 °C. This makes it possible to avoid a distortion of the analysis that could be due to the features of the low CVD temperatures (less than 300 °C) which are known to affect the silica structure and porosity significantly. Eventually, these data give us an opportunity to select specific film features, which are used in the next chapters to formulate the structures of BPSG films, explaining the reasons of observed differences in BPSG film properties.

12.1. Technique for Analysis of Moisture Effects in Thin Glass Films

Moisture absorption effects in thin films can be studied qualitatively by implementing a FTIR spectroscopy technique. This is widely used in research practice and integrated circuit production. This non-destructive analytical technique shows changes of the initial level of additives and residual moisture in a spectrum that happen during the film exposure to ambient. As a result of the literature analysis, a large number of moisture-related species found in FTIR spectra of boron- and phosphorus-contained glass thin films is presented in Table 2.3. Despite a wide variety of peaks in spectra, the interaction of films with moisture is frequently described in a more simple way. It is assumed to take into account just a few peaks corresponding to the following bonds: the B – OH bond at 3200 – 3300 cm^{-1}, the Si – OH bond at 3650 cm^{-1}, and the H – O – H bond at 3300 – 3500 cm^{-1}. Normally, areas of these peaks are summarized and called either "hydroxyl" or "water (moisture)-related" areas (peaks). We will use this simple below, but in most of cases it is a sufficient approach to characterize the difference in the moisture interaction of films, deposited using different techniques.

Depending on the film composition and film properties, the moisture absorption effects can be seen in FTIR spectra in a few days of the film exposure to the clean-room air. However, in most cases these effects are slow enough (for instance, it is true for samples thermally annealed at elevated temperatures). Sometimes it takes up to a month to distinguish differences between the film samples. To accelerate and to pronounce the interaction of films with moisture, autoclave testing, also called "Pressure-Cooker testing" (PCT), is used. PCT can be performed, for instance, at the following conditions particularly used in our studies: 121 °C, 1.2 atm pressure and 100 % humidity for 168 hours time. After such a PCT treatment, films can be also evaluated using the FTIR technique to see how strong the differences in moisture-related areas are. In addition, the decrease in phosphorus concentration during the PCT, for example, measured before and after the PCT with FTIR or XRF spectroscopy, can be considered to be caused by the phosphorus leaching from the film.

Application of SIMS for the film analysis, including hydrogen analysis, was demonstrated by Kern et al [50]. Later, Yoshimaru et al [178] used SIMS in order to determine the degree of water penetration into the film. He fixed profiles of hydrogen and deuterium species in BPSG films before and after PCT with excess pressure of either H_2O or D_2O. Using SIMS profiles and the amount of absorbed water calculated after weighting the films with a microbalance before and after PCT, the increase in film weight after PCT was considered caused by water absorption. Dyer [217] used SIMS analysis to measure the hydrogen and hydroxyl depth profiles in the films to determine the extent of moisture penetration. The author also implemented the results of quantitative SIMS analysis of hydrogen content in BPSG films deposited by Method 3, in order to perform a calibration of water content in the films and to propose possible origin of moisture species in the films.

To study water adsorption sites in the films, water desorption from films between room temperature and 700 °C was also measured using a Thermal Desorption Spectroscopy (TDS) technique [178,217].

12.2. Film-Moisture Interaction

Moisture absorption effects in FTIR spectra of the films. The absorption and reaction of moisture with SiO_2, binary and ternary glass films as well as the effect of post-deposition film thermal annealing, have been studied extensively. Most of the data have been obtained for particular glass films and a pre-defined CVD method, as well as thermal treated conditions. A summary of the peaks according to the literature analysis is presented in Table 2.3. Basic trends are illustrated below.

Figure 2.24 shows the results of FTIR monitoring of SACVD BPSG with 4 wt % [B] and 4 wt % [P] for a total of 9 days storage [186]. A gradual weak increase in the intensity of the moisture-related areas around 3200 – 3600 cm^{-1} and a gradual decrease of the peak at 1330 cm^{-1} attributed to the P = O bond can be seen. However, few changes were observed in the FTIR moisture-peaks after 9 days of storage for the similar SACVD BPSG films thermally annealed at 850 °C and 900 °C. Similar results for annealed films can be found in other references proving the fact that densified films are less prone to interact with ambient moisture. Summarized changes in FTIR spectra of as-deposited and annealed BPSG films are shown in Figure 2.25. It is clearly seen that prolonged anneal at temperatures above 750 °C reduces the film absorption capability with respect to the moisture causing sufficient film densification.

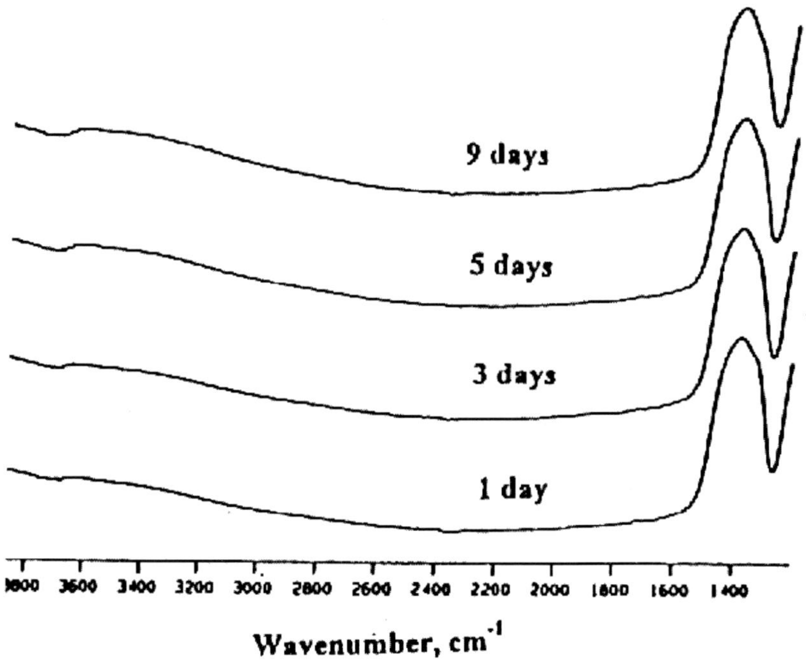

Figure 2.24. FTIR spectra of as-deposited 4 wt % [B] – 4 wt % [P] BPSG film deposed using Method 5 after 1-9 days exposure to ambient moisture. Presented at 3rd Intern. Dielectric for ULSI Multilevel Interconnection Conf. (DUMIC), 1997 [186].

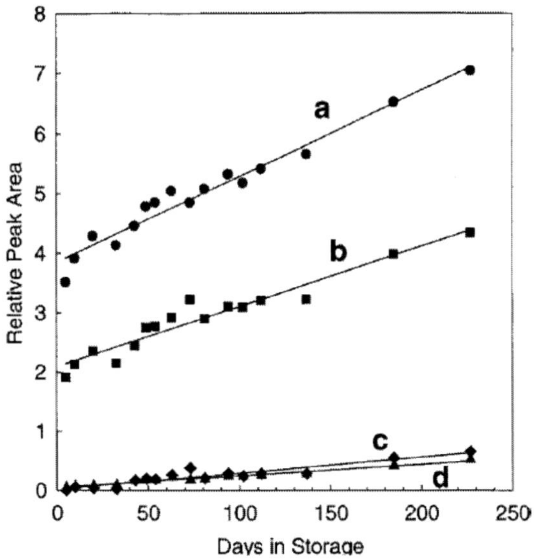

Figure 2.25. Integrated transmission FTIR peak areas for the hydroxyl envelope of BPSG/TEOS film stacks as a function of storage time in a clean-room [271]. Definitions: (a) as deposited, area range 2898 – 3754 cm^{-1}; (b) 500 °C annealed, area range 2897 – 3768 cm^{-1}; (c) 750 °C annealed, area range 3101 – 3703 cm^{-1}; (d) 900 °C annealed, area range 3217 – 3689 cm^{-1}. Reproduced by permission of ECS – The Electrochemical Society.

Another feature of BPSG films is a strong relation between film composition and moisture interaction. This can be clearly seen in films with a high additive content. For instance, as-deposited SACVD TEOS-ozone BPSG films (Method 5) with similar boron and phosphorus total content but different ratios of phosphorus to boron were exposed to clean-room air for 6 days. After a few days of the film exposure to clean-room air, FTIR spectra show a clear correlation of moisture absorption with the additive ratio in films, see Figure 2.26. Obviously, the moisture absorption effect becomes stronger with the increase of the boron-to-phosphorus ratio in BPSG films. We concluded that the strong increase of moisture-related areas in FTIR spectra goes along with the increase of boron concentration in BPSG films when the latter exceeds a level, defined by an empirical equation [B] < 0.4 [P].

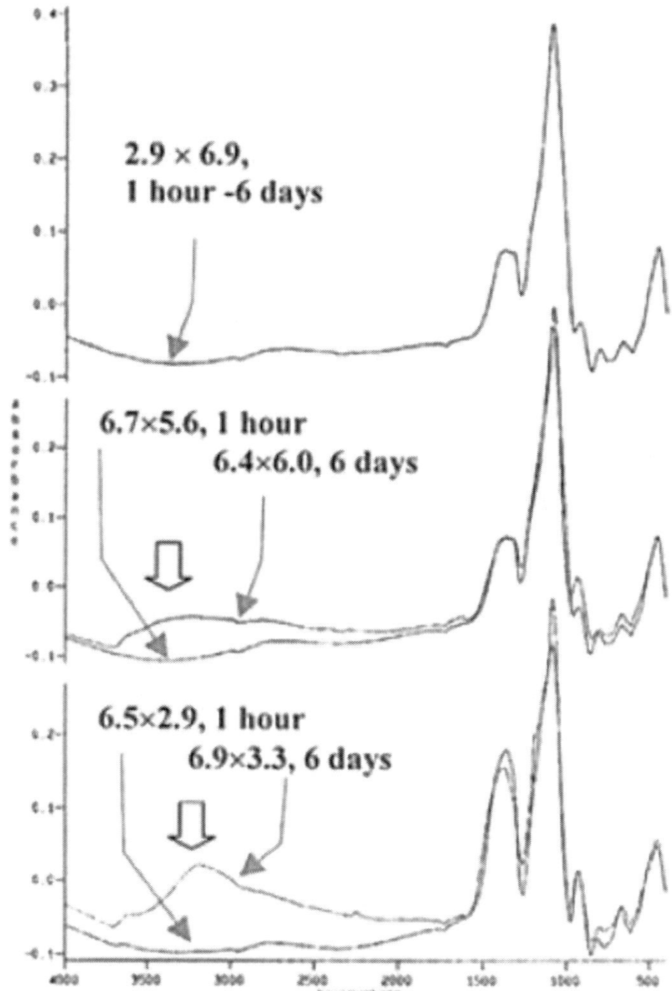

Figure 2.26. FTIR spectra of as-deposited SACVD BPSG films (Method 5). Additive content and a number of days of sample storage in ambient air are shown on the spectra [224].Reproduced by permission of ECS – The Electrochemical Society. Block arrows show the moisture-related areas in FTIR spectra.

However, BPSG films revealed some differences with respect to the interaction with moisture. Considering changes in FTIR spectra of films after an air exposure or, especially, after PCT, we found that the quantity of absorbed moisture depends on the gas-phase deposition method [186,224]. To clarify this, we analyzed and compared SACVD BPSG films and PECVD BPSG (deposited using Method 5 and Method 8, respectively) with total additive content less than 9 wt. % and an additive ratio close to one.

The worst case of moisture absorption was observed for TEOS-ozone-deposited BPSG films (Methods 4 and 5). The best results were found for PECVD and, especially, for HDP-CVD deposited films (CVD Methods 8 and 9, respectively). For instance, a dramatic change in the FTIR spectra around 3200 – 3600 cm^{-1} was seen after PCT of annealed films, indicating a strong water absorption by SACVD BPSG films during this treatment, as shown in Figure 2.27(a). Changes of FTIR spectra after PCT were found to be correlated with the strong changes of additive distribution in upper film thickness area observed by SIMS analysis. The additive profiles across the as-deposited and annealed SACVD BPSG films were nearly flat, but in contrast, profiles were found to be dramatically changed after PCT, as shown in Figure 2.27(b). A significantly smaller change in the FTIR spectrum of PECVD BPSG films (Method 8) was found after PCT, as shown in Figure 2.27(a). Moreover, we did not observe significant changes of additive distribution in PCT PECVD BPSG films that is illustrated in Figure 2.27(b). Thus, these data clearly show that the additive leaching effect is much more pronounced for the films deposited by Method 5.

BPSG composition and additive leaching from the films. It was reported that phosphorus, if it exceeds a certain concentration in the PSG film, leaches out from the film. Interacting with water, phosphorus oxide becomes phosphoric acid that attacks and corrodes aluminum metallization [8,9,16]. In contrast, leached boron becomes weak boric acid and this seems to be a reason why almost no film-related boron leaching data have been published compared to those of phosphorus leaching. Numbers characterizing phosphorus leaching from films are different. For instance, Nagasima et al [16] showed that in PSG films with less than 8 wt % [P], the insoluble silica matrix protects phosphorus from attack by absorbed moisture. However, at higher concentrations phosphorus leaches out, since oxides such as P_2O_5 are water soluble. BPSG films deposited using the PECVD method from hydrides with nitrous oxide [51] revealed phosphorus leaching after 4 wt %.

To illustrate the details about the phosphorus leaching effect, we use data obtained by Yoshimaru et al [178]. PSG and BPSG films deposited using Method 1 exceeding 9 wt % phosphorus showed high phosphorus leaching. Figure 2.28(a) illustrates the dependence of the phosphorus leaching ratio for PSG and BPSG films on the phosphorus concentration after 120 hours in the PCT (120°C, 100 % relative humidity).

The phosphorus leaching ratio (%) is defined as the ratio of the phosphorus concentration difference before and after PCT divided by the phosphorus concentration before PCT. Figure 2.28(b) shows the dependence of the phosphorus leaching ratio for PSG and BPSG and on the PCT time with the same other conditions, respectively. Phosphorus in PSG with 10 wt % [P] leached after 64 hours. The phosphorus leaching ratio apparently saturates after 120 hours at about 25 %. Adding boron to the film makes phosphorus leach more easily. The PCT time required for phosphorus leaching decreases, and the phosphorus leaching ratio becomes saturated at a higher ratio. Phosphorus in BPSG with a 4 wt % [B] and 10 wt % [P] showedphosphorus leaching after only 8 hours in the PCT. The phosphorus leaching ratio became saturated at 50 % after 24 hours in the PCT. PSG and BPSG with 9 wt % [P] or less, however, showed almost no phosphorus leaching after a PCT continuing up to 150 hours. Finally, the authors concluded that a certain limit in the phosphorus concentration near 9 wt % [P] is not affected by boron concentration. Boron only shortens the PCT time required for BPSG phosphorus leaching and decreases the residual phosphorus concentration after a long time in the PCT.

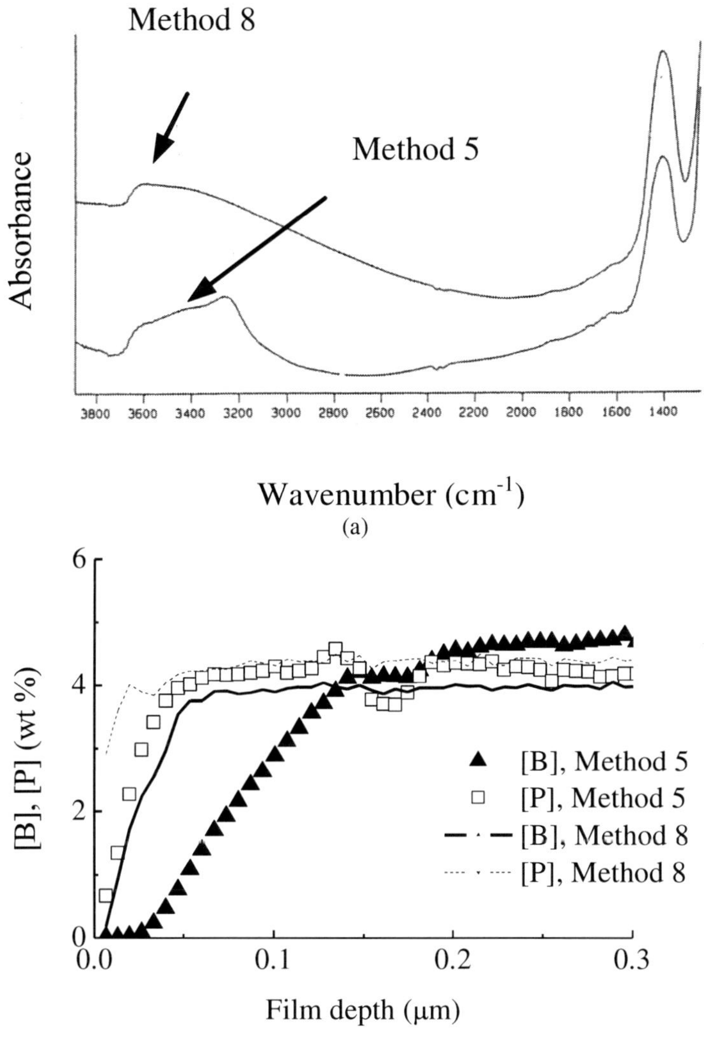

Figure 2.27. FTIR spectra of moisture-related area for autoclave treated annealed BPSG films deposited using Method 5 and Method 8 (a) and SIMS profiles of boron and phosphorus in BPSG films deposited by Method 5 and Method 8 after autoclave testing (b).Presented at 3rd Intern. Dielectric for ULSI Multilevel Interconnection Conf. (DUMIC), 1997 [186] (a) and at 4th Intern. Dielectric for ULSI Multilevel Interconnection Conf. (DUMIC), 1998 [208].

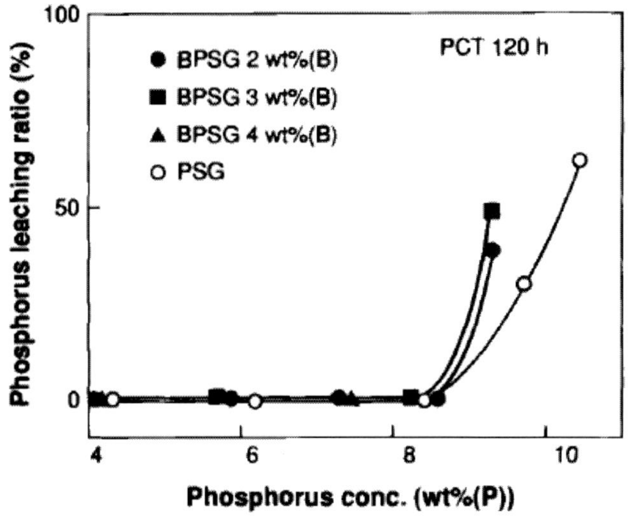

(a)

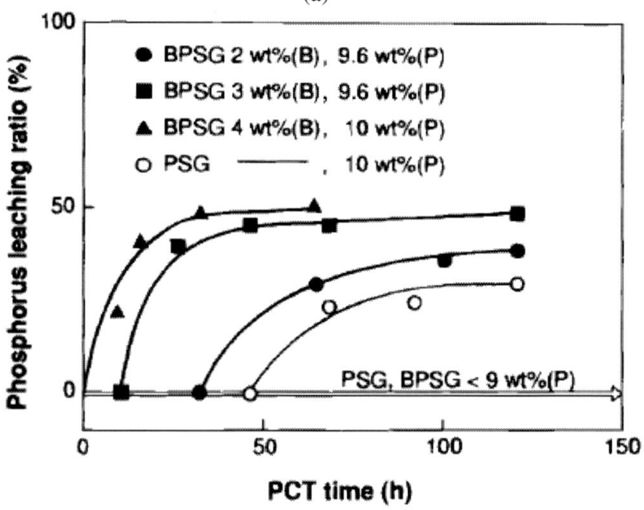

(b)

Figure 2.28. Dependence of the phosphorus leaching ratio for PSG and BPSG on the phosphorus concentration after 120 h in the PCT (120 °C, 100 % relative humidity) (a) and dependence of the phosphorus leaching ratio for PSG and BPSG on the PCT (20 °C, 100 % relative humidity) time (b) [178]. Reproduced by permission of ECS – The Electrochemical Society.

An obvious physical consequence of the additive leaching effect during the glass film moisture absorption/interaction is a liquid droplet formation on the surface of films. This was easy to see under the optical microscope as long as the size of droplets in SACVD BPSG films was found to be up to 30 μm. The size and density of liquid droplets were proportional to the total additive content in films. Liquid drops were also found on the surface of PECVD films after PCT, but their size and density was found to be much smaller than those in SACVD BPSG films [186]. The content of these droplets is supposed to be water soluble phosphoric and boric acids.

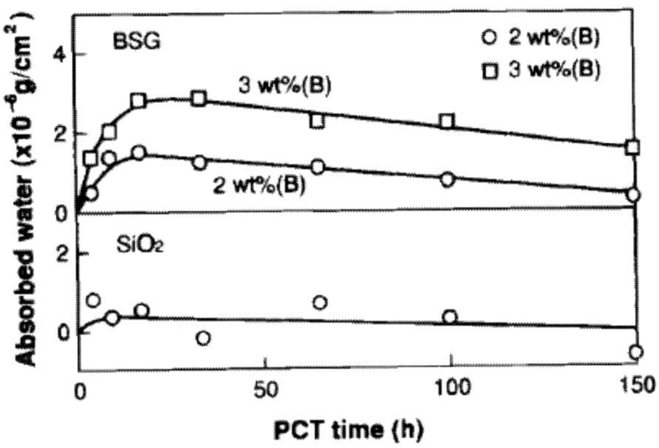

(a)

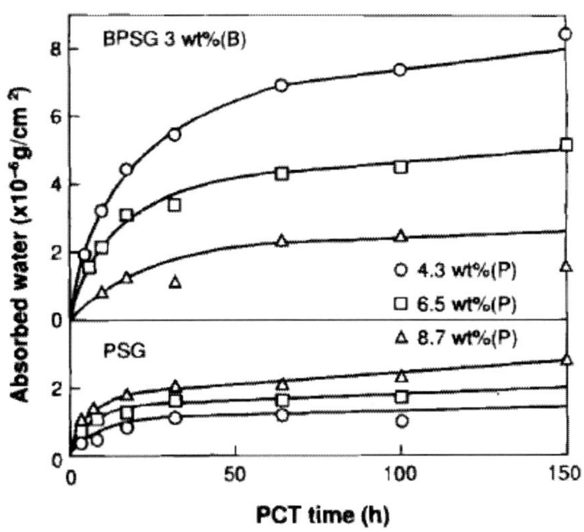

(b)

Figure 2.29. Dependence of absorbed water in SiO2 and BSG on the PCT time (a) for boron concentrations in BPSG films are 2 and 3 wt % and dependence of absorbed water in PSG and PSG on the PCT lime for phosphorus concentrations in films 4.3, 6.5, and 8.7 wt% and the boron concentration in BPSG is 3 wt % (b) [178]. Reproduced by permission of ECS – The Electrochemical Society.

A quantity of moisture absorbed by films. The capacity of boron- and phosphorus-contained films (with 9 wt % [P] and less, i.e., films that showed no phosphorus leaching) to absorb moisture was evaluated by Yoshimaru et al by the weight increase during PCT [178]. Films were weighted with a microbalance before and after PCT and the difference was attributed to the absorbed water content in the film. Figure 2.29(a) shows the dependence on the PCT time of absorbed water in SiO_2 and BSG films.

BSG film absorbed more water than SiO_2.The amount of absorbed water increased with an increase in the boron concentration. BSG film with 3 wt % [B] contained three times more water than SiO_2. Water absorption saturated after a short PCT time, and then decreased. The number of water adsorption sites in the film may decrease after a long time in the PCT. Figure 2.29(b) shows the dependence of absorbed water in PSG (lower half) and BPSG (upper half, 3 wt % [B]) on the PCT time as a parameter of phosphorus concentration. BPSG absorbed more water than PSG at the same phosphorus concentration. Both BPSG and PSG, however, absorbed a similar water content at the highest phosphorus concentration. Water absorbed by BPSG decreased with increasing phosphorus concentration. By contrast, water absorbed by PSG increased with an increased phosphorus concentration. The data suggest that increased boron and phosphorus enhanced water adsorption if they exist alone in silicate glass film, but do not necessarily enhance water absorption in BPSG films.

Instrumental analysis of hydrogen-contained species in the films. Hydrogen-contained species were studied in a few reports. Xie et al [184] measured the hydrogen content of phosphorus-contained silica and SiO_2using nuclear reaction profiling. He showed that the concentration of hydroxyl groups and the rate that water was absorbed within the films depended on the deposition method, additive concentration, and the post-annealing temperature. It was found that the H content was $10^{21} - 10^{22}$ atoms/cm^3 in as-deposited films. After approximately 150 hours of storage, 2 wt % [P] PSG films absorbed water, but films with 8 wt % of [P] did not. Annealing decreased the H content whereas leaving the films in a room environment increased it. They related the electrical properties of the films to the H content.

The hydrogen concentration in SACVD BPSG films was studied in [224]. In as-deposited 4.5 wt % [B] – 4 wet % [P] SACVD BPSG film capped with PECVD TEOS-oxygen silicon dioxide layer immediately after deposition was found to be as low as 3.2×10^{20} atoms/cm^3. This was found to be not higher than that of PECVD silicon dioxide film (about 6×10^{20} atoms/cm^3), or HDP-CVD silicon dioxide film (about 4×10^{20} atoms/cm^3). For SACVD BPSG films, the hydrogen concentration was found to be about 1.3×10^{20} atoms/cm^3 after RTA at 850 °C for 30 seconds. In all cases, hydrogen concentration was uniform across the films. Hydrogen concentration in as-deposited and non-capped SABPSG films stored in clean-room conditions for 3 weeks was found to be increased up to about 3×10^{21} atoms/cm^3 slightly decreasing from the surface to the film depth. This increase of hydrogen content corresponds to the adequate changes in FTIR spectrum moisture-related areas.

Dyer [217] studied the SIMS depth profiles for hydrogen and hydroxyl for the RTA and FA heated BPSG films deposited using Method 4 with 1.6 wt % [B] and 3.2 wt % [P]. The chemical source for both of these secondary ions was taken to be water-related (e.g., H_2O and OH). Data showed that the extent of water penetration in the films was limited and that the depth of penetration was a clear function of the anneal conditions used. Among the RTA films, the dependence of the penetration depth on the RTA parameters was similar to that of the other measures of stability. The penetration was deeper for lower RTA temperatures and shorter RTA times with the effect of temperature being dominant.

Yoshimaru et al [178] studied SIMS profiles in BPSG with 2 wt % [B] and 6.5 wt % [P] after 32 hours in the PCT. He used either H_2O or D_2O as the moisture source and Figure 2.30(a,b) shows the results for H and O – H profiles using H_2O (a) and for D and O – D profiles using D_2O (b). Profiles before the PCT are shown by dashed lines. The H profile before the PCT was found to be not flat. As could be expected, the D profile before the PCT was flat and the ion intensity was about two orders of magnitude lower than for the H profile. These results may be due to the effects of residual H_2O in the SIMS system or H_2O adsorbed at the film surface. The change in D profiles before and after the PCT was found to be larger and more reproducible than that of the H profile. The authors concluded that more reliable information on water penetration into the film could be obtained through the use of D_2O rather than H_2O.

Figure 2.31 shows D and OD profiles for SiO_2, BSG, and PSG after 32 hours in the PCT. Water was found to penetrate SiO_2 and BSG completely after the PCT. The water penetration of PSG was not complete, being only to a depth of about 0.4 μm. Therefore, water penetrated SiO_2 and BSG rapidly, but slowly for PSG and BPSG, which contain phosphorus. The D ion intensity in BSG, PSG, and BPSG in the water-penetration region were found to be higher than that in SiO_2. It was proposed that both boron and phosphorus increase water concentration in the films, but phosphorus appeared to suppress the water penetration depth in silicate glass differently from boron.

To summarize the water penetration effect in more detail, Yoshimaru et al calculated the water concentration profile in film from D profiles and the amount of absorbed water. Figure 2.32 shows the typical water profiles in BPSG films (2 wt % [B] and 6.5 wt % [P]) as a function of the PCT time. These profiles do not exhibit a simple exponential behavior, but consist of the following three regions: (1), near the surface, where the profile was rather flat and the water concentration was high; (2), in the middle of the film, where water concentration profiles decreased sharply; and (3), near the interface between BPSG and Si, where the profile was flat and had a low background intensity. Water concentration in (I) near the surface was termed surface water concentration (the surface water concentration apparently indicates the maximum water concentration that films contain). The water penetration depth in the film after PCT was defined as the depth of the point where the water concentration was 10^{20} molecules of H_2O in 1 cm^3 because the point was located in the middle of the region (2) and it was easy to define the depth from the water concentration profiles.

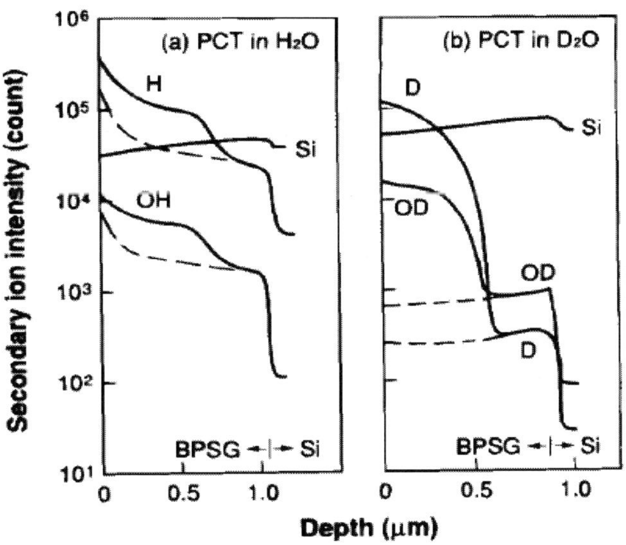

Figure 2.30. Typical SIMS profiles of water in BPSG (2 wt % [B] and 6.5 wt % [P]) after 32 h in the PCT. At left are H and O – H profiles with H_2O as the moisture source. At right are D and O – D profiles with D_2O as the moisture source. Dashed lines show profiles before the PCT [178]. Reproduced by permission of ECS – The Electrochemical Society.

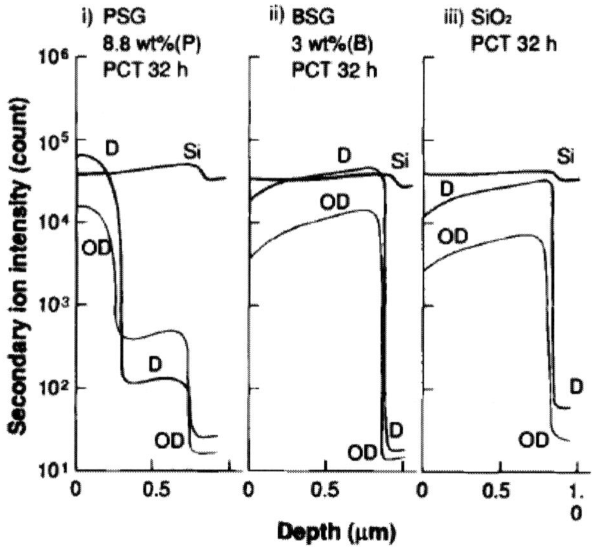

Figure 2.31. Deuterium profiles for PSG (8.8 wt % [P]) (i), BSG (3 wt % [B]) (ii) and SiO_2, (iii) after 32 hours in the PCT [178]. Reproduced by permission of ECS – The Electrochemical Society.

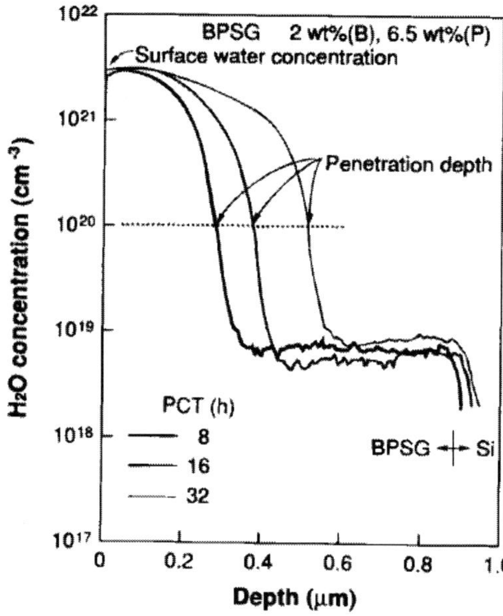

Figure 2.32. Typical water profiles in BPSG (2 wt % [B], 6.5 wt % [P]) as a function of PCT time (from 8 hours to 32 hours) [178]. Reproduced by permission of ECS – The Electrochemical Society.

Figure 2.33(a,b) demonstrates the dependence of the surface water concentration on film phosphorus concentration and the dependence of water penetration depth on film phosphorus concentration after 16 hours in the PCT at a different boron concentration, respectively.

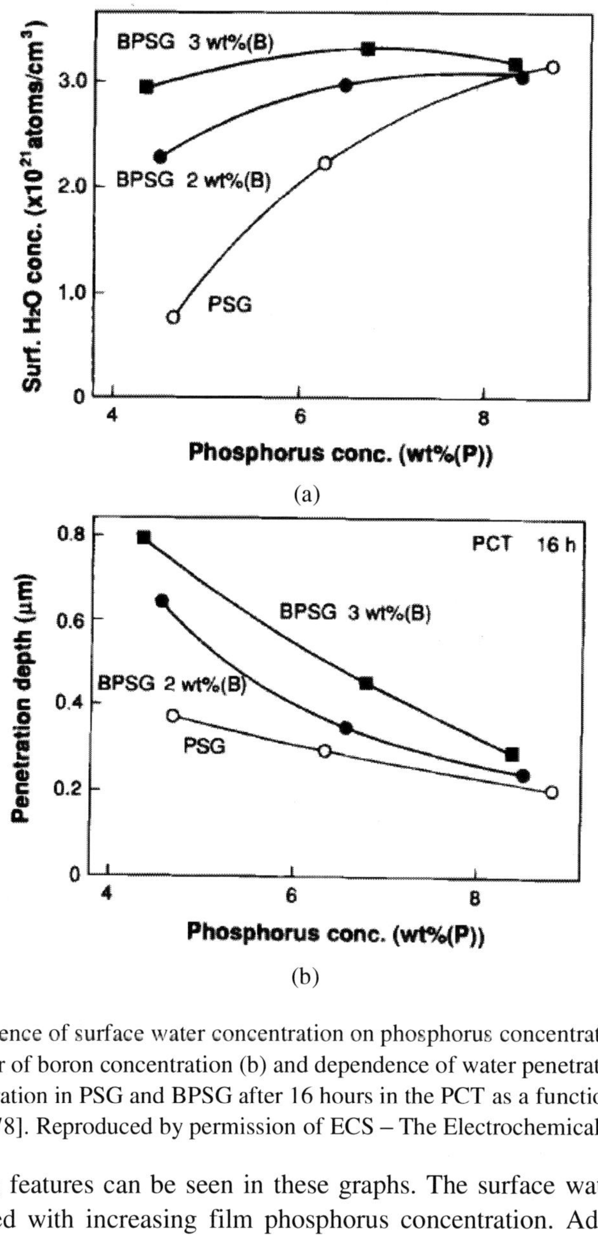

Figure 2.33. Dependence of surface water concentration on phosphorus concentration in PSG and BPSG as a parameter of boron concentration (b) and dependence of water penetration depth on phosphorus concentration in PSG and BPSG after 16 hours in the PCT as a function of boron concentration (b) [178]. Reproduced by permission of ECS – The Electrochemical Society.

The following features can be seen in these graphs. The surface water concentration of PSG film increased with increasing film phosphorus concentration. Adding boron to PSG film increased the water concentration in film with a low phosphorus concentration. Films with high phosphorus concentrations changed very little with boron addition, however. Boron increased the surface water concentration in BPSG film, but had less of an effect with increasing phosphorus concentration. The penetration depth of PSG film decreased with increasing phosphorus concentration. Adding boron to PSG film increased the water penetration depth. For film with about 4.5 wt % [P], the penetration depth of PSG film was about 0.4 μm and that of BPSG with 3 wt % [B] was about 0.8 μm. However, for film with 8.5 wt % [P], the BPSG penetration depth was almost the same as that of PSG film. The authors concluded that phosphorus appeared to suppress water penetration, and boron enhanced it. But boron had less of an effect with increasing phosphorus concentration. The dependence of the water penetration depth on boron and the phosphorus concentration was found to be similar to that observed for surface water concentration.

The final dependence of the water penetration depth of PSG and BPSG films with 2 wt % [B] on the PCT time is presented in Figure 2.34. The penetration depth logarithm increases linearly with the logarithm of the PCT time. This relationship is expressed in the author's original equation $Y = At^b$ [178], where Y is the penetration depth, t is the PCT time, A and b are constants. The constant B of PSG and BPSG is about 0.4.

Thermal Desorption Spectroscopy (TDS) analysis. The TDS technique allows clarifying the status of water species in films. Yoshimaru et al [178] studied SiO_2, BSG, PSG and BPSG films after 100 hours PCT. Figure 2.35(a) shows the water desorption from SiO_2 and BSG with 3 wt % [B] reflecting similar water desorption profiles as long as both films demonstrate a sharp water desorption peak at about 280 °C.

Water desorbed at low temperatures is assumed to be that physically adsorbed in the film pore structure. According to the authors [178], boron apparently increases the number of film pores because the amount of water desorbed from BSG film at low temperatures (280 °C) is larger than that for SiO_2 film. Figure 2.35(b) shows the water desorption from PSG with 8.5 wt % [P] and BPSG film with high and low phosphorus concentrations. PSG film did not show a sharp water desorption profile, but had a broad desorption profile between 250 and 700 °C. PSG film could have fewer pores than SiO_2 and BSG films. The water desorption profile for BPSG film with a low phosphorus concentration was similar to that of BSG and SiO_2 films. The film showed a sharp water desorption peak at about 280 °C. The water desorption profile of BPSG film with a higher phosphorus concentration was similar to that of PSG film, showing only a broad profile from 250 to 700 °C. BPSG film with a high phosphorus concentration presumably had a smaller number of pores than the BPSG film with a low phosphorus concentration. According to the author's conclusion, the pore structure may influence the water penetration path in silicate glass films, and the water penetration depth was determined by the change in pore structure caused by boron and phosphorus.

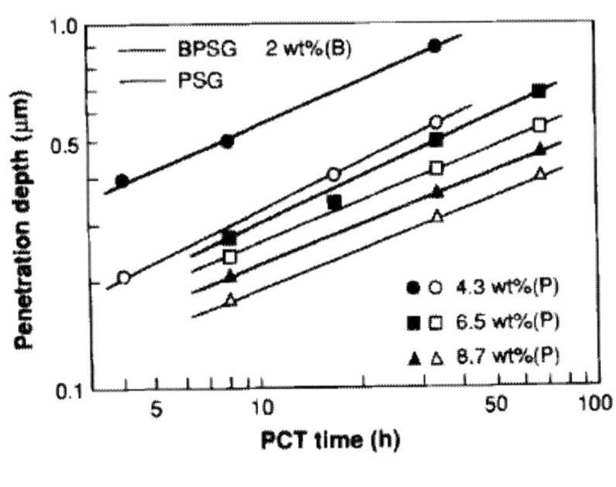

Figure 2.34. Dependence of water penetration depth of PSG and BPSG films with 2 wt % [B] on PCT time [178]. Reproduced by permission of ECS – The Electrochemical Society.

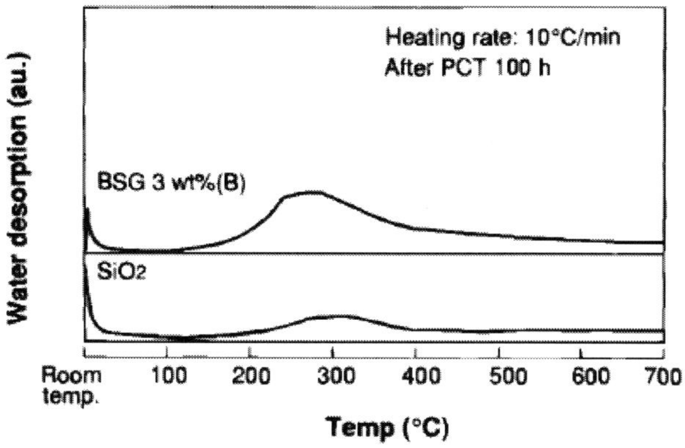

(a)

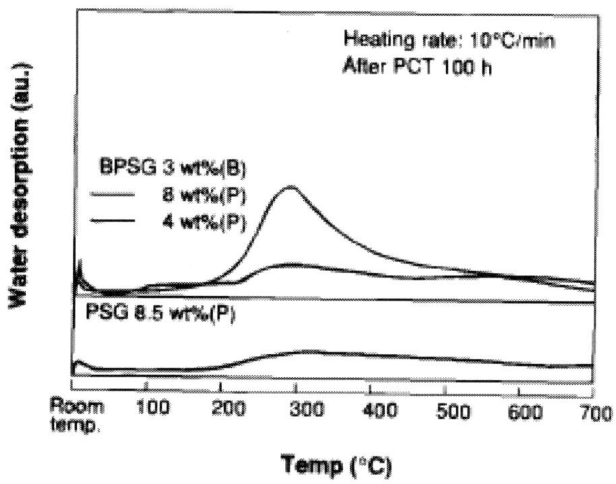

(b)

Figure 2.35. Water desorption from SiO2 and BSG with 3 wt % [B] alter 100 hours in the PCT (a) and water desorption from PSG with 8.5 wt % [P] and BPSG with high and low phosphorus concentrations after 100 hours in the PCT (b) [178]. Reproduced by permission of ECS – The Electrochemical Society.

Moisture absorption and film properties. There were very few publications about changes in the film properties during long time film exposure to ambient air. Changes in the film properties over the exposure time to the ambient seem to be a feature of the as-deposited films at long air-exposure time. Long-time thermally annealed, i.e. significantly densified films revealed very moderate interaction with the ambient moisture and stability of the film properties on time. For instance, Dyer [217] showed that the residual stress exponentially decreased and the thickness of as-deposited films exponentially increased over time due to the moisture instability whereas these properties were stable in furnace annealed films. Films treated under various RTA conditions (temperatures from 700 to 900 °C and times from 10 to 50 sec) were chemically similar to each other but had different physical properties depending on the specific conditions of the anneal process.

To monitor net charge values in as-deposited BPSG films during long time exposure to the clean-room ambient, the author used the Surface Charge Analysis (SCA) technique [214].

Charge values in both studied BPSG film types (deposited using Method 5 and Method 8) revealed stable values up to 250 hours after film deposition with a little decrease, as shown in Figure 2.36(a).

Normally, the storage of as-deposited BPSG films for several days caused the formation of so-called "circular BPSG defects" on the film surface (see Chapter 13 for details). The complete removal of these defects (which have a liquid nature) from the surface of glass film can be achieved by vacuum pumping in helium flow at 440 ºC for 2 minutes that is equal to the first step of the film deposition recipe. Vacuum pumping caused a dramatic change of SCA values and Figure 2.36(b) shows the decrease of SCA values after vacuum pumping for BPSG films as a function of the total additive concentration. However, we did not observe serious changes of the net charge for thermally annealed and long time exposed to clean-room ambient BPSG films after pumping. This clearly indicates the difference between the film's density and the film's ability to absorb moisture. Charge changes after pumping in as-deposited (and supposed to be porous) BPSG films were explained by the removal of moisture and products of additive-moisture interaction from the film.

The proposed reason for the observed physical and net charge changes was moisture absorption affecting mostly as-deposited films. However, as-deposited BPSG films are not in use in IC device technology that makes a study of their property changes rather meaningless. Based on common sense, the time delay after BPSG film deposition and following processing should be minimized. For all of the anneal conditions, the post anneal film density is proposed to be a consistently good predictor of the subsequent film property stability during time.

12.3. Description of Moisture Absorption Effects in BPSG Films

Film composition and moisture absorption activity. Hydrogen-contained species in the glass films are normally attributed to the water molecules. However, the author distinguished three hydrogen sources in the 4.5 wt % [B] – 4 wt % [P] borophosphosilicate glass film samples deposited using Method 5 [270]. Approximately 20 % of the initial hydrogen (3.2×10^{20} atoms/cm^3) amount was supposed to be present in alkoxy groups ($-OC_2H_5$). These groups are residuals of incomplete oxidation of the initial alkoxy compounds of silicon, boron, and phosphorus upon oxidation reaction. Approximately 80% of the initial hydrogen amount can be in the form of water as the product of oxidation reactions of organic esters. The initial absolute water concentration in the studied SiO_2 capped 4.5 wt % [B] – 4 wt % [P] BPSG film was estimated to be about < 0.15 wt %. Here the thermodesorption data [178] were used for the SIMS calibration. The water absorbed by glasses from ambient air is also the hydrogen source. It was found to be increased up to about 3×10^{21} atoms/cm^3 in a 3 week period when the samples were stored in a clean-room wafer box. According to the SIMS data on the hydrogen content, the water concentration after such exposure was estimated to be about 1.8 wt % and this was purely absorbed moisture by the films. For the glasses produced by PECVD, the water content in these glasses even upon prolonged exposure to air did not exceed tenths of a weight percent.

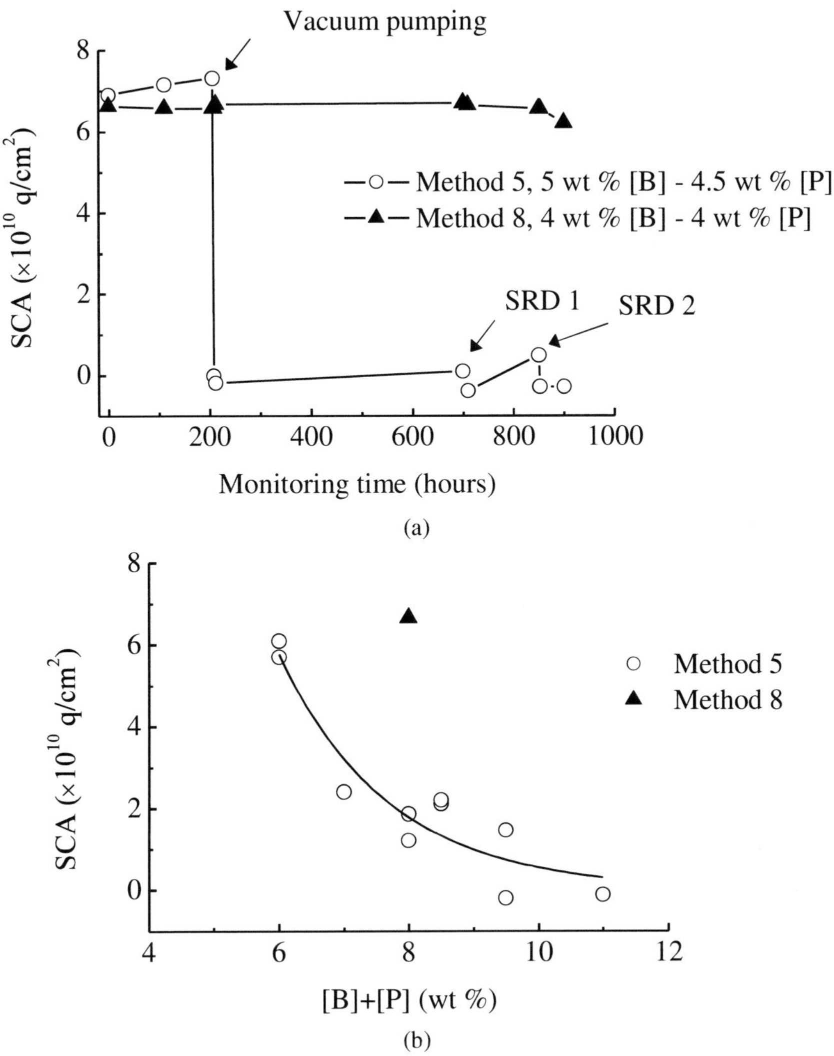

Figure 2.36. Monitoring of as-deposited BPSG films with different additive content during long time storage followed by vacuum pumping (a) and SCA values vs. total additive content in films after pumping. Presented at 4th Intern. Dielectric for ULSI Multilevel Interconnection Conf. (DUMIC), 1998 [214].

As can be seen in Figure 2.26, as-deposited boron-enriched films are prone to absorb significantly more moisture compared to the films with the other compositions. Undoubtedly, this strong absorption confirms a deep moisture penetration in the film. In fact, Yoshimaru et al [178] concluded that boron in BPSG films enhances the water penetration into the depth of the film, but the phosphorus was prone to fix moisture on the film surface, suppressing moisture penetration into the film depth. Figure 2.37 shows the dependence of the surface water concentration on the phosphorus concentration at different boron concentrations. The data are the same as those in Figure 2.33(a), except that the unit on the abscissa is changed from the phosphorus weight percentage to atom/cm^3.

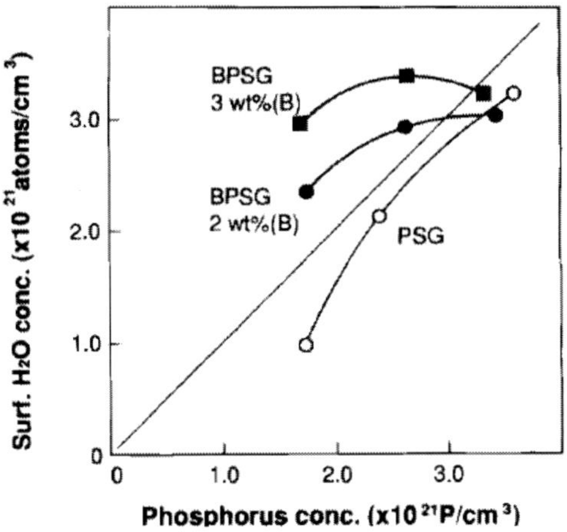

Figure 2.37. Dependence of surface water concentration on phosphorus concentration as a function of boron concentration. The 45° angle straight line indicates a case where all phosphorus in the film combines with water molecules [178]. Reproduced by permission of ECS – The Electrochemical Society.

The common explanation of the interaction of phosphorus-contained glass films with moisture is that P_2O_5 in silicate glass interacts with water according to a scheme (2.4):

$$(\text{-Si-O-})_3P{=}O + H_2O \rightarrow (\text{-Si-O-})_3P(OH)_2 \qquad (2.4)$$

This scheme assumes that each phosphorus atom site in the film combined with one molecule of water, i.e. the water concentration (H_2O/cm^3) becomes the same as the phosphorus concentration (P/cm^3). Thus, in the case where all absorbed water combines with all phosphorus in the film, it could be expected that the data points in Figure 2.37 would fall on a straight 45° angle line. In Figure 2.37, the data points for PSG are plotted near and parallel to the 45° angle line, i.e. almost all water in PSG film combines with P_2O_5.

Data points of BPSG films do not follow this line. Water concentration in BPSG film with a high phosphorus concentration falls near the line and data points for water in BPSG film with low phosphorus fall above the line. According to [178], this result suggests that boron increases the number of additional water absorption sites in BPSG film, but highly contented phosphorus reduces the number of additional water absorption sites.

As shown above, the common explanations of the interaction of the phosphorus-contained glass film with the water molecules are usually done using the double P = O bond in accordance with the scheme (9.1) [178]. Along with the increase of peaks intensity in the moisture-related areas, this interaction is normally associated with the loss of P = O intensity in FTIR spectra. For instance, Figure 2.38 shows Dyer's data [217] where a direct correlation of absorbed moisture and P = O bond loss can be seen.

After thermal anneal, all phosphorus-contained films revealed a decrease in moisture-related peaks, and an increase in the P = O peak intensity. This change reflects water desorption from the film, which is commonly described as follows (2.5-2.7):

$$H_2O_{\text{(absorbed)}} \rightarrow H_2O_{\text{(gas)}}, \qquad (2.5)$$

$$2(\equiv SiOH) \rightarrow \equiv Si - O - Si \equiv + H_2O, \quad (2.6)$$

$$\equiv P(OH)_2 \rightarrow \equiv P = O + H_2O \quad (2.7)$$

Based on schemes (2.4) and (2.7), one could propose the formation in the BPSG films the structural groups with 5 equal phosphorus-oxygen bonds, for instance, $(\equiv Si - O -)_2(OH)_2P - O - B(- O - Si -)_3$. However, as long as the phosphorus is known to save $P = O$ bond in oxygen-contained compounds, the interaction of the phosphorus-contained films with the moisture can be rather explained using the following scheme:

$$(\equiv Si\text{-}O\text{-})_3OP + H_2O \rightarrow (\equiv Si\text{-}O\text{-})_2OP\text{-}OH + \equiv Si\text{-}OH \quad (2.8)$$

We believe the scheme (2.8) adequately explains the well-known phosphorus leaching effect from the phosphorus-contained films discussed above, because the scheme (2.4) does not give such an opportunity. In fact, further interaction of the reaction product $(\equiv Si\text{-}O\text{-})_2OP\text{-}OH$ in scheme (2.8) with ambient moisture H_2O should produce $(\equiv Si\text{-}O\text{-})OP(OH)_2 + \equiv Si\text{-}OH$. Following this logic, after reaction with moisture $(\equiv Si\text{-}O\text{-})OP(OH)_2$ it should produce phosphoric acid $OP(OH)_3 + \equiv Si\text{-}OH$. In reality, the presence of droplets of phosphoric acid on the phosphorus-contained films is a well-known feature of these films.

The explanation of boron effects is slightly more complicated. Explaining high activity of as-deposited boron-rich BPSG films with respect to the moisture (see Figure 2.26), a simple scheme (2.9) similar to (2.8) can be used:

$$(\equiv Si\text{-}O\text{-})_3B + H_2O \rightarrow (\equiv Si\text{-}O\text{-})_2B\text{-}OH + \equiv Si\text{-}OH \quad (2.9)$$

Similarly to phosphorus leaching, the following interaction of the reaction product with water should produce boric acid $B(OH)_3$. However, the boron leaching effect during the glass thin film interaction with the moisture is not as strong as the phosphorus one. It has never been clarified in the literature. For instance, Yoshimaru et al showed that the B-O absorption peak (1380 cm^{-1}) of BSG film with 3 wt % [B] is stable during 100 hours in the PCT. This indicates that B_2O_3 in annealed BSG film does not react with water. The change in the B-O peak in BPSG is not clear, because the B-O peak in BPSG overlaps with the P=O (1300 cm^{-1}) peak and the P=O peak decreases with PCT time. The Si-OH (3650 cm^{-1}) and H-O-H (3300-3500 cm^{-1}) peaks in BPSG film after PCT, however, are smaller than that of BSG film. This indicates that B_2O_3 in BPSG scarcely interacts with water and, therefore, B_2O_3 in the film apparently does not act as a water absorption site in BPSG. In other words, this result suggests that boron increases the number of additional water absorption sites in BPSG film without interacting with water.

We believe the following explanations of accumulated data and existing discrepancies can be done. Using the discussion in Chapter 9, the activity of the boron-rich films to the moisture can be explained by threefold boron coordination with respect to the oxygen in these films and its ability to change coordination to a fourfold one. Boron oxide is known to be a Lewis acid. Thus, one can propose that besides the interaction according to the scheme (2.9), it can be an additional path of interaction like forming a group $(\equiv Si\text{-}O\text{-})_3BOH_2$ with a fourfold boron coordination. However, we believe that this interaction is supposed to be even weaker than that, producing a group with a weak water bonding [(≡Si-O-)$_3$B]•H_2O. This type of bonding is supposed to help the water molecules move easily into the depth of the film. The validity of this conclusion can be proved by the following simple comparison. The size of a water molecule (distance between hydrogen atoms is about 1.5 Å) is less than the distance between planar layers in B_2O_3 structure (4.3 Å) and less than middle-range order (2.8 – 5.5 Å) in glasses.

Gradually adding P_2O_5 to the B_2O_3 or SiO_2-B_2O_3 mixture resulted in the change of boron coordination from threefold to fourfold with respect to oxygen atoms. Therefore, the share of threefold coordinated boron atoms decreases and the film structure becomes more ordered. As a result, the film becomes less active with respect to the moisture interaction, particularly, in terms of moisture penetration into the film depth. As for the phosphorus, it still remains an active agent with respect to the moisture and fixes water molecules closer to the film surface.

Additional features of BPSG film structure. Phosphorus in CVD glass films was assumed to be randomly substituted in the silica network preferably in the form of P_4O_{10} or, as commonly designated, P_2O_5.This means that all the phosphorus atoms are defined to be in the form of P^{5+}. According to considerations made in Chapter 9 in these glass films all glass-forming oxide species are supposed to exist as tetrahedra for silicon, phosphorus and partly, boron, and triangles and boroxol groups for boron too. Assuming the film consists of tetraherdra only, the structural groups (≡Si-O-)$_2$PO-O-, (≡Si-O)$_2$B-O-, (≡Si-O-)$_2$OP-O-B(-O-Si≡)$_3$ can be proposed.

However, based on the leaching studies, Yoshimaru et al [178] proposed that part of the phosphorus in BPSG films deposited using Method 1 with a concentration higher than 9 wt % [P] may form P_2O_5 clusters. The clustered phosphoric oxides may leach out during the PCT, leaving the porous silica structure. This was observed by using Transmission Electron Microscopy observation using the replica method that the film surface of phosphorus-leached areas was extremely porous.

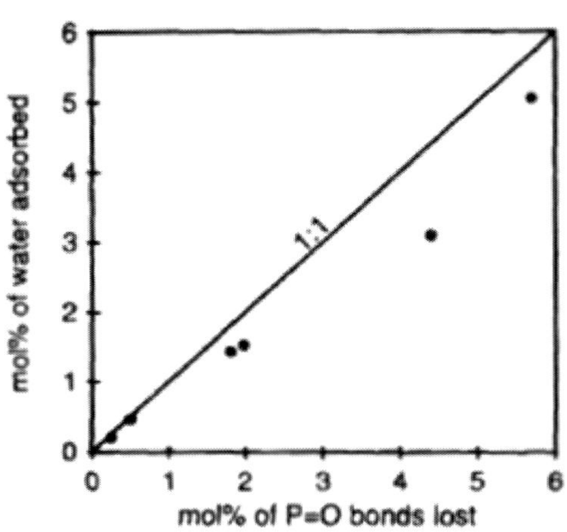

Figure 2.38. Stoichiometric relationship between absorbed water and lost P = O bonds for the PSG films with various 6.2 mol % P [217]. Reproduced by permission of ECS – The Electrochemical Society.

Chapter 13

DEFECT FORMATION PHENOMENON IN BPSG FILMS

13.1. INTRODUCTION: PROBLEM DESCRIPTION

Despite the active implementation of BPSG films in IC device technology since the mid 1980s, there was a serious challenge in the practical use of these films. It has been a very complicated matter with a long history, since the first detailed BPSG paper was published by Kern et al. in 1982 [50].

This challenge is known as a phenomenon of defect formation on the surface of BPSG films after the film deposition or its high temperature thermal treatment. The term "BPSG defect" can be defined as any film imperfection observed on/in BPSG thin film. BPSG films after an optimized CVD process reveal surface (or micro-particle) defect density (DD, cm^{-2}) less than about 1 cm^{-2} with defect size less than about ten microns. These defects are due to the equipment maintenance quality and chemical reaction features.

However, additional defects appear on the BPSG films after the film exposure to ambient air and/or after post-deposition thermal anneal of BPSG films. Until recently, a few publications presented images of different BPSG defect shapes. Summaries of BPSG defect studies were published in reviews by the author [93,207,226,246]. These papers concluded that the key of this phenomenon is the film composition. The addition of phosphorus and, especially, boron oxides into silicon dioxide glass matrix causes changes in the film structure and properties and, therefore, new issues. The positive side of BPSG film implementation is a much better ability to flow comparing to PSG films. The negative side is a decrease of film stability with the increase of additive, especially – boron, content in the glass.

A brief conclusive summary of published BPSG defect data are presented in Table 2.10. In this table, additive concentrations are presented in elemental weight percents and shown as [B] and [P]. Analytical methods used for studies are designated as OM (optical microscopy), SIMS, SEM, LS (Laser scanning), AES (Auger Electron Spectroscopy), and TEM (Transmission Electron Microscopy), AFM, EDS (Energy-Dispersive Spectroscopy). Images of BPSG defects are presented in Figure 2.39(a-n), 2.40(a-e), Figure 2.41(a,b).

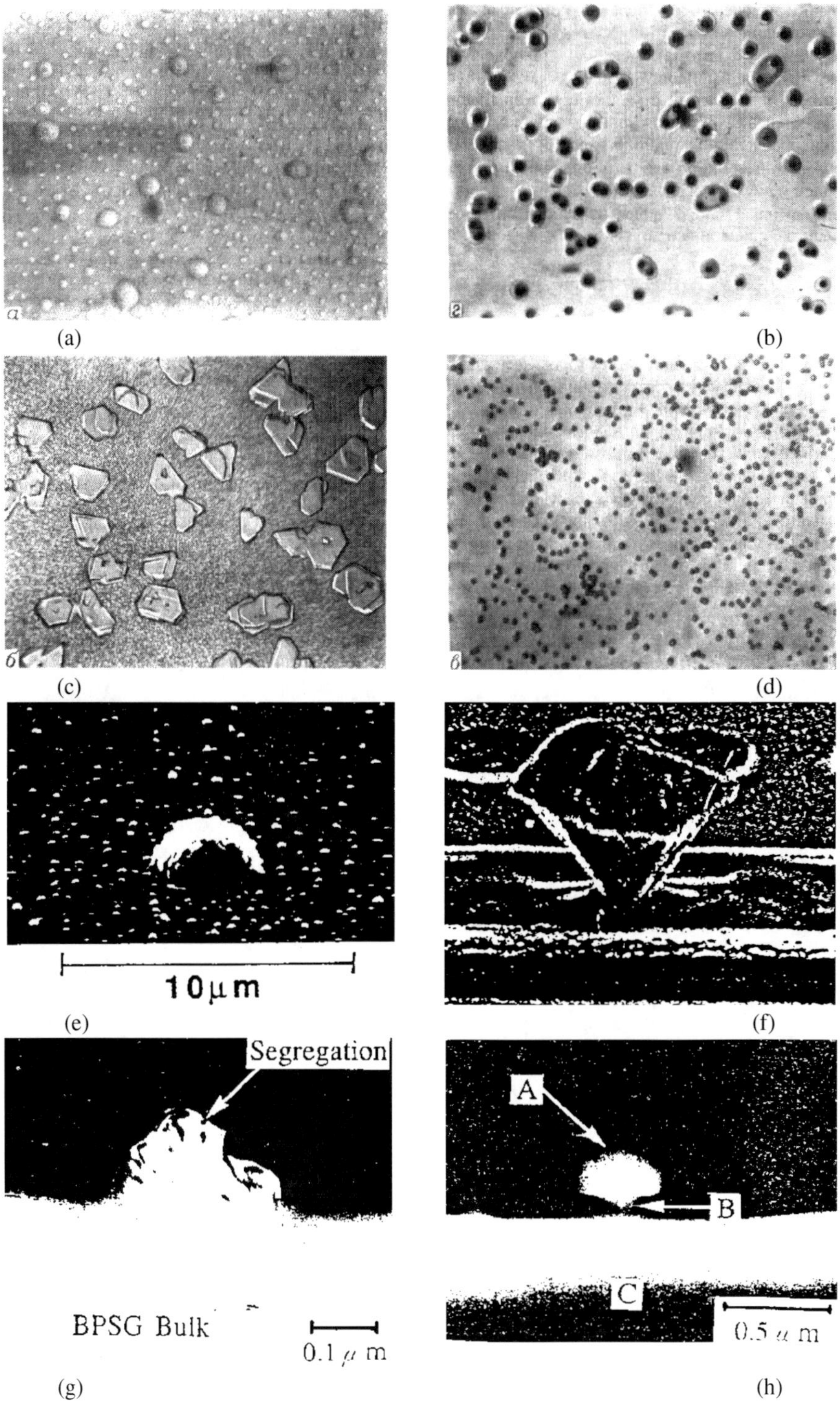

Figure 2.39. (Continued).

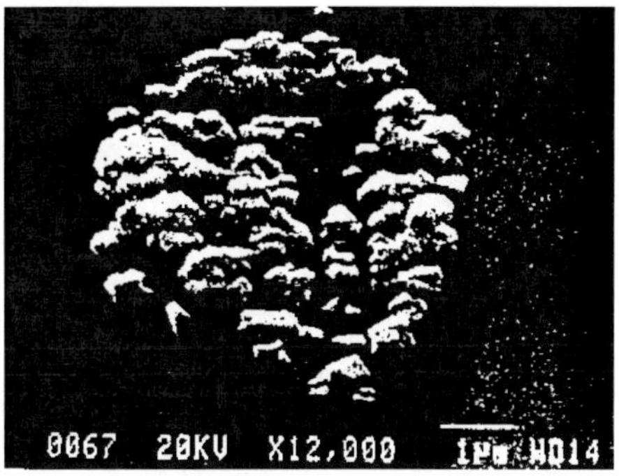

(i)

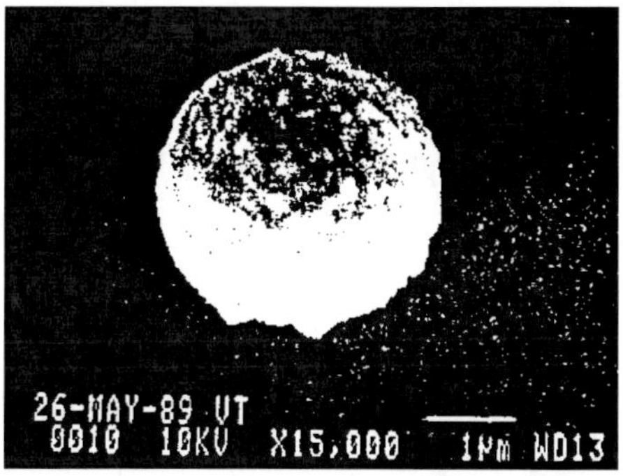

(j)

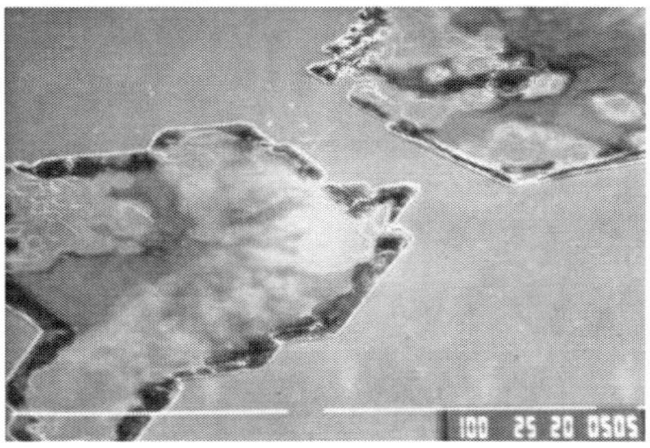

(k)

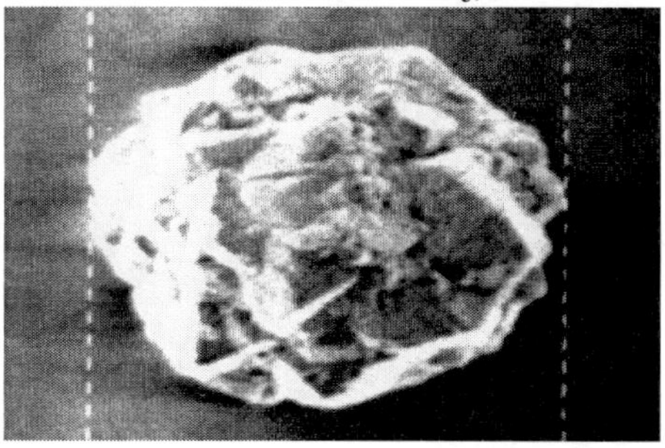

(l)

Figure 2.39. Images of observed defects in BPSG films: (a-c) [89]; (d) [82] and (e) [102] (Reproduced by permission of ECS – The Electrochemical Society); (f) [130] (Reproduced by permission of Cambridge University Press, Copyright 1993). (g,h) [113] (Reprinted with permission from American Institute of Physics, Copyright 1992; (i,j) [112] (Reprinted with permission from American Vacuum Society, Copyright 1992); (k) the author unpublished data, (l) [163] (Reproduced by permission of ECS – The Electrochemical Society).

It can be seen that a variety of BPSG defect types were observed on the film surface and this confirms that BPSG defect formation was a serious issue for the IC technology. Defect appearance undoubtedly depends on the additive concentrations in the films. Some attempts to optimize additive content in BPSG films were done along with the tightening of device gap-fill and planarization requirements. The Chemical Mechanical Polishing (CMP) technique is now used for BPSG planarization purposes. As a result, the actuality of BPSG film planarization on the device steps using post-deposition high-temperature anneal has been decreased. In contrast, along with the shrinkage of device elements, an issue of void-free gap-fill for narrow device gaps has arisen. This is consistent with requirements to lower BPSG film post-deposition anneal temperature and time to provide new silicide technology schemes. Thus, an increase of additive concentrations in BPSG film and optimization of post-deposition anneal either at isothermal furnace (FA), or Rapid Thermal Anneal (RTA) conditions has still been considered as a good option for advanced ULSI technology. This trend keeps the BPSG defect formation effects under control and research. Among other studies, the author's studies on this topic have been performed on a systematic basis [186,187,190,211,226]. This has allowed some important conclusions and the drawing of a mechanism of BPSG defect formation.

Table 2.10. Summary of BPSG deposition/anneal methods and condition, observed defects

Ref N	Deposition condition, (anneal condition)	Additives (wt %)	Method used	Defect characterization (in original author's terminology)
[50]	SiH_4-O_2; 450 °C APCVD (750–950 °C)	[B]=4–6 [P]=3–4	OM SIMS	"Boric acid crystals"; their removal damaged the film surface
[61]	SiH_4-O_2; 400 °C LPCVD	[B]=<6 [P]=5–7	OM	"Hygroscopic films" at [B]>5%, crystal formation at air exposure
[62]	SiH_4-O_2; 450 °C; APCVD; (750-900 °C; PH_3+O_2+N_2)	[B]=3.6 [P]=4	SEM	"BPO_4 particles", ~ 10 μm, on BPSG films with PH_3 during anneal; no defects without PH_3.
[69]	SiH_4-O_2; 450 °C APCVD	[B]=< 8 [P]=< 10	OM	"Surface crystallization" at [B]>10%; "hygroscopic glass" at [B]>6% and [P]>6%.
[78]	TEOS-O_2; 650 °C LPCVD	[B]=<10.8[P]=<5.6	OM	"Boric acid crystals" at [B] 5-10.8% at air exposure
[82]	TEOS-O_2; 680 °C; LPCVD; (900–950 °C; steam; PBr_3; O_2)	[B]=1.1–4.6 [P]=3.5–5.7	OM SEM	"Microcrystal phase separation" in PBr_3; round shape; ~ 2-4 μm; rich with [P]; no defects in steam.
[89]	SiH_4-O_2; 420 °C; APCVD; LPCVD; (800-950 °C; Ar; O_2)	[B]=2–8 [P]=2–10	OM SEM	"4 defect types" at total dopants > ~10%: "a"-round ~10 μm ([B]≤[P]); "b"-crystals ~30 μm ([B]≥P]); "c"–round ~1 μm; "d"–round double and triple ~5 μm ([B]≈[P]).
[102]	APCVD; (740–850 °C; N_2)	[B]=4 [P]=5.5	SEM	"Crystal BPO_4 precipitates", ~ 4 μm
[109]	SiH_4-O_2; 420°C; LPCVD (700-850 °C)	[B]=4–9.5 [P]=3–6	SEM OM LS	"Boric acid crystallites"; ~ 0,5 μm; suppressed by top "undoped" oxide
[112]	SiH_4-O_2; 400 °C; LPCVD 900 °C; steam-N_2; O_2/N_2; O_2; (treatment in water carrier box or water rinsing before anneal)	[B]=2–4 [P]=6–7	OM SEM AES	"Grapelike crystalline cluster defects": "a"- ~ <1 μm, segregated in circles (short exposure time); "b" – transparent boric acid crystals; ~ <10 μm (long exposure time). Water soluble. No defects found in steam.
[113]	SiH_4-O_2; 370 °C; APCVD; (900 °C; N_2)	[B]=3.76 [P]=7.05	SEM TEM	"Surface segregates"; ~0,5 μm; proposedBPO$_4$ compound
[120]	TEOS-O_3; 400 °C; APCVD (750–850 °C)	[B]=3.9 [P]=4;	SEM	"Dopant precipitates" at total dopants >~10%.
[126] [128]	SiH_4-O_2; 450 °C; APCVD; TEOS-O_2; 410 °C; PECVD; (800 °C, steam, 900 °C, N_2)	[B]=4.5 [P]=4.5	SEM LS	"Thermo-induced defects"; ~ 4 μm.
[130]	SiH_4-O_2; 450 °C; APCVD (920°C; /$POCl_3$/O_2)	[P]=6-8	SEM LS	"BPO_4 mushrooms"; ~ 2-10 μm;

Ref N	Deposition condition, (anneal condition)	Additives (wt %)	Method used	Defect characterization (in original author's terminology)
[146]	SiH_4-O_2; 450 °C; APCVD (RTA 120 sec at 900-1000 °C)	No data	SEM OM LS	"Rainbow defects";~ 1 μm; "borate defects";~300 μm.
[153] [154]	SiH_4-O_2; 450 °C; APCVD (RTA 45 sec at 900-1100 °C, N_2;O_2)	No data	SEM TEM	Defects defined as "flowers", "mushrooms", "precipitates" with submicron size, all were attributed to BPO_4 particles. Their surface concentration depended on deposition and anneal temperature
[163]	SiH_4-O_2; 450 °C; APCVD (850-950 °C; N_2;O_2; steam)	[B]=2.7-5 [P]=5-8	SEM	"BPO_4 microcrystals"; ~ 4,4 μm; Concentration is lower at steam, and high gas flow, higher in N_2.
[179]	TEOS-O_3; 400 °C; APCVD (850 °C)	[B]=4-7 [P]=3-10	SEM	"BPO_4 precipitates"; < 0,1μm inside BPSG film
[170]	APCVD; (900°C; N_2)	[B]=5-7 [P]=6-8	SEM	"Three types of BPO_4 crystals", ~ 1 μm, on the surface and inside film
[186] [187] [190] [211] [226]	TEOS-O_3; 440 °C; SACVD; TEOS-O_2; 410 °C; PECVD; (850–950 °C; N_2; O_2)	[B]=3-10 [P]=3-7	LS AFM OM SEM	Round liquid defects> 0,2 μm; merged liquid defects; crystals, dendritic crystals with size up to a few millimeters; solid defects with size about a few μm
[288] [290]	SiH_4-N_2O; 400 °C PECVD (950 °C)	[B]=1-4.9 [P]=3.4–9.7	OM SEM EDS	Particles <0.4 μm in as-grown films, disappeared in air; particles 1–2 μm in annealed films; amorphous, polymer/-cristalloid, proposed BPO_4

13.2. Methodology of BPSG Thin Film Defect Studies

Defect formation phenomenon is a common feature of BPSG films. Some authors reported that defects in BPSG films could be stimulated by underlying layers. In order to study BPSG film defects, the films were deposited on bare silicon wafer after standard chemical treatment. The film thickness used for defect observation is normally in the range of 0.5 – 1 micron.

There was a difference seen in the behavior in defect density and speed of defect appearance on the film surface for films deposited by different CVD techniques. Therefore, some types of BPSG films could not considered being good for investigation. For instance, this conclusion was done due to the long period of defect appearance and the low efficiency of studies with microscopy techniques for some types of BPSG films. According to the author's studies performed using different types of BPSG films [89,186,187,190,211,226], the films deposited by Method 5 (i.e. TEOS-ozone SACVD BPSG films) were considered to serve as an "ideal" object for the investigation of defect formation phenomenon. This was true because these films were found to be more pronounced to form surface defects. It means that at similar additive concentrations in the films deposited by different methods, the time of defect appearance on TEOS-ozone BPSG film surface was shorter and the speed of defect formation was higher. Eventually, surface defect density was much higher and convenient for observations.

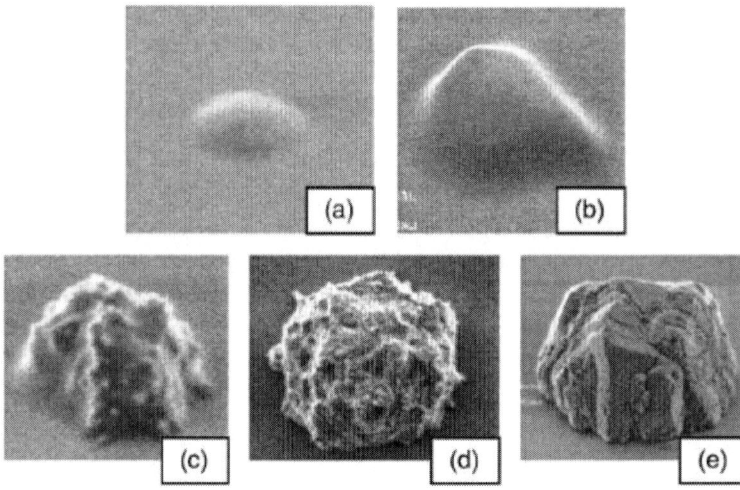

Figure 2.40. SEM images of the particles present on the BPSG surface after high-temperature annealing [288]: (a,b) amorphous; (c,d,e) polymer/crystalloid. Reproduced by permission of ECS – The Electrochemical Society.

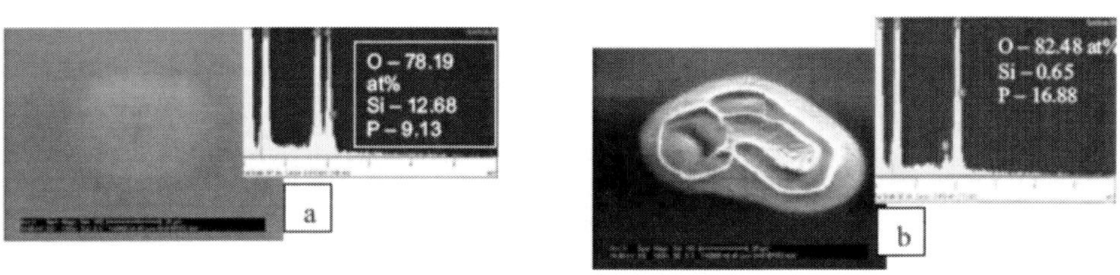

Figure 2.41. SEM image of the typical as-grown defects that are generated at an extremely low SiH_4 gas flow (0.12 slpm) and, consequently, with high boron incorporation of 6 wt %. Defect type (a) differs from defect type (b) by Si/O/P ratios (EDS spectrometer does not detect B and H atoms) [290]. Reproduced by permission of ECS – The Electrochemical Society.

Considering the differences in our approach for BPSG defect investigation as compared to the other publications, it is necessary to highlight the following. First, defect data presented in Table 2.10 and in Figure 2.39 - Figure 2.41 were published without indications of the time since the beginning of film exposure to the ambient air after the film deposition or its thermal anneal. This complicates the analysis of the published results. However, it was shown that the time between the completion of the CVD film process or the film anneal and the time of the glass exposure to ambient air (t_e) is very important. The ambient air condition was chosen to be a standard clean-room condition, namely: 22 °C and 40 % of moisture content. For these conditions, depending on the additive content in the films, defects in BPSG films could appear either faster, or slower. In Figure 2.42, two boundary lines corresponding to 1-hour and 24-hours defect-free areas in as-deposited silane-based BPSG films ([69] and [89], respectively) are presented. Other data points (triangular points) represent defects for BPSG compositions shown in Table 2.10 without indication of exposure time to air. Lines for BPO_4 and crystallization area are the same as shown in Figure 2.13.

Secondly, to perform comprehensive BPSG film defect studies, the author used a combination of non-destructive optical methods, such as laser scanning (LS, Figure 2.3), high-resolution optical microscopy (OM, Figure 2.4), AFM (Figure 2.5), and SEM (Figure 2.7). Using the LS method, the films were monitored immediately after film deposition or thermal anneal. This allowed observing the defect appearance and growth from the beginning of these processes. These approaches of monitoring allowed us to reveal the early stages of defect formation and defect growth, and the relationships between different types of observed defects. Eventually, these analytical methods together allowed covering a wide range of surface concentration of BPSG defects (DD, cm^{-2}) and the defect size, as shown in Figure 2.43.

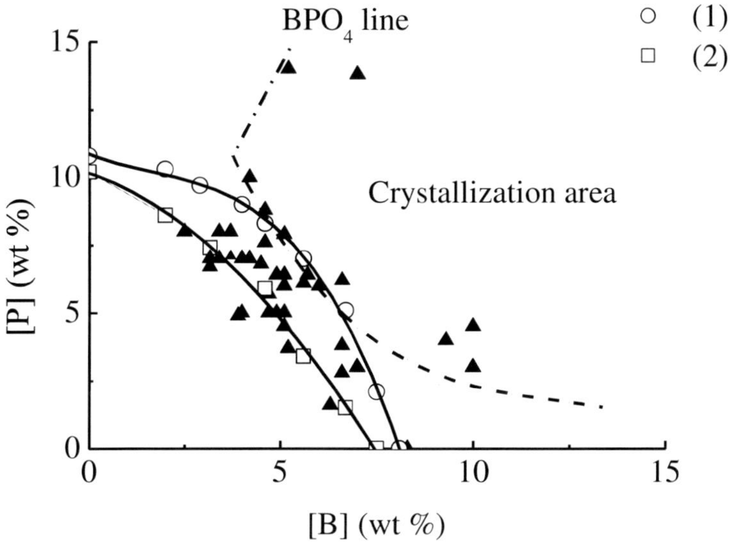

Figure 2.42. Boundary lines correspond to 1-hour (1) and 24-hours (2) defect-free areas in as-deposited silane-based BPSG films (re-calculated form data [69] and [89], respectively). Triangle data points represent the BPSG film compositions, which were indicated to have defects, but without clarification the time after the beginning of the film exposure to the ambient air.

Defect studies were performed in the following sequence. After the film deposition on the pre-cleaned silicon substrate with diameter 200 mm, or after the film thermal anneal, the LS method was used immediately to see the initial defect density and to localize the place of defect appearance on the substrate. The films were then monitored with the LS method for some time (t_m) in order to detect changes in the areas of defect appearance on the wafer and to obtain a distribution of defect size, see Figure 2.3(a-c). Monitoring was performed till the DD value exceeded about 7 cm^{-2}. This value was found to be a certain upper limit of the laser scanning tool that was used. Along with LS monitoring and, especially, after exceeding the LS tool limit, the films were studied using OM and AFM.

It is necessary to mention that the OM technique was used for defect monitoring in the center of wafers. This was chosen because there was a significant difference in SACVD BPSG defect density and defect size within a 200 mm silicon wafer. This non-uniformity did not depend on the thickness or additive non-uniformities within wafers, which were evaluated to be about ±1.5 % and ±0.2 wt %, respectively. For this type of film, a difference in the film porosity was proposed to be the only factor, which effects SABPSG defects distribution within across the wafer.

13.3. Kinetics of BPSG Film Defect Growth during the Film Exposure to Ambient Air

Kinetics of defect density changes in BPSG films studied with a laser scanning method were presented in detail in our paper [187]. SACVD BPSG films were studied with similar boron and phosphorus concentrations and total additives content in the range of 6 – 11 wt %. At the beginning of the registration, defects were found to have a round shape and a size above 0.2 μm that was due to the limit of the LS tool resolution that was used.

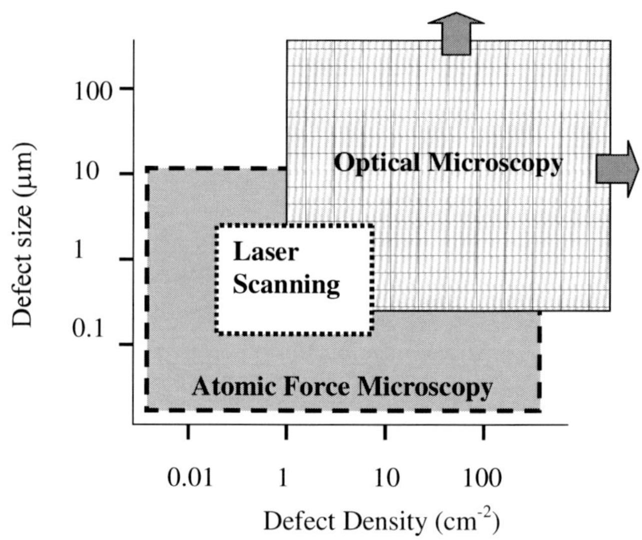

Figure 2.43. Diagram of studied BPSG film defect size and defect density ranges with the use of optical microscopy, laser scanning and Atomic Force Microscopy methods.

We found principal differences between defect appearance and growth in as-deposited and annealed SABPSG films. The appearance of defects on the top of as-deposited SACVD BPSG film started in some wafer areas after several hours of film exposure to the clean-room air (see Figure 2.3(a-c)). A certain incubation period (depicted by block arrow) in defect appearance in as-deposited BPSG films was found. The beginning of defect registration depended on additive concentrations, as shown in Figure 2.44(a). As can be seen, the time of incubation period had a reverse dependence on the total additive concentration in the film.

After the beginning of defect appearance, the rates of defect increment (DD/t_m) increased with an increasing of additive concentrations. There was a growth followed by a saturation trend of $\log(DD/t_m)$ observed in as-deposited SACVD BPSG films in the range of total additive concentration 8–11 wt % (see Figure 2.44(c)). From the beginning of defect registration, a rate of total defect density increment (DD/t_m) was estimated to be in the range of 0.05 – 0.5 cm^{-2} per hour (see Table 2.11). The rate value was found to be dependent on the film composition and thickness of the silicon dioxide cap layer on the top of BPSG film.

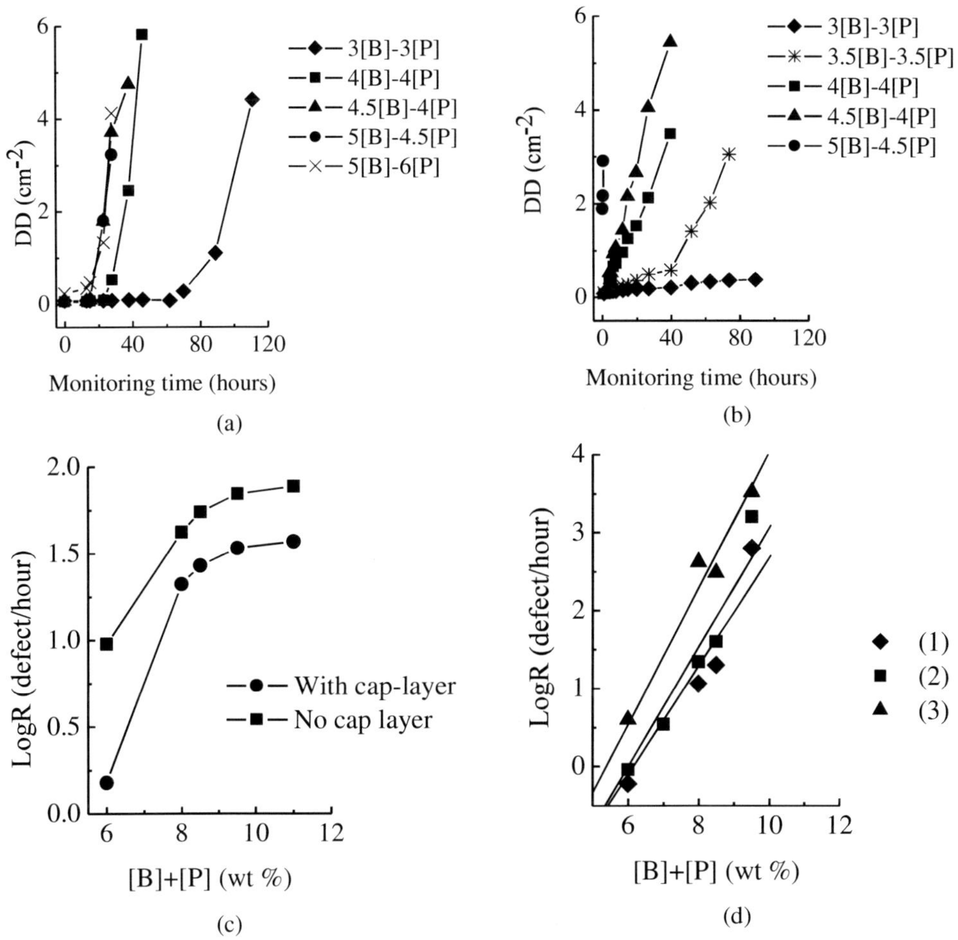

Figure 2.44. SACVD BPSG defect density growth trends in as-deposited (a) and densified (b) films, and the rate of defect increment vs. total additive content in as-deposited (c) and densified (b) films. Presented at 3rd Intern. Dielectric for ULSI Multilevel Interconnection Conf. (DUMIC), 1997 [187].

In contrast, the formation of defects in thermally annealed (densified) films was observed immediately after unloading the wafers from the densification furnace, i.e. after starting the film exposure to the clean-room ambient(see Figure 2.44(b)). For annealed SABPSG films, the concentration dependencies of $\log(DD/t_m)$ were in the range of 0.1–3.5. These dependencies were approximated with the empirical linear equation: $\log(DD/t_m) \sim \gamma \times ([B]+[P]-6)$. The value of coefficient γ was empirically found to be in the range of 0.6–0.9, depending on anneal conditions (see Figure 2.44(d)). For isothermal anneal conditions, temperature 900 °C and total additive content about 8.5 wt %, the change of coefficient γ value with anneal time (t_a, minutes) was roughly approximated with empirical equation $\gamma \sim 0.17 \times \sqrt{t_a}$, as shown in Table 2.11. The latter dependence indicates that diffusion limitations are responsible for the rate of defect formation after the film anneal. Yoshimaru et al [163] proposed that defect formation after BPSG film anneal is a result of vaporized P_2O_5 interaction with B_2O_3 on the film surface during a post-anneal film cooling stage. We believe that our observation directly confirms that the defect formation process in BPSG films after FA is a consequence of additive out-diffusion from the film depth to the film surface followed by evaporation to the space between wafers placed in the furnace. The particular rate value was found to be dependent on the film composition, thickness of cap layer, anneal temperature and anneal time.

The estimation of the defect size distribution during the period of film exposure to clean-room air was carried out by the quantitative LS measurement of defect parameters. LS data gave an average size in the range from 50 to above 100 arbitrary units (the increasing of this arbitrary size from 50 up to above 100 shows the increasing of defect size, see Figure 2.3). For both as-deposited and annealed films, the defects were as small as 0.2 μm in the beginning of their registration. The number of defects gradually increased with time. It started to decrease at a certain time when the number of large defects started to increase.

This situation is shown in Figure 2.45(a,b) for the total number of defects and defects with defined arbitrary sizes. Figures.2.45(c,d) show the changes in a normalized manner, namely as shares of defects with defined arbitrary sizes. One can see that for both as-deposited and thermally-annealed BPSG films. There were similar processes of defect size growth along with a decrease of the small defect shares. This effect was found to be more pronounced for as-deposited SACVD BPSG films as compared to the other types of BPSG films.

Table 2.11. Summarized kinetic data for SACVD BPSG defect appearance and growth

Characteristic	BPSG film state	
	As-deposited	**Annealed**
"Incubation" period	Yes, depends on [B],[P]	No
Rate of total defect density increment DD/t_m ($cm^{-2} \times hour^{-1}$)	0.05–0.50 depends on [B],[P]	0.01–4.66 depends on [B],[P] and t_a
Concentration dependence for a rate of total defect density increment	$\log(DD/t_m) \sim \gamma$ ([B]+[P]–6), where $\gamma \sim 0.80$ at [B]+[P]<8%, and $\gamma \sim 0.05$ at [B]+[P]>8%	$\log(DD/t_m) \sim \gamma$ ([B]+[P]–6), where $\gamma \sim 0.85$
Approximate dependence $\gamma = f(t_a)$	-	$\gamma \sim 0.17 \times \sqrt{t_a}$ (900 °C)

Here γ is a coefficient of proportionality.

Defect growth was observed also with the use of the AFM technique (see Figure 2.5(a,b)). Using AFM data of the change in average particle diameter with the time of storage after densification for 5 wt % [B] – 4.5 wt % [P] APCVD BPSG film, we can clear see that defect size gradually grows. This was an average evaluation taken for a few defects while monitoring their size during film exposure to ambient.

A trend of gradual saturation of defect quantity and defect size in annealed films was observed clearly after a few days of the film exposure in air. With a long exposure time, the defect density of annealed SACVD BPSG films with about 4 wt % of each additive was estimated to be about 10^8 cm^{-2}, and the defect size was estimated to be in the range of 0.5–1.5 μm. By contrast, an appearance of new defects and defect growth in as-deposited films with a similar additive content was found to take place much longer as compared to that in annealed films, even up to a few months. Accordingly, defect density was found to be significantly higher and the distribution of defect size was found to be significantly wider compared to annealed films.

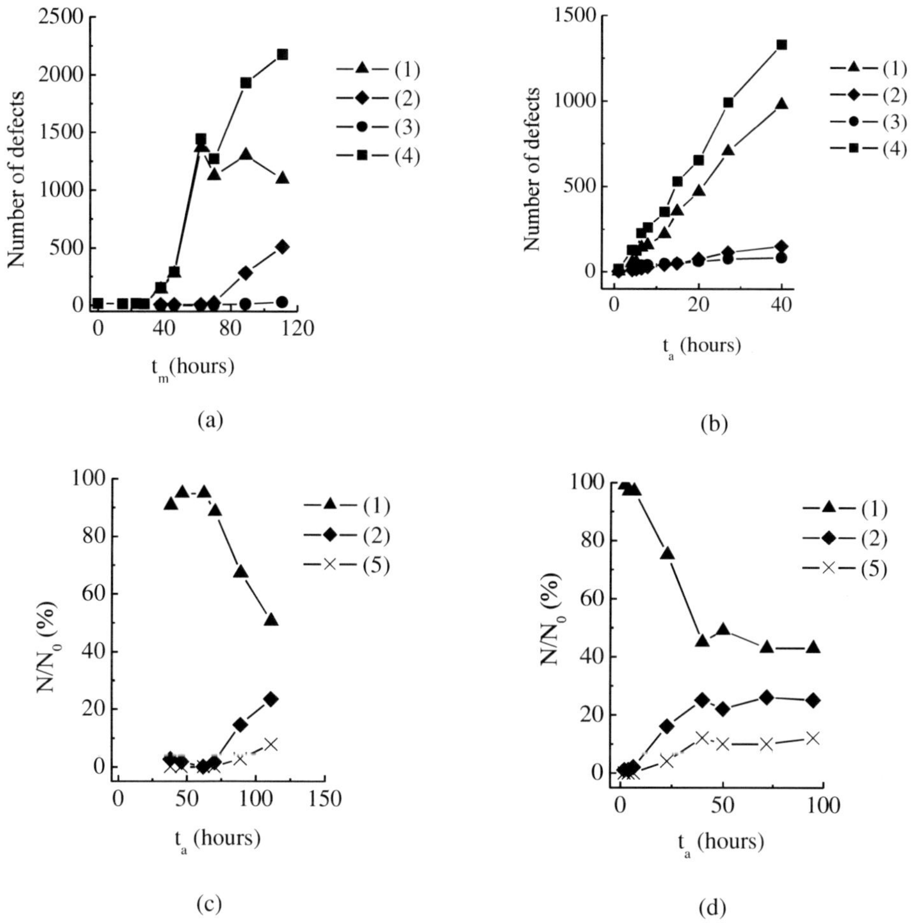

Figure 2.45.Changes in defect number with different arbitrary size defined by LS method for as-deposited (a) and thermally-annealed (b) films and shares of defects with defined arbitrary sizes for as-deposited (c) and thermally-annealed (d) films. Presented at 3rd Intern. Dielectric for ULSI Multilevel Interconnection Conf. (DUMIC), 1997 [187].

The generalized trend of change in defect appearance and growth in as-deposited BPSG films is shown in Figure 2.47 and it can be characterized as a defect evolution process. It reflects the following stages of the defect phenomenon in as-deposited BPSG films: induction, appearance, growth, and saturation. The latter stage, namely crystallization, is discussed below. The particular time for the selected evolution stage is a function of the total additive content and boron share in the total additive content.

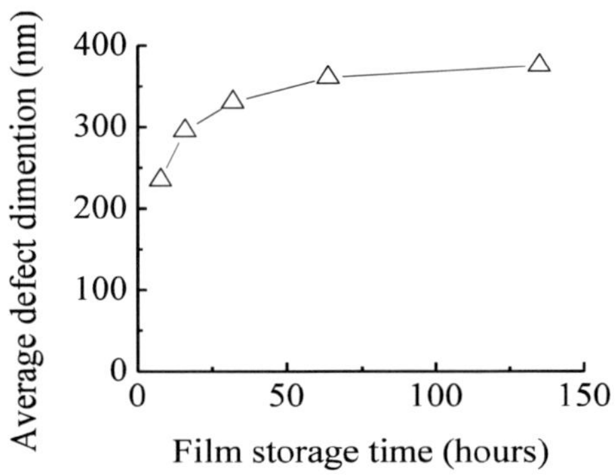

Figure 2.46. A plot calculated using AFM data of the change in average particle diameter with the time of storage after densification for 5 wt % [B] – 4.5 wt % [P] APCVD BPSG film (re-drawn from Ref. [225]).

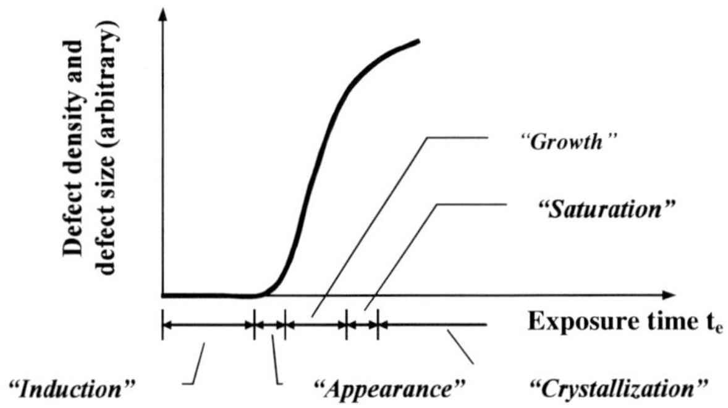

Figure 2.47. A curve of generalized changes in defect concentration and their size during film exposure to air and selected stages of the defect evolution process in as-deposited BPSG films.

13.4. Classification of Observed BPSG Defect Types

Despite the differences in shape, size and conditions of appearance, defects presented in Figure 2.39, Figure 2.40, and Figure 2.41 were primarily considered by researchers as either solid BPO_4 particles, or solid crystals of boric acid for films with high boron concentration. The line for BPO_4 compound presented in Figure 12.2(b) was drawn using the data in Figure 12.2(a) obtained for the annealed bulk ternary system SiO_2-B_2O_3-P_2O_5[333-335]. In fact, some studied BPSG films with high additive content were very close to (or even inside) the area drawn for the BPO_4 compound (see Figure 2.13). However, most of studied BPSG compositions were far enough from BPO_4 boundary. However, these low concentrations did not allow the use of the basic BPSG film advantage, e.g. flow capability. Based on common sense, it is necessary to keep the additive content as close as possible to the curve shown in Figure 2.42 as the 24-hours defect-free boundary line.

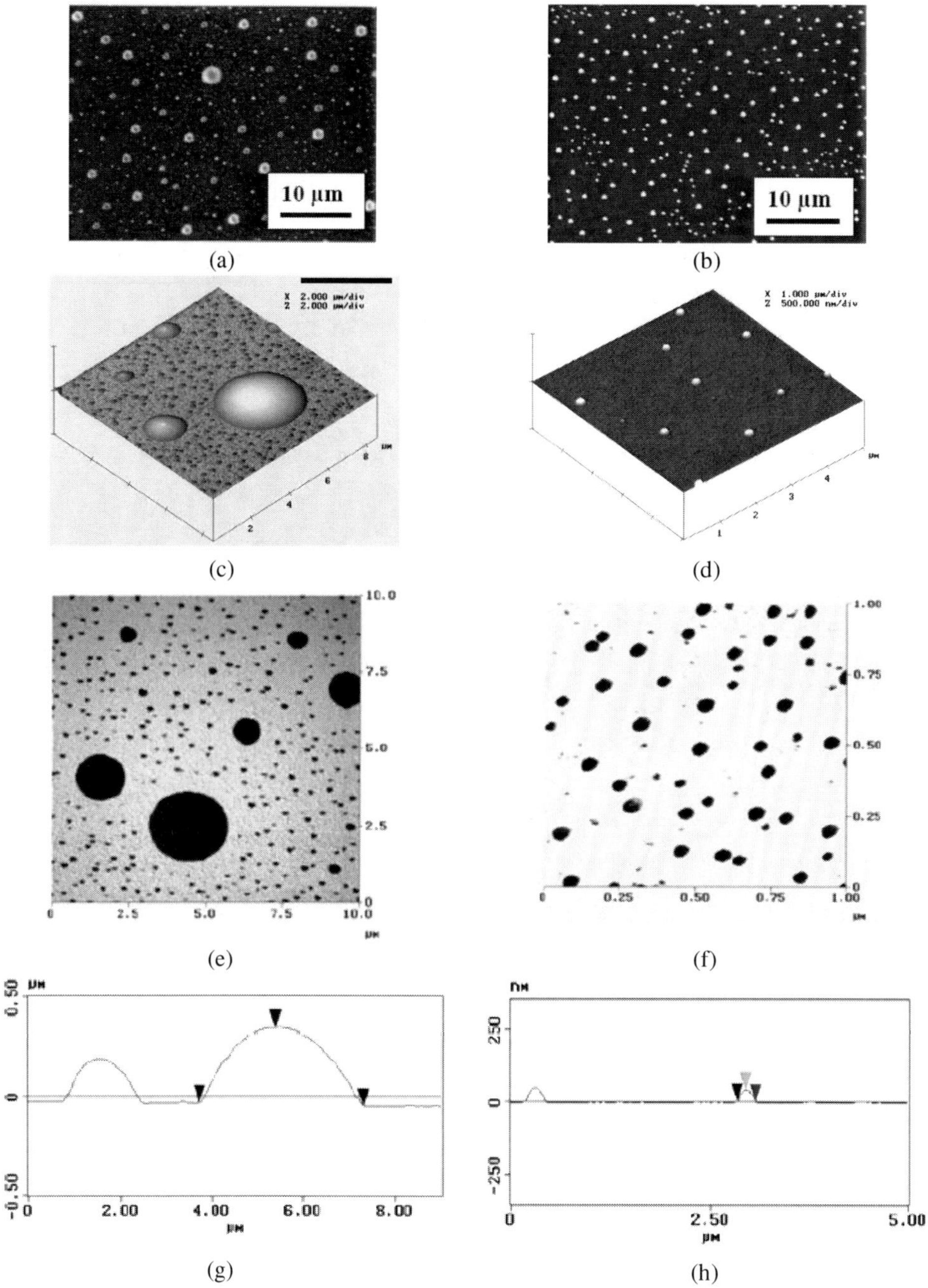

Figure 2.48. Liquid state defects after film deposition LD_I (left) and after film anneal LD_{IT} (right) studied with OM (a,b) and AFM (c-h) methods. Presented at 3rd Intern. Dielectric for ULSI Multilevel Interconnection Conf. (DUMIC), 1997 [187].

Depending on the chosen boron- and phosphorus additive content in BPSG films, a variety of defect types were observed and characterized earlier. It was done during periodic monitoring of different BPSG film compositions using a combination of OM and AFM methods. The types of defects were found to depend on the additive concentration in the film, post-deposition exposure time to air, annealing conditions (temperature, time, ambient) and the post-anneal exposure time to air. The author selected observed stable forms of BPSG thin film defects into a few different groups [89,186,187,190,211,226].

Liquid state defects. The first group of typical BPSG defects included round surface defects (sometimes called also "droplets") with a size ranging from 0.2 μm to a few microns (see Figure 2.4(a,b), and Figure 2.5(a,b,c)). These defects were observed for all studied BPSG film compositions. The most important conclusion of our studies [187,190] was a liquid nature of these rounded defects at the beginning of their appearance on the film surface. Some typical images of these defects are shown in Figure 2.48(a-h). They were defined as liquid state defects LD_l for as-deposited and LD_{lT} for thermally annealed films, accordingly.

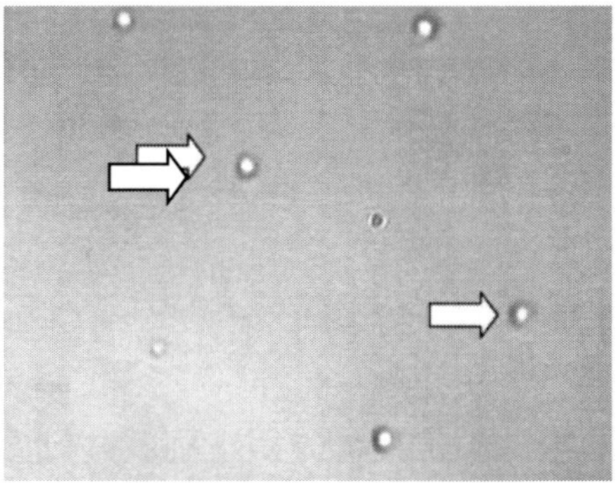

(a)

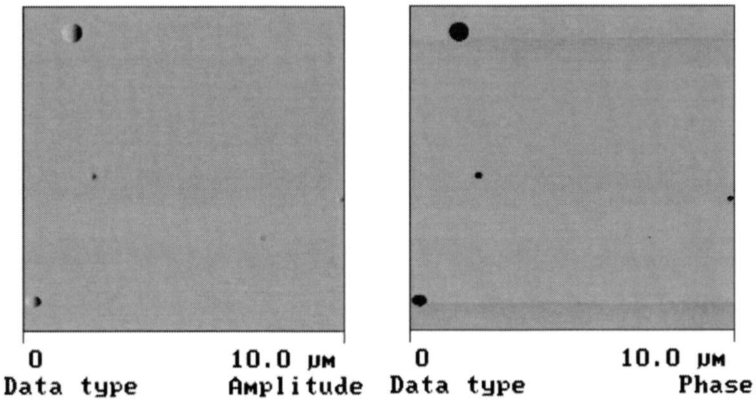

(b)

Figure 2.49. Liquid state defects LD_l (highlighted by arrows) in as-deposited phosphosilicate glass films with 4 wt % phosphorus content:(a) BPTEOS films, OM method and (b) HDP-CVD films, AFM in amplitude mode (left) and AFM phase mode (right). Presented at 8th Intern. Symposium on IC Technology, Systems and Application, ISIC-1999 (a) [298] and Intern. Symp. Interconnects and Contact Metallization for ULSI at 196th Meeting of Electrochemical Society, 1999 (b,c) [300].

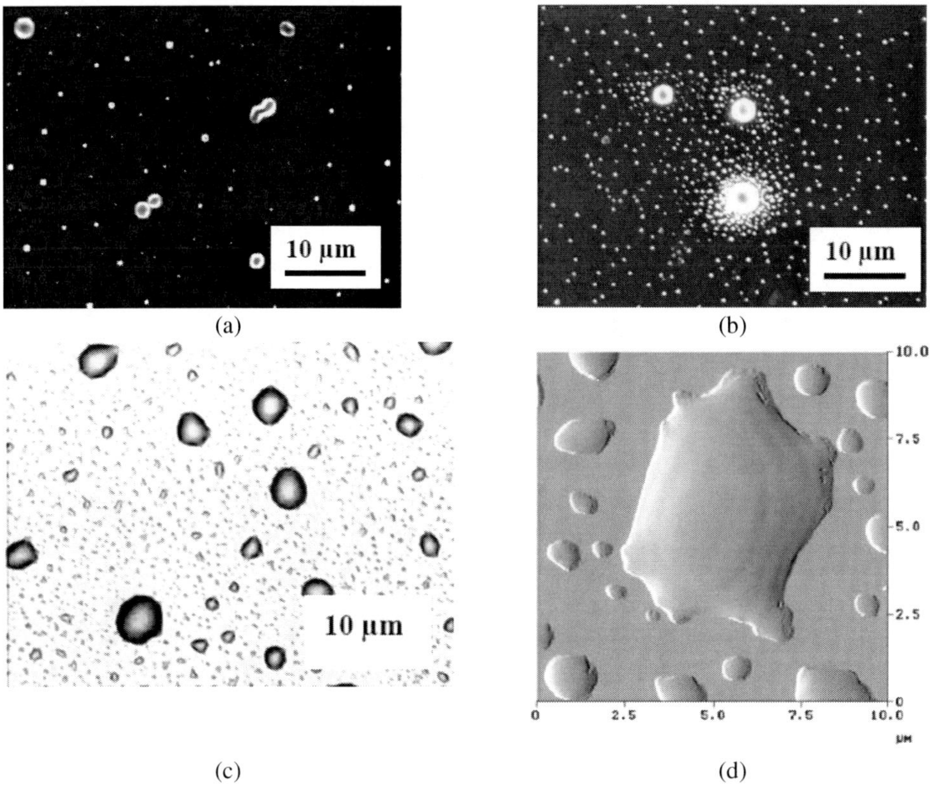

Figure 2.50. Images of liquid state defects LD_2 (a,b) and LD_4 (c,d). Analytical methods used: OM dark field (a,b) and OM bright field (c), amplitude mode of AFM (d). Presented at 4th Intern. Dielectric for ULSI Multilevel Interconnection Conf. (DUMIC), 1998 [211].

The difference in defect density and defect size in as-deposited and thermally annealed films can be clearly seen. A common feature of these defects was their rounded shape, particularly a ratio of diameter to altitude in the range of 6–8. The liquid nature of these defects was confirmed with the phase scanning mode of AFM technique (see Figure 2.5(c)). Figures 2.48(c-f) shows three-dimensional mode and phase mode AFM images of the liquid state defects. The difference between defect and film surface material was very clear from the phase mode images. These defects were always the first defect type observed on the BPSG film surface during film monitoring irrespective of the type of BPSG films. Therefore, these defects were recognized to be the key for understanding of all defect formation processes on/in BPSG films.

Note, that this type of defect was similar to that observed on the surface of PSG films exposed to clean-room air. For instance, images of rounded defects found on the surface of PECVD and HDP-CVD PSG films are shown in Figure 2.49(a-c). It is important to note, that film surface roughness affect the shape of the liquid defects. For instance, if the film roughness was too high, the shape of liquid defects was not rounded. It was found to depend on the surface, as shown in detail AFM images earlier, see Figure 2.6(a-d) [225]. Other stable types of defects with a liquid nature were also classified depending on the shape and the size. In as-deposited and annealed BPSG films, non-symmetric defects LD_2 and LD_{2T}, accordingly, were found. They were observed at a similar additive content (see Figure 2.39(d) and Figure

2.50(a,b)). They clearly existed in a process of single droplet merge. Relatively large symmetric defects LD_3 and LD_{3T} were mostly observed at high phosphorus content in BPSG films (see Figure 2.39(a)).

Large non-symmetric defects LD_4 and LD_{4T} (as-deposited and annealed films, accordingly) were observed at a relatively high total, but with similar additive content (see Figure 12.50(c,d)).

Crystalline state defects. In addition to liquid defects, crystal defects were also observed on the surface of BPSG films after some exposure time to ambient air. It is important to note that there was a direct link found between the liquid and crystal defects. First, an example of a crystal clearly surrounded by liquid is shown in Figure 2.51(a). Second, a sequence of liquid droplet growth and transformation into crystals in as-deposited films is shown in Figure 2.51(b) (here the exposure time to air is presented qualitatively by numbers from 1 to 9).

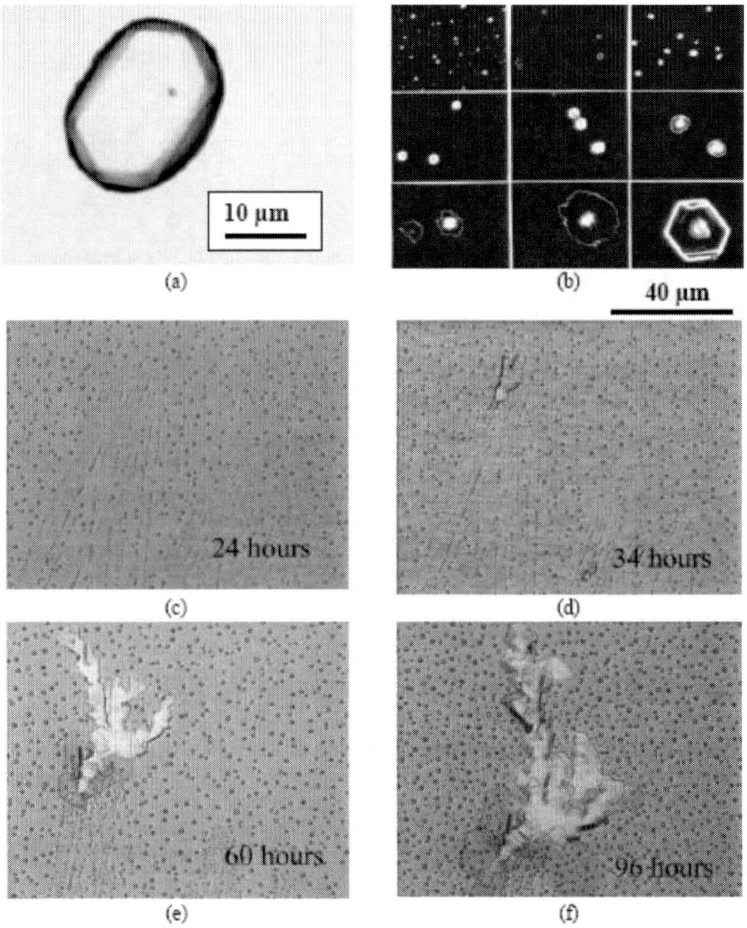

Figure 2.51. An example of defect formation stages in SACVD BPSG films: a crystal inside a liquid state defects (a), a sequence (from left to right in top followed by middle and followed by bottom raws) of crystal formation from small liquid droplets in 5 wt % [B] – 6 wt % [P] BPSG film (b), and some steps of crystal growth in time (c – f) initiated by touching of the film surface by metal tweezers (see traces) in 6 wt % [B] – 6 wt % [P] BPSG film. Images (a,b) were presented at 4th Intern. Dielectric for ULSI Multilevel Interconnection Conf. (DUMIC), 1998 [211]).

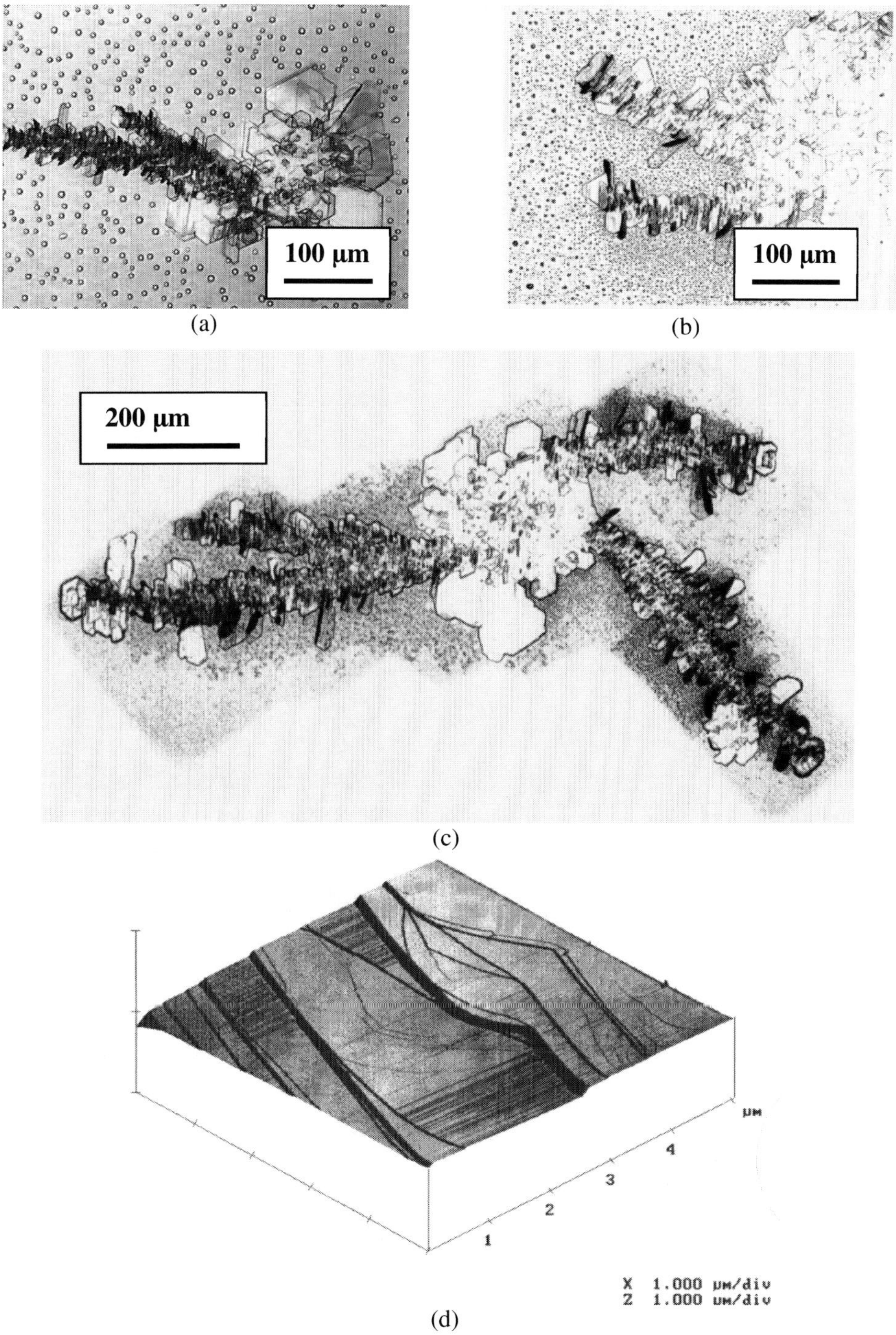

Figure 2.52. Crystal BPSG defects: fragments of defect CD_1 in as-deposited (a) and annealed (b) 6.3 wt % [B] – 6 wt % [P] SACVD BPSG film, an example of a dendritic crystal in annealed film (c) and an AFM image clearly demonstrating layers on the crystal surface (d). Presented at 4th Intern. Dielectric for ULSI Multilevel Interconnection Conf. (DUMIC), 1998 [211].

A few stages of the growth of a crystal in an as-deposited BPSG film are shown in Figure 2.51(c-f). In this case, initial crystallization nuclei has been created on the surface of the film artificially by a metal tool touching it. This second group of BPSG surface defects includes the following crystal defect types, defined depending on the shape and size as CD_1, CD_2 and CD_3 in as-deposited, and as CD_{1T}, CD_{2T} and CD_{3T} in thermally annealed films. Representative OM and AFM images of selected defect types are shown in Figure 2.52, Figure 2.53, and Figure 2.54. Generally, the shape and the size of crystals depend on the total additive content and the ratio of additives in the film.

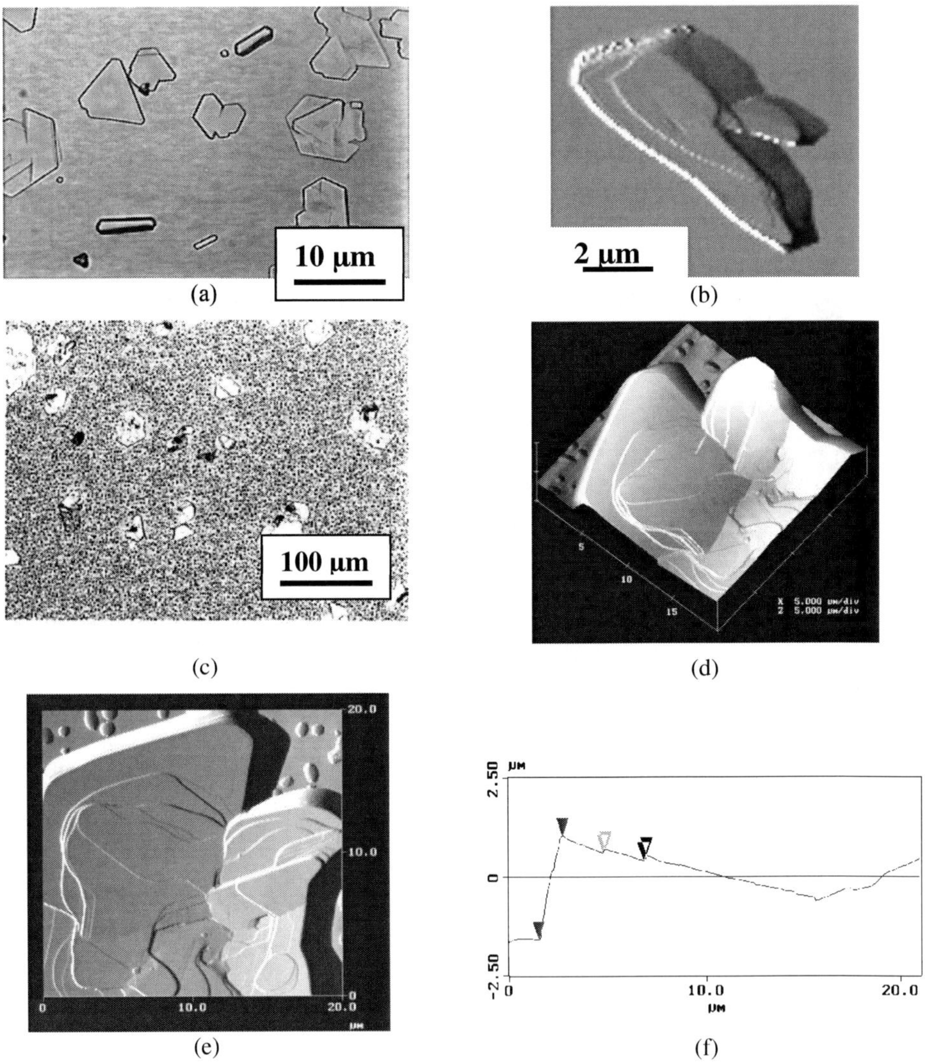

Figure 2.53. Examples of crystalline defects CD_2 in as-deposited BPSG films (a – OM method, b- AFM method), co-existed with liquid defects after BPSG film anneal, (c – OM method, d,e – AFM method, highlighted by block arrows). Figure (d,e,f) show AFM images of a crystal defect: 3-d view, top view and cross-section view, respectively. Presented at 4th Intern. Dielectric for ULSI Multilevel Interconnection Conf. (DUMIC), 1998 [211].

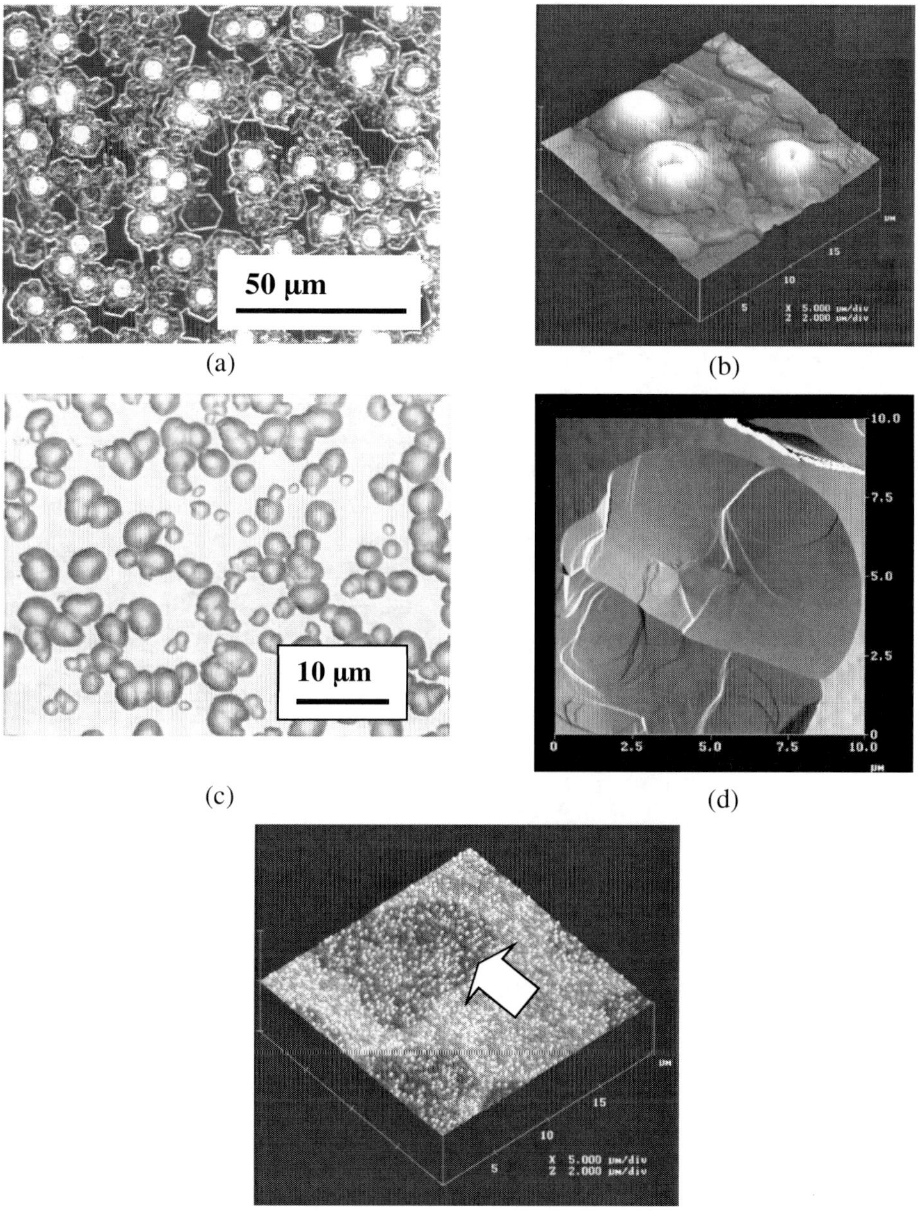

Figure 2.54. Images of crystalline defects CD3 in boron-rich BPSG films with the total additive content more than 14 wt %: after film deposition (a,b), after film anneal (c,d), after defect removal (e). Images (a,c) obtained using OM method, (b,d) using AFM method. Presented at 4th Intern. Dielectric for ULSI Multilevel Interconnection Conf. (DUMIC), 1998 [211].

Crystals CD_1 and, especially, CD_{1T} were very large, stretching up to several hundred microns and were shaped in the dendritic form, as can be seen in Figure 2.53(a-c). Such crystals grew to more than 200 μm in size in less than 60 hours after deposition. These defects were usually seen with a high total additive content and slightly boron-reached BPSG film compositions. The precipitation of these crystals were believed to be due to the contact with the liquid droplets on the BPSG surface, saturated with boron- or phosphorus-contained compounds, which were leeched out from the film by air moisture. The liquid droplets always surrounded the crystals. AFM pictures of the crystal surface revealed very fine contoured layers on the surface of these crystals (see Figure 2.52(d)).

However, small crystals CD_2 and CD_{2T} were also typically shaped (see Figure 2.53(a-d)). These crystals can be also seen together with dendrites, or on other areas of the wafer. As compared to dendrites, these small crystals were typical for boron-reached compositions. The co-existent liquid droplets were normally seen only in annealed films, as shown in Figure 2.53(c,d,e), but not in as-deposited films where the surface was normally free of droplets. AFM scans of the small crystals are shown in Figure 2.53(d,e,f) and reveal very fine contoured layers on the surface of these crystals.

The last type of BPSG film crystal defects was observed for films with significant excess of boron (for instance, 10 wt % [B] – 3 wt % [P]). This defect type was defined as a defect CD_3. Optical and AFM images of the surface of as-deposited films are shown in the Figs.2.54(a,b). These defects consisted of a thin layer of hexagonal type crystal, which was faceted and appeared over a largely connected network on the film surface. Inside every hexagonal crystal, a large semi-hemispherical shaped nuclei can be seen. Some of these semi-hemispherical defects had deep craters in the center which makes it look somewhat like a microscopic "volcano" (see Figure 2.54(b)). After film anneal in oxygen at a temperature as high as 900 °C, CD_{3T} defects found on the surface of the film were slightly different from those found in the previously controlled as-deposited BPSG films (see Figure 2.54(c,d)). Only semi-hemispherical like defects were seen distributed throughout the surface of the film. The sizes of these defects range from about 5 μm to about 20 μm in diameter. As seen by the AFM image in Figure 2.54(d), these defects appeared to be in the form of layers. After the defects were removed from the film surface in deionized water, a slight depression on areas once occupied by the large semi-hemispheric like crystal defects can be seen (see Figure 2.54(e)), highlighted by a block arrow. Inside these highly-boron enriched annealed films, sub-surface defects with size about 0.2 – 0.3 μm were found using the SEM cross-sectional technique (as shown in Figure 2.7(b)).

After storage of the films with crystal defects types CD_1 and CD_2 for about one month, small liquid droplets were found on the top of the dendrites, as indicated by block arrows in Figure 2.55(a,b). Liquid was also found to have accumulated on the side of these crystals, as shown by block arrow in Figure 2.55(c). Further storage of the annealed films for about 3 months led to the change in the structure of the crystals into large liquid droplets (see Figure 2.55(d-f)) followed by black round bodies formation (shown in Figure 2.55(g)). At the same time, the long term storage of liquid droplets co-existent with crystals led to the formation of small round defects (as shown in Figure 2.55(h)).

Solid state defects. The third type of BPSG defects in BPSG films thermally annealed at temperatures above 900 °C was identified as solid state defects (SD_T). These small defects were found to be either separated or clustered (segregated) in a shape of a "tiger paw" or a ring (as shown in Figure 2.56(a-d)). The diameter of such rings was found to be close to the diameter of the liquid state defects that occupied the surface of the film before anneal, (compare Figure 2.56(a) and Figure 2.56(b)). After some time, liquid appearance around these small solid state defects was observed.

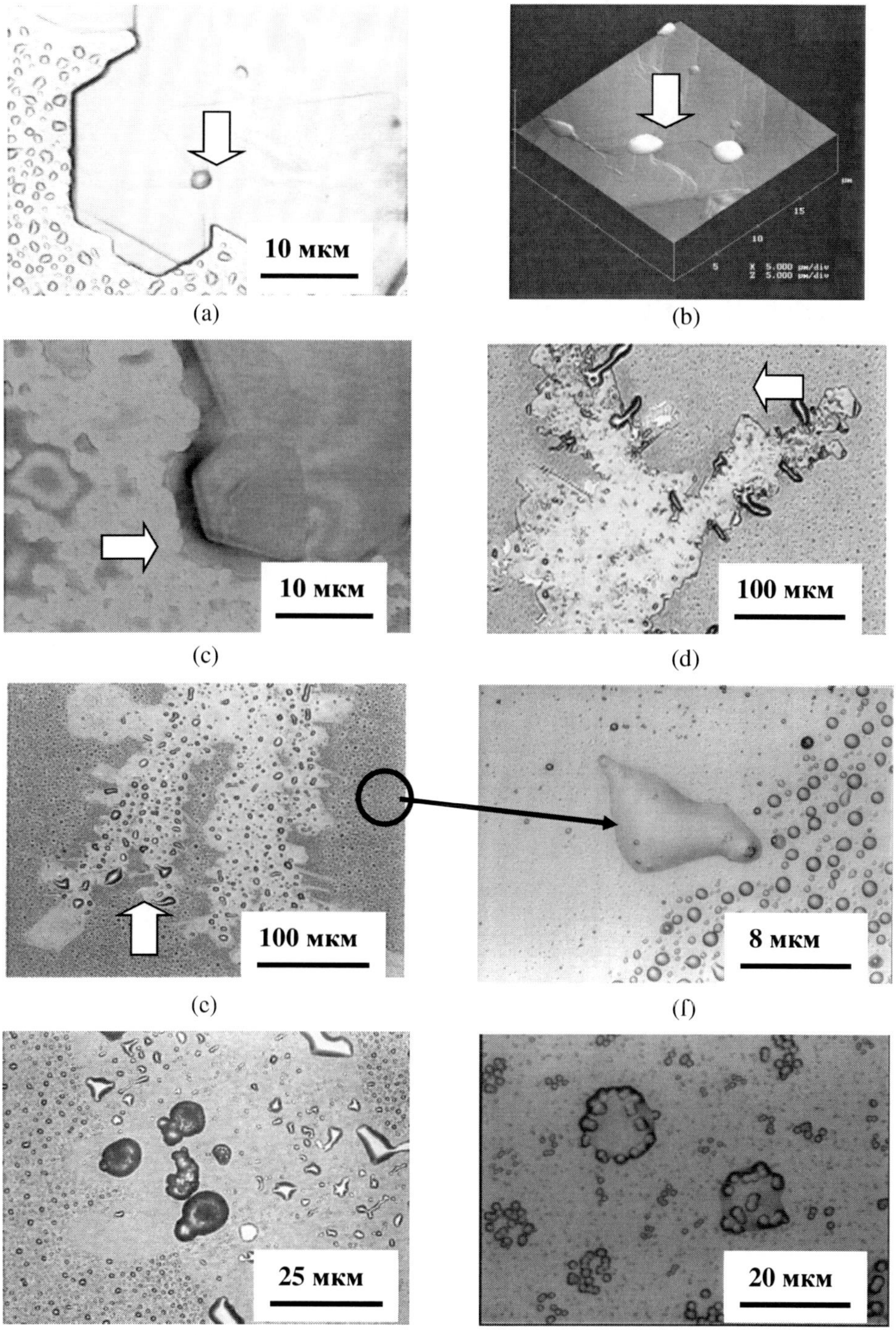

Figure 2.55. Optical (a,c-h) and AFM (b) images of the processes on BPSG film surfaces during long time exposure to air (see text for details). Presented at 4th Intern. Dielectric for ULSI Multilevel Interconnection Conf. (DUMIC), 1998 [211].

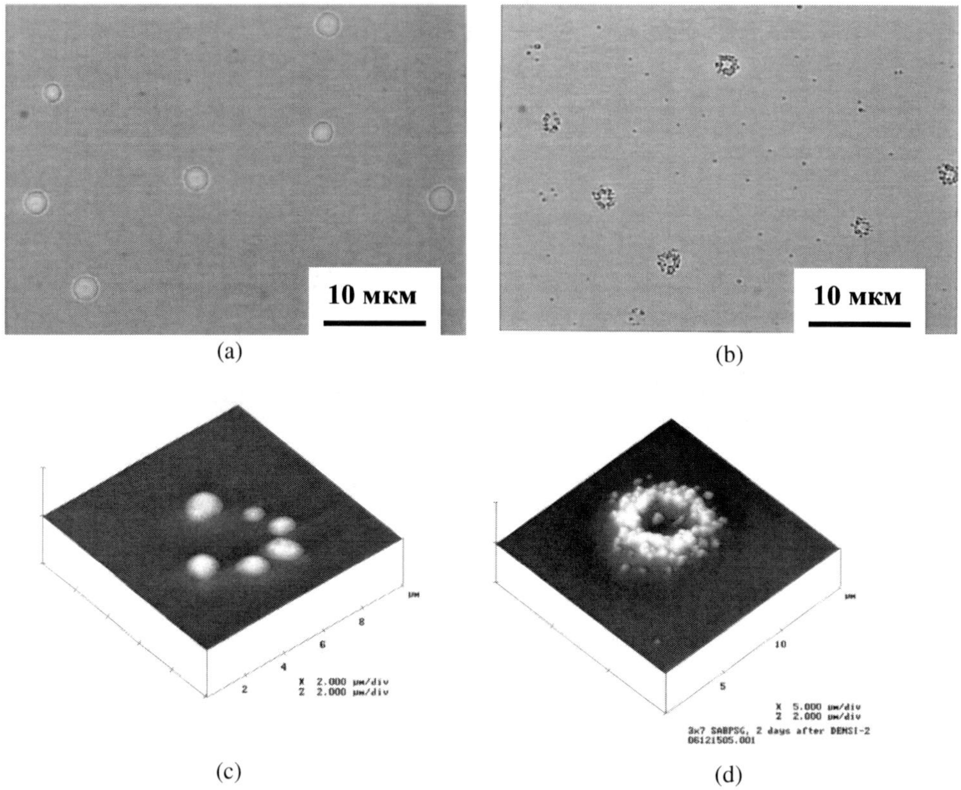

Figure 2.56. Optical images of liquid state defects in 3 wt 5 [B] – 7 wt % [P] SACVD BPSG films after deposition (a) and optical (b) or AFM (c,d) images of separated or segregated solid state defects in the same films after thermal anneal at 900 °C for 12 minutes in oxygen. Presented at 4th Intern. Dielectric for ULSI Multilevel Interconnection Conf. (DUMIC), 1998 [211].

According to most researchers, the solid state defects were BPO_4 particles. They form only after high temperature (higher than about 950 °C) anneal. This conclusion was clearly confirmed by Yoshimaru et al [163], who compared structures of such types of BPSG defects with BPO_4 tetragonal crystals and found their coincidences (see Table 2.12).

Table 2.12. Spacing of lattice planes for the particle and for BPO_4 quoted from the Joint committee on Powder Diffraction Standards [163]. Reproduced by permission of ECS – The Electrochemical Society

Particle	BPO_4 (tetragonal)	
d (nm)	d (nm)	Lattice plane
0.3667	0.36351	(101)
0.2242	0.22546	(112)
0.1855	0.18641	(211)
0.1822	0.18175	(202)
0.1474	0.14697	(213)
0.1219	0.12114	(303)
0.1122	0.11272	(224)

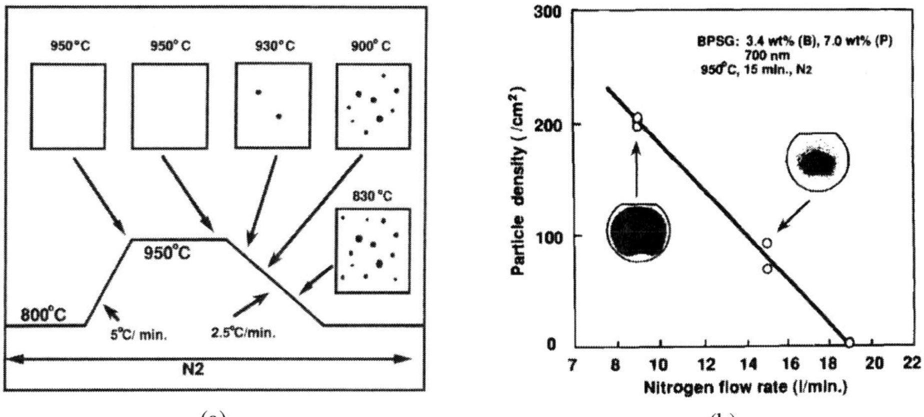

Figure 2.57. Particle growth density change during anneal cycle for 4 wt % [B] – 7 wt % [P] BPSG film observed by microscope (a) and effect of nitrogen flow rate on the particle growth density of BPSG film (b) [163]. Reproduced by permission of ECS – The Electrochemical Society.

The authors of [163] also monitored the defect density on the BPSG surface in-situ using an anneal chamber with a build-up microscope. They showed that formation of BPO_4 particles began from the saturated gas-phase during the post-anneal cooling stage (as shown in Figure 2.57(a)).

Further investigation showed that the defects density had a reverse order with respect to the gas flow rate during the film anneal, as shown in Figure 2.57(b). To prevent particle formation during and after the film anneal, the author discovered the necessity to exchange additive-rich vapor from the space near the wafers. Dry nitrogen anneal ambient gave the worst case of BPO_4 particle density while oxygen-steam ambient was found to be the best one. According to classical chemistry, BPO_4 can exist either in a non-hydrated form (dissolves very slowly in water), or in a hydrated form with 2, 4 or 6 molecules of water. These hydrated forms are easily soluble in water. Based on these BPO_4 facts, it can be proposed that the non-hydrated form was the result of anneal in dry ambient, but the soluble form was due to its formation in the steam ambient.

Finally, a comprehensive summary of additive concentration ranges corresponding to observed BPSG defect types is presented in Table 2.13 and in Figure 2.58 for BPSG films after a long term exposure to ambient that means a steady state of the defects.

13.5. BPSG Defect Formation Schemes

A few schemes of BPSG defect formation during/after the film anneal have been discussed in the literature. The first scheme, proposed by Susa et al [62], explained a formation of large BPO_4 particles during high-temperature BPSG film anneal in phosphine-oxygen mixture. According to the scheme, during the film anneal there was an interaction of boron in the film with gas-phase P_4O_{10} molecules formed as a result of phosphine oxidation.

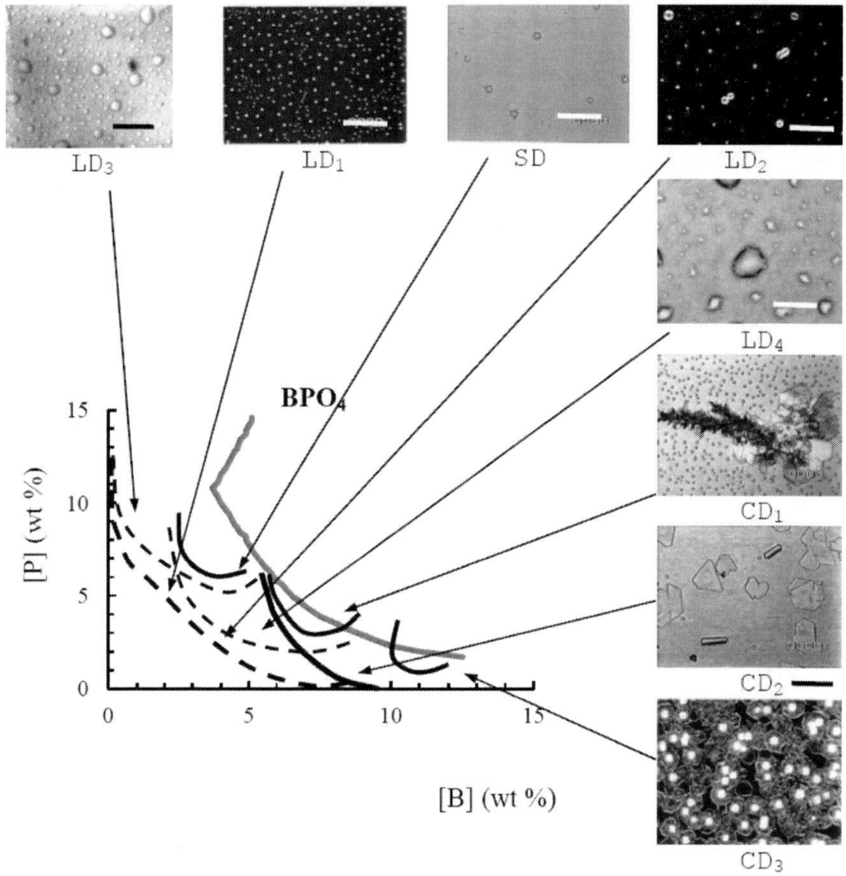

Figure 2.58. Approximate concentration ranges of classified BPSG film defect types [liquid defects (- - -), solid state and crystalline defects (-)] and their typical images made by optical and AFM techniques. Length of the lines on images corresponds to 10 μm [224].Reproduced by permission of ECS – The Electrochemical Society.

Ahmed et al proposed a two-stage scheme [112]. In the first stage, boron and phosphorus oxides assumed to react with ambient moisture forming acids followed by soluble amorphous form $BPO_4•H_2O$. The latter transforms to a non-soluble crystalline BPO_4.after the film anneal in nitrogen or oxygen at temperatures higher than 900 °C. In the case of steam anneal, there were no such BPO_4 particles found and this phenomenon was explained as follows. Elimination of BPO_4 crystal development by steam ambient may be due to the suppression of both phosphoric and boric acids formation, which, according to the cited authors, were the BPO_4 predecessor compounds. Instead of acids, hydrated oxides were assumed to form, which lost their water and, then, formed the corresponding oxides B_2O_3 and P_2O_5. The three-stage scheme was proposed by Imai et al [113]. The scheme stages were as follows: 1) formation of sub-surface additive-rich areas; 2) phosphorus evaporation from the film with volatile P_4O_{10} formation; 3) interaction of P_4O_{10} with boron in an additive-rich area with the formation of amorphous BPO_4 followed by its crystallization during film cooling. The last scheme proposed by Yoshimaru et al [163] assumed only phosphorus out-diffusion in nitrogen or oxygen ambient followed by volatile P_4O_{10} formation. This oxide reacted with film surface B_2O_3 to form BPO_4. However, according to this model, at steam ambient of both boron and phosphorus out-diffused from the film. Hence, BPO_4 particle growth on the surface was suppressed because there were no nucleation sites on the surface. It is easy to conclude that all cited schemes were devoted to the explanation of BPO_4 particles formation after the film anneal or, according to our classification, the solid state defects SD_T.

Despite these reasonable explanations of BPO_4 formation, there were no detailed studies and no understanding of "the whole picture" of BPSG defect formation phenomenon. In the author's previous studies, it was shown that defect formation phenomenon was linked to the film structure and its features. Structures for BPSG films were proposed for the first time in [214,224] and they are presented in Figure 2.59(a,b). Among them is a scheme of "ideal" non-porous BPSG film with randomly distributed additive atoms. This scheme is mostly applicable to the films deposited by PECVD methods or deposited at high-temperature CVD conditions, as well as to films after high temperature anneal. Another scheme of porous film with partly randomly distributed additives in the film matrix and partly accumulated in embedded clusters in the form of phosphoric oxide [nP_4O_{10}] or phosphoric-boron oxide [$nP_4O_{10}{\times}mB_2O_3$]was developed to explain low-temperature TEOS-ozone BPSG film features. The additive-rich clusters that were supposed to be built into the porous film were believed to be more reactive to moisture than just phosphorus in the glass matrix. These two schemes allowed us to explain most of observed features in BPSG films.

Table 2.13. An approximate correlation of defect types after thermal anneal with film evolution step before anneal

Total additive (wt %), [B]-[P] ratio	Thermal anneal	Correlation of BPSG defect types after anneal with the stage of glass evolution after deposition, see Figure 2.46.			
		Induction	Appearance – Growth	Saturation	Crystallization
< ~ 6 Any	Grown	-	-	-	-
	Annealed	-	-	-	LD_{1T}
~ 6 – ~ 9 Any	Grown	-	LD_1	LD_2	-
	Annealed	LD_{1T}	LD_{1T}, LD_{2T}	LD_{1T}, LD_{2T}	LD_{2T}
> ~ 9 [B] >> [P]	Grown	-	LD_1	LD_2	CD_3
	Annealed	LD_{1T}	LD_{1T}	LD_{2T}, CD_{2T}	CD_{3T}
> ~ 9 [B] > [P]	Grown	-	LD_1	LD_2	LD_2, CD_2
	Annealed	LD_{1T}, LD_{2T}	LD_{1T}, LD_{2T}	LD_{2T}, LD_{4T}, CD_{1T}-CD_{2T}	LD_{2T}, LD_{4T}, CD_{2T}
> ~ 9 [B] ≈ [P]	Grown	-	LD_1, LD_2	LD_1, LD_2, LD_4	LD_2, LD_4, CD_1, CD_2
	Annealed	LD_{1T}, LD_{2T}, SD_T*	LD_{1T}, LD_{2T}, SD_T*	LD_{1T}, LD_{2T}, SD_T*, CD_1-CD_2	LD_{2T}, LD_{4T}, CD_{1T}, CD_{2T}, SD_T*
> ~ 9 [B] < [P]	Grown	-	LD_1, LD_2	LD_1, LD_2, LD_4	No
	Annealed	LD_{1T}, LD_{2T}, SD_T*	LD_{1T}, LD_{2T}, SD_T*	LD_{1T}, LD_{2T}, SD_T*	LD_{2T}, LD_{4T}, CD_{1T}, CD_{2T}, SD_T*
> ~ 9 [B] << [P]	Grown	-	LD_1, LD_2	LD_1, LD_2, LD_3	No
	Annealed	LD_{1T}, LD_{2T}	LD_{1T}, LD_{2T}	LD_{2T}, LD_{3T}	LD_{2T}, LD_{3T}

Note: * depending also on the anneal conditions.

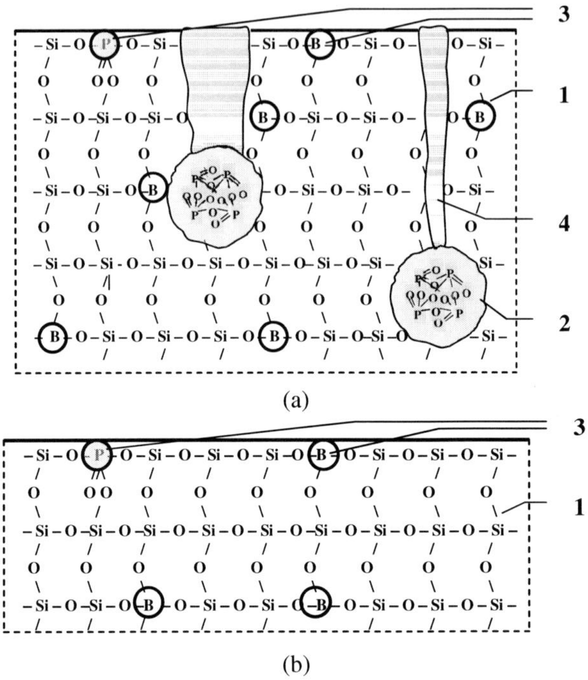

Figure 2.59. Schemes of BPSG film structures: porous with randomly distributed additive-contain clusters and randomly distributed additives (a) and non-porous with randomly distributed additives (b). Definitions: 1 – glass matrix; 2 – cluster; 3 – surface additive atoms; 4 – pore with solution [208].

To generalize defect formation data and to create a defect formation mechanism for any type of BPSG film, we introduced a term “a specific surface site” (S*) [224]. This S* was proposed to be an area of appearance and to further the development of the defect formation process. It was assumed to be the result of the interaction of moisture and additives inside the film. This definition allows us to link all of the observed film features and to create a mechanism of defect formation in as-deposited and thermally annealed films. Depending on the additive concentration, step of defect evolution, and film anneal conditions, different types of defects can be formed on this site, as summarized in Table 2.14 and in schemes presented in Figure 2.60(a-h).

To explain the defect formation mechanism, the author proposed the following logic and simplified the steps of the scheme. Moisture absorption and reaction of moisture with the additives in the porous as-deposited film resulted in the accumulation of boron and phosphorus acids inside the film pores. Next, out-diffusion of additives to the film surface and formation of the liquid defect nucleus on the film surface took place. Liquid defects interacted with moisture and grew due to the merging with neighboring defects (as shown in Figure 2.60(c)). Their shape and size depended on the additive concentration and time of exposure in ambient. Due to the continued processes of acids out-diffusing to the film surface, defects gradually saturated with additives. After that, crystallization processes took place and different types of crystalline defects formed depending on the additive concentration of the film. Crystal growth in as-deposited films was mainly due to the out-diffusion of additives to the film surface from the porous film.

Table 2.14. Steps of defect formation mechanism in as-deposited and annealed BPSG films

Film	Selected evolution stage in film	Basic steps of defect formation schemes	Factors affecting defect formation
BPSG film after deposition	1. Induction	Interaction of ambient moisture with surface additives; moisture absorption; moisture penetration into the film; formation of acids in glass pores and their diffusion to the film surface.	Total additive; [B]/[P]; film porosity; time
	2. Appearance	Formation of liquid surface nuclei with acids initially in the largest pores; acceleration of moisture absorption by nuclei due to direct contact with ambient; growth and formation of LD_1.	Same as 1
	3. Growth	LD_1 merging into LD_2 and LD_3. The beginning of S* formation on and under defect places.	Same as item 2
	4. Saturation	LD_2 merging and LD_4 formation; LD_3 growth; formation of saturated mixtures in largest LD_2 and LD_4; their interaction with film surface; accumulation of additives in S*.	Same as item 3
	5. Crystallization	Solid nuclei formation inside saturated LD_2 and LD_4; crystallization and growth of $CD_1 - CD_3$.	Same as item 4
BPSG film after thermal anneal	6. Thermal treatment of the glass film	Glass softening; film structure improvement; decrease and disappearance of the glass pores; additive diffusion to the gas ambient and formation of additive-rich vapor with the concentration of P_4O_{10} and H_3BO_3 depending on additive content, moisture content in glass, gas ambient and the rate of gas exchange, temperature and time. Evaporation of the liquid defects and formation of boron oxide rich S*; partial evaporation of existent crystals.	Same as item 5; additive content in gas phase; gas ambient; gas exchange rate
	7. Glass cooling	Gas-phase P_4O_{10} cluster formation during temperature ramp-down and their sedimentation on the film surface; interaction of cluster with boron oxide rich S* with the formation of SD_{1T} – BPO_4.Possible crystallization of boric components as anα- B_2O_3 and /or metaboric acid.	Same as item 6, and the rate of cooling
	8. Post-annealed exposure in ambient moisture	Growth of LD_{1T} due to the absorption of moisture by clusters; their growth and merging into LD_{2T}-LD_{4T}; interaction of BPO_4 with moisture; fast lateral growth of CD_{1T} and growth of CD_{2T} due to the interaction with LD_{2T} andLD_{4T}.	Same as item 7

During the high temperature anneal that was needed for film flow, the out-diffusion of additives from the film to the gas phase, mainly phosphorus, took place (as shown in Figure 2.60(e)). On the cooling stage, phosphorus oxide clusters nucleated in the vapor phase and deposited on the film surface creating surface defect nucleus. At high phosphorus content there was a possibility of BPO_4 particle formation due to the reaction of condensed P_4O_{10} with boron oxide on the surface. However, if the additive concentration was low enough, immediately after initial exposure of annealed film to ambient, liquid state defects started to form because of the interaction of moisture with highly active phosphorus oxide nucleus. As for as-deposited films, the growth of defects in annealed films was due to the merging of the neighbor defects (as shown in Figure 2.60(g)). Intensive lateral crystallization in annealed films was due to the interaction of crystals and surface droplets because of elimination of pores during film anneal.

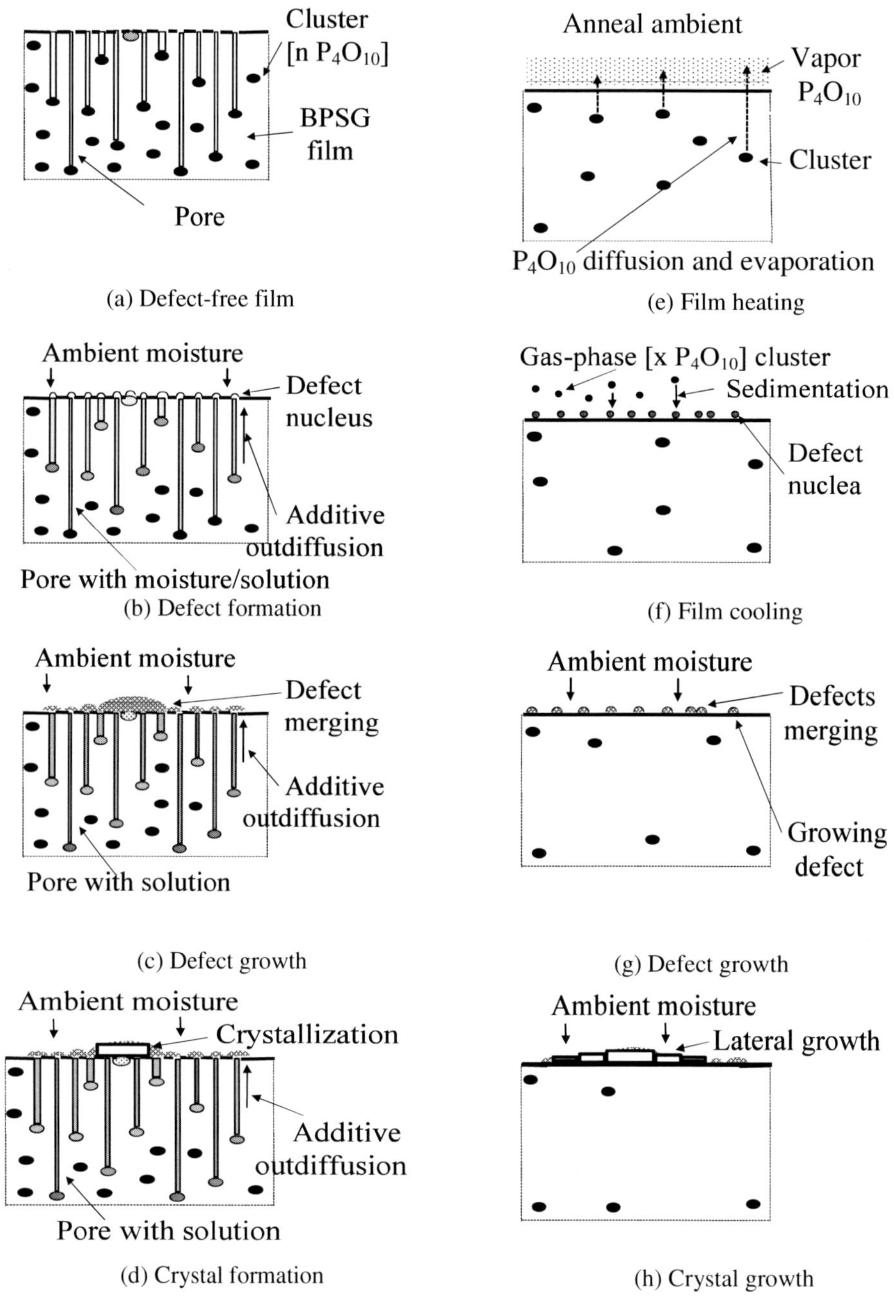

Figure 2.60. Schematics of BPSG surface defect formation and growth in as-deposited (left side) and annealed (right side) films [224]. Reproduced by permission of ECS – The Electrochemical Society.

13.6. Quantitative Definition of Defect-Free BPSG Thin Film Compositions

The analysis above shows the importance of additive concentration levels in BPSG films for defect formation processes. In general, all types of BPSG film revealed similar behavior that allows us to summarize data irrespective of the film deposition method. Summarized BPSG film defect data show that the total additive content in BPSG films cannot exceed a certain limit, which can be chosen taking into account device and technology needs. On the other hand, additive concentration must be sufficient enough in order to use film advantages, such as the film flow capability. Approaches for characterization of BPSG films discussed in [89,93] allowed us to define the optimized boundary of additive concentrations. This predicts good BPSG film quality without serious moisture absorption and, therefore, defect problems, during the selected time of film exposure to ambient. For convenience, this time was chosen to be 24 hours. The optimized defect-free area boundary was defined using an empirical equation (2.10):

$$[P] < -0.17[B]^2 - 0.19[B] + 9.6 \quad (2.10)$$

where additive concentrations are presented in elemental wt %. It corresponds to the 24-hour defect-free curve in Figure 2.42. For TEOS-ozone BPSG films, which revealed more pronounced defect formation effects, Eq.(2.10) and defect-free curve in Figure 2.42 correspond to about 18 hours after film deposition, instead of 24-hours for the other types of BPSG films.

The increase of the boron concentration in BPSG films was found to enhance the moisture penetration depth into the film [178]. At the same time, the increase of phosphorus concentration provided a limit of water on the film surface. These data gave us an opportunity to compare defect formation and moisture absorption phenomena quantitatively in terms of boron share [B]/([B]+[P]) in the total additive concentration in BPSG films. Thus, a correlation of 24-hours defect-free area boundary from Figure 2.42 and the moisture penetration depth into the film during PCT (calculated by Vassiliev [213] using experimental data [178]) is presented in Figure 2.61. These data show that the increase of the boron share [B]/([B]+[P]) above ~ 30 % causes a sharp increase in moisture penetration depth during PCT (see open symbols). At the same time, above this boron to phosphorus ratio we can see a gradual decrease of the defect formation boundary to the lower total additive content. Therefore, in this area defect formation effects become more pronounced.

This summarized graph with data, based on different research approaches and published by different research groups, confirms that the boron to phosphorus ratio ~ 0.4 revealed the basic BPSG material feature, namely, complete structure ordering. This case can be expressed numerically as follows (2.11):

$$[B] \leq\sim 0.4[P] \quad (2.11)$$

Equation (2.11) can be used as an additional limit to enhance film stability with respect to the moisture absorption.

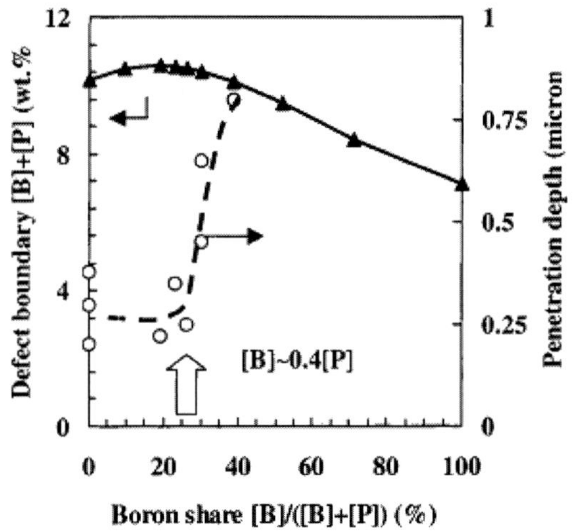

Figure 2.61. Water penetration depth (re-calculated using data from [178] and a boundary of 24 h defect-free area defined in [89] vs. boron share in BPSG films [269].Reproduced by permission of ECS – The Electrochemical Society.

13.7. BPSG Film Defect Prevention and Removal

Generally, any measures leading to the decrease of the concentration of "a specific surface site" (S*) on the film surface are helpful to reduce defect formation processes in BPSG films. For as-deposited BPSG films, these measures include: (1) a general decrease of additive concentration in BPSG film; (2) an enhancement of film density; (3) deposition of BPSG film as a stack with different additive concentrations in subsequent layers; (4) an increase of silicon dioxide cap-layer thickness on the top of BPSG film and (5) removal of additives from the subsurface of BPSG film by water rinsing after film deposition, etc.

The following additional measures can be used in order to prevent defect formation in BPSG films during and after high temperature anneal: (1) a decrease of anneal temperature and anneal time; (2) an improvement of gas exchange conditions during anneal and (3) capping of BPSG film after anneal with silicon dioxide layers. The effectiveness of defect removal from the film surface depends on the nature of defects and, therefore, on the additive concentrations in the film. Liquid defects, containing acid solutions, can be effectively removed by water rinsing. Removal of boric acid crystals were done, for instance, using methanol rinsing [50]. Solid state BPO_4 defects are mostly non-removable because of the very low solubility of this compound in water.

Summary

Chapter 13 summarizes the understanding of defect formation processes in as-deposited and thermally annealed BPSG films based on a comprehensive analysis of experimental data irrespective of the BPSG film deposition method and related fundamental data. Defect growth kinetics and evolution stages after the film exposure to clean-room air is characterized by taking into account the total additive concentrations and additive ratio in the film, ambient exposure time, and anneal conditions such as temperature, time, gas ambient. A classification of defect types observed on BPSG films is performed. Concentration ranges for classified defect types are defined and the chemical nature of defects is discussed. A correlation of defect types after thermal anneal with the step of BPSG film evolution before anneal is identified. An approach to explain most of the experimental data and observations is presented and an optimized boundary of additive concentrations for BPSG films is outlined as an empirical equation.

It is necessary to note that the methodology of BPSG film defect analysis that is presented is believed to be applicable to more complex film compositions, for instance, germanium-boron-phosphorus-contained silicate glasses. At the moment, there is no information regarding defect formation processes in such films. However, considering their very complex composition, as well as the required high total additive content for further film flow improvement (i.e. lowering of anneal temperature), significantly more pronounced defect formation problems can be predicted.

Chapter 14

INTERRELATION OF BPSG FILM DEPOSITION, COMPOSITION, STRUCTURE, AND PROPERTIES

Thin silicon dioxide and glass films were deposited using a variety of gas-phase deposition methods and chemical compound combinations. Despite a number of published survey papers describing thin film CVD processes and film properties [7-14, 27-34], they cannot be considered to be adequate. For instance, there were no monographs or focused summaries devoted to the correlation of glass film composition, structure, properties with gas-phase deposition features. To a certain extent, the author attempted to perform such summaries earlier [2,3,224,269,270].

Chemical bonding and structures of thin CVD vitreous films were found to be close enough to those of bulk glasses with similar glass composition. Therefore, knowledge of classical chemistry of the vitreous materials can be applied for the explanation of structural features and differences in thin glass film properties and structures. The major common feature of the CVD thin films is that the formation of the glass structure occurs directly during thin film synthesis. Therefore, the glass film composition and properties should be strongly defined by the compound types used and the deposition conditions. The film deposition temperatures for IC technology applications are normally high enough (> 350 °C), excluding the formation of amorphous polymeric compounds. Therefore, we can reasonably assume that as-deposited glass thin film structures can be considered to be fixed after CVD and relatively stable. This conclusion was confirmed by FTIR spectra of different as-deposited glass films. In fact, all types of BPSG films deposited at temperatures in the range cited already contained the same peaks for Si – O, B – O, Si – O – B, P = O bonds. As long as practically used PSG, BSG, and BPSG films preferably contained a SiO_2 component, these complex vitreous materials can be analyzed using a silica structural network base where the other oxides are assumed to be additives. Changes in FTIR spectra after the post-deposition film anneal were attributed to a certain glass structure improvement, but not glass structure forming. At the same time, the temperature of post-deposition film anneal was normally lower (< ~ 900 °C) than that of the glass crystallization temperature. This fact always causes some concerns regarding the film structure and its stability.

This chapter is devoted to the analysis of the relationships between CVD features and silicon dioxide and silicate glass film properties and structures.

14.1. A Brief Summary of Consolidated Material

Glass film deposition kinetics. Detailed summary of glass thin film deposition kinetics and gas-phase deposition processes can be found in Part 1. CVD processes were realized at APPCVD, SACVD, LPCVD, PECVD, HDP-CVD conditions. It was shown that silicon dioxide CVD processes can occur according to surface or gas-phase reaction limited mechanisms depending on the chemical compounds used. It was found that all studied phosphorus compounds accelerate oxidation reactions of the silicon compounds, indicating the common character of this phenomenon. This effect was explained by a development of chain-brain, i.e. free-radical reactions, when phosphorus compounds were added to the silicon compound and strong oxidizers. In other words, all phosphorus-contained glass film CVD processes seem to have a similar type of reaction mechanism. Boron compounds, such as boric acid ethers, were found to accelerate oxidation reactions of silicon organic compounds, such as TEOS. This was also an indication of a change of the SiO_2 CVD reaction mechanism. An opposite type of CVD reaction mechanism change was that the boron ethers were prone to decrease the deposition rate of the silane-oxygen reaction, which was explained by the inhibition of chain-brain oxidation reactions.

It was concluded that the film deposition rate could be considered the only important gas-phase deposition process parameter. It allows calculating the so-called "effective constant of the deposition process, k_{eff}", which has been used for an effective quantitative evaluation and comparison of the gas-phase film CVD (see more details in Chapter 5). Generally, a lower k_{eff} value corresponds to a slower gas-phase deposition process and vice versa.

Reaction schemes and classification of CVD processes. Observed thin glass film deposition kinetic features were explained with a simplified multi-route and multi-step kinetic scheme of the silicon-based thin film gas-phase deposition processes (see Chapter 5 for details). The proposed reaction scheme included three reaction routes: a heterogeneous route (I), a two-step reaction route (II) of thin-film formation, and a route of particle (aerosol) formation in the gas phase (route III). This simplified reaction scheme has been capable of explaining all of the basic experimental data on CVD glass deposition kinetics. According to this scheme, the reaction route differed depending on the compound combination and the deposition conditions. For instance, CVD features of the TEOS pyrolysis reaction can be explained using the heterogeneous reaction route. We have also chosen the heterogeneous route to explain very specific so-called "surface-sensitive" TEOS-ozone CVD process. In fact, despite some differences, very good film thickness uniformity as well as excellent step coverage and gap fill were the common features of this TEOS-based deposition process.

By contrast, typical so-called gas-phase reactions (such as silane APCVD/LPCVD oxidation and PECVD processes) were explained with the two-step reaction route including two consecutive reactions. Differences in the observed CVD kinetic features were explained by the limiting step of the deposition process. The latter was assumed to be either a step of homogeneous formation of intermediate product (1st step), or a step of its heterogeneous consumption (2nd step).

To find links and relationships between the observed gas-phase deposition process features, we have made an attempt to classify those features (see Table 1.3). A certain fixed reaction route or a limiting step of the kinetic scheme was also attributed to every selected CVD process group. Changes of the deposition kinetic features with deposition conditions or gas-phase additives, like phosphorus- or boron compounds, meant that a certain deposition process "shifts" from one group to another group, as shown by arrows in Table 1.3. This was explained by the change of the reaction route and/or the limiting step of the CVD reaction.

Glass film properties review. We have investigated the thin glass film deposition temperature range from about 350 °C to about 520 °C; the most preferred temperature is about 400 °C. In order to analyze the experimental data in more detail, we can divide the accumulated data into two sections: the film property commonalities and the film property differences.

Despite the differences in the film deposition techniques and compound combinations, a set of thin film parameters was found to be similar to all film types deposited at the optimized CVD conditions. Among these parameters are the following: refractive index values, P^{+5} state of phosphorus in the film (that indicates the full phosphorus oxidation with the formation of phosphorus oxide P_4O_{10} or P_2O_5), uniform phosphorus and boron distribution in the film depth, low film defect density and low impurities content (such as carbon and hydrogen) in the film.

The basic differences in as-deposited films were found to be a surface roughness and a wet etch rate WER (or wet etch rate ratio WERR - a wet etch rate of the film divided by that of thermal silicon dioxide). The latter indirectly indicated the differences in the film density. A film anneal at relatively high temperatures and a film exposure to a wet ambient allowed obtaining greater differences in the glass film properties, such as film shrinkage and a film ability to interact with ambient water vapor. The highest film wet etch rate ratio and film shrinkage (up to 7 %) were registered for CVD TEOS-ozone silicon dioxide and BSG films that went along with so-called "surface sensitivity" effect. The lowest film etch rates and shrinkage (less than 0.1%) were registered for PECVD and, especially, for HDP-CVD films. For BPSG films deposited by the same method, higher density of the phosphorus-rich films comparing to the boron-rich films was confirmed by the fact that the shrinkage of the former was found to be about two times higher.

For PSG and ternary BPSG films, most of the features were related to the total phosphorus and boron concentrations, as well as to their ratio in the films. Generally, FTIR spectra of the as-deposited and annealed BPSG films reflect a known set of bonding characteristics of the films. The interaction of a BPSG film with air water vapor caused the appearance of so-called "hydroxyl" groups, i.e. the peaks of the total bonds – OH in the glass film. We showed that for similar PSG and BPSG film compositions, the quantity of absorbed moisture depends on the CVD method. The worst case of moisture absorption was observed for CVD TEOS-ozone BPSG films deposited by Method 4 and Method 5. The best results were found for PECVD and especially so for HDP-CVD deposited silicon dioxide and glass films. After thermal anneal, all these films revealed a decrease in moisture-related peaks, and a slight increase in the Si – O peak and P = O peak.

Another feature of BPSG films was a strong relation between the film composition and moisture interaction. We observed that the strong increase of moisture-related areas in FTIR spectra went along with the increase of the boron concentration in BPSG films deposited by Methods 4 and 5, when boron content exceeded a level, defined by an empirical equation (2.11). This observation correlated with the data obtained by Yoshimaru et al for APCVD autoclave-treated silane-based BPSG films (Method 1) [178].

Increase of k_{eff} →

SiO_2

CVD process with heterogeneous reaction mechanism (HTG 1) or heterogeneous limiting step of 2-step model (HTG 2)

CVD process changes

- gas-phase particles ↑
- step coverage ↓
- gap-fill ↓

Film structure

Decrease of thin glass film structure disordering

Film properties:

- porosity ↓
- shrinkage ↓
- density ↑
- wet etch rate ↓
- moisture absorption

Glass films

Gas-phase reactions with homogeneous limiting step (HMG)

Figure 2.62. Simplified diagram of relationship between an effective constant of the deposition rate. (Signes ↓ and ↑) reflect decrease or increase of process characteristic, respectively.

14.2. Relationships between a Glass Film Structure and the CVD Features

The following trend was detected when the gas-phase deposition processes were characterized using k_{eff} values. The films deposited using fast gas-phase deposition processes with k_{eff} ~ 1.5 cm/sec (like chain-brain oxidation CVD or PECVD) were normally known to have low shrinkage and a low wet etch rate ratio. Therefore, they had a relatively high film density. To date, the densest CVD films are known to be the HDP-CVD films. In fact, for unbiased HDP-CVD processes (i.e., for the film deposition without simultaneous argon sputtering) k_{eff} values were estimated to be as high as ~ 10 cm/sec and above. In contrast, the films deposited using very slow low-temperature CVD TEOS-ozone process with k_{eff} ~ 0.3 cm/sec (and which kinetic features were explained with heterogeneous reaction mechanism) are known to have high shrinkage and high WERR values, indicating relatively high film porosity. Other gas-phase deposition processes with intermediate k_{eff} values seem to be between these two poles and the respected films revealed close enough density-related properties.

We explain the observed film features with a term “structure disordering”. In a scheme in Figure 2.62, the only low-temperature deposition processes (400 – 450 °C) are considered. The TEOS-ozone CVD process can be placed on the left side of the scheme, reflecting an example of a very slow process. It produces the glass film with the disordered structure.

On the other pole of the scheme in Figure 2.62, we place high-speed CVD processes, conventional PECVD and HDP-CVD processes. The kinetic scheme of these processes assumes intermediate products formation in the gas phase followed by their diffusion to the wafer surface. This type of deposition process provided high film density. Therefore, these films were significantly less disordered as compared to the films deposited by CVD processes with a heterogeneous reaction mechanism. Here it is necessary to mention that studied phosphorus compounds enhance the rate of CVD oxidation processes. Therefore, according to the diagram in Figure 2.62, this effect additionally improves the structure ordering. All the rest of the CVD processes can be placed between these two selected poles, which can explain the relatively small difference between these glass film properties. Thermal anneal allows improvement of the structure ordering. This can be seen as changes in shrinkage values, etch rate ratio, FTIR spectra, etc. However, this effect depends strongly on the anneal temperature, which is intentionally lowered more and more with every further generation of ULSI devices.

To confirm the adequacy of the term ''structure disordering", it is useful to mention the following information from classical glass chemistry. In terms of structure, these materials were characterized using so-called short-range order, which was a consequence of the chemical bonding between atoms. The length of short-range order (in terms of its radius) was estimated by Nemilov [338] to be about 1.6 – 2.8 Å. However, the structures of real glass materials are far enough from perfection. For instance, besides an "ideal" tetrahedral silica structure, it can be a number of the other polyhedra structures. This type of structures was characterized using the so-called medium-range order. This term characterizes the structural disordering, showing the reason for porosity of a vitreous material. The radius of medium-range order was estimated to be 2.8 – 5.5 Å.

Among the other conclusions taken from the scheme in Figure 2.62, one is the most important for glass film IC device applications. This is about the structural disordering changes, which are opposite the film step coverage and, therefore, the film gap fill. This generally indicates that slow low-temperature CVD processes would produce a conformal, but inherently disordered glass film that is very important for the practical semiconductor device technology applications.

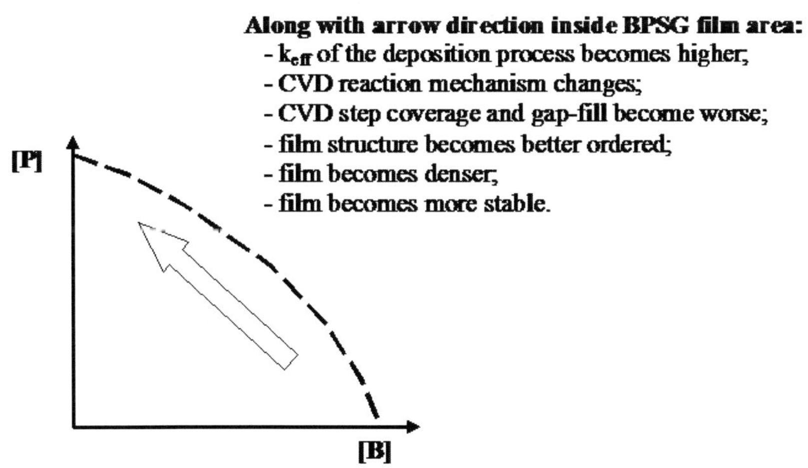

Figure 2.63. A summary of CVD process film features expected to be along with a change of BPSG film composition inside a "defect-free" BPSG film area.

The summary of basic relationships between thin glass film CVD processes and BPSG film features is presented in Figure 2.63. An area of "potentially good BPSG film" is outlined by the dashed line. Inside this area, the block arrow indicates a change of BPSG film composition and summarizes adequate changes of basic CVD processes and film properties.

14.3. Correlation of the Structure Disordering and the Phosphorus- and Boron Silicate Glass Film Composition

Previous considerations have been done mostly for vitreous silica (SiO_2) films. However, the structures of silicate glasses are more complex compared to that of silica. The main conclusion of the film property studies was that there were serious differences in the physical and chemical properties of BPSG films deposited by different CVD techniques. To explain the possible reasons for such differences, it was essential to analyze the information about the film structure. As long as phosphorus and boron oxides structures differ from SiO_2 structure, the addition of phosphorus oxide and, especially, the boron oxide to the silica network enhances the glass film structural disordering.

The glass structure of boron-rich compositions consists of triangles and boroxol groups. As mentioned in Chapter 9, in both cases boron reveals threefold coordination with respect to oxygen atoms. In this case, strong structural differences between elementary structure groups reflected as tetrahedra $SiO_{4/2}$ and boroxol $2(BO_{3/2})$ groups lead to the glass structure disordering. Threefold boron coordination explains the well-known high activity of the boron-enriched BSG films with respect to the moisture, for instance, producing a group with a weak water bonding $[(\equiv Si - O -)_3B]\bullet H_2O$. CVD method features also affect this disordering. For example, the TEOS-ozone BSG CVD process (performed in the presence of boron ethers) was found to reveal a "surface sensitivity" effect [122]. This was explained previously in terms of heterogeneous reaction mechanism. Therefore, the density of these films is lower as compared to the other types of BSG films. In fact, by contrast, PECVD BSG films are significantly denser.

Along with the increase of the ratio of phosphorus compound concentration to the boron compound concentration in the gas phase, the heterogeneous reaction mechanism of BSG process TEOS-ozone CVD process changes to gas-phase BPSG chain-brain reaction mechanism. Adding P_2O_5 to boron-rich compositions resulted in the decrease of the boron share in BPSG film and changes in the glass structure. Namely, the boron oxide network becomes broken and the boron changes coordination from threefold to fourfold. Following the previous discussion, it means that more and more boron atoms in the film network become fourfold coordinated with oxygen, therefore improving structure ordering. This certainly increases the film density and decreases the ability of BPSG film to react with moisture. The lowest capability of BPSG film to react with water vapor was found to correspond to the ratio $[B] \sim 0.4\,[P]$, where additives concentrations are presented in elemental wt %. Re-calculation of this value shows that such BPSG films contain the equal quantity of B_2O_3 and P_2O_5 (oxide) molecules that is represented in Figure 2.14 by a straight dashed line. Eventually, the corresponding boron phosphate BPO_4 glass structure was characterized as a tetrahedral PO_4 and BO_4 based network. Therefore, in these glass films all glass-forming oxide species are supposed to exist as tetrahedra SiO_4, BO_4, and PO_4. They form the structural groups $(\equiv Si - O -)_2OP - O - B(- O - Si \equiv)_3$ confirming significant improvement of glass film structural ordering. Further, at $[B] << 0.4\,[P]$, the glass structure consists primarily of SiO_4 and PO_4 tetrahedra and the quantity of BO_4 tetrahedra can be ignored. Figure 2.14 represents a summary of the structural features of BPSG films, expressed using Myuller's structural group symbols, clearly demonstrating the differences in the BPSG film structures caused by coordination of the glass-forming atoms.

14.4. Correlation of the BPSG Structure Disordering and Moisture Absorbance and Defect Formation Processes

Along with the worsening of the BPSG film structure ordering, the increase of the boron concentration in BPSG films was found to enhance the moisture penetration depth in the film [178]. At the same time, the increase of phosphorus concentration provides a limit of water on the film surface. These data give us a possibility to compare defect formation and moisture absorption phenomena quantitatively in terms of boron share [B]/([B]+[P]) in the total additive concentration in BPSG films. Thus, a correlation of 24-hours defect-free area boundary from Figure 2.42 and the moisture penetration depth [178] was shown in Figure 2.61. The data show that the increase of boron share above ~ 30 % ([B] ~0.4 [P]) causes a sharp increase in moisture penetration depth during PCT (see open symbols). At the same time, above this boron to phosphorus ratio we can see a gradual decrease of the defect formation boundary to the lower total additive content. The latter means the enhancement of defect formation processes at higher boron content in BPSG films.

This summarized graph with data, taken from totally different research groups and from different research approaches clearly confirms that boron to phosphorus ratio ~ 0.4 revealed the basic BPSG material feature, namely, complete structure ordering.

14.5. Specific Additive Features in BPSG Film Structure

Taking all of the above into consideration, in this section we discuss two schemes of BPSG film, which generalized all accumulated information regarding the film features. These schemes were introduced for the first time by Vassiliev et al in Ref. 214. It was assumed that additive concentrations were equal, for instance, close to 4 wt % of each element that was close enough to the BPSG film composition mainly used in VLSI and ULSI technology. To equalize the additive content, the author excluded BPSG structure differences related to the difference in the film composition. Therefore, it allowed to consider the only differences in BPSG films, which could be due to the CVD processes features, or thermal anneal features. According to the author's long-term research, the most interesting set of BPSG film properties was obtained for TEOS-ozone BPSG films deposited either at atmospheric, or subatmospheric CVD conditions, i.e. by Methods 4 and 5. The scheme of CVD processes presented in Chapter 5 shows that the TEOS-ozone based CVD process is the most complicated one due to the differences in TEOS-ozone oxidation reaction (assumed to have a heterogeneous reaction mechanism) and phosphorus compound oxidation (assumed to have a chain-brain reaction mechanism). Thus, it was believed that these TEOS-ozone BPSG films can be taken as the best example of specific structural features of the film.

An "ideal", i.e. highly-ordered BPSG thin film can be defined as a glass matrix with typical bonds: Si – O – Si, Si – O – B, Si – O – P and P = O. This glass structure is assumed to have "ideal" glass properties, including an equal dependence of the film glass viscosity on the both boron and phosphorus additives (in terms of their oxide weight percent). The real properties of BPSG films differ from the ideal scheme because of differences in compounds, oxidizers and deposition conditions used. The main reason for the difference of the glass property seems to lie in the additive state of the film. For instance, there were found to be at least four principal differences in TEOS-ozone SACVD BPSG film properties as compared to other types of BPSG films.

First, there was a higher deviation of additive concentration in the film depth indicated by the SIMS profiles (see Chapter 8, Figure 2.10(a,b)).

Second, the observed non-uniformity of the defect appearance on the film surface as well as a high wet etch rate and net charge deviation data were explained by high film porosity.

Third, a higher moisture absorption and additive out-diffusion of PCT treated TEOS-ozone BPSG films was observed as compared to PECVD BPSG films (see Figure 2.27(a,b)). Large liquid droplet formation on the surface of BPSG and high phosphorus content PSG films, obviously show the role of phosphorus in inducting these surface defects. In general, SACVD BPSG films, as well as TEOS-ozone APCVD films were prone to form surface defects at total concentrations as low as 8 wt %.

Fourth, there were differences in flow properties between SACVD BPSG and other types of BPSG films (detailed consideration will be presented in Part 3). It was shown that SACVD BPSG film flow capability revealed significantly higher additive concentration dependence with respect to that of the other types of BPSG films deposited either at APCVD or LPCVD conditions [223]. It was proposed that this could indicate that a significant share of the measured phosphoric oxide content was not included in the glass matrix and thus, probably existed as local additive-rich clusters in film.

The first scheme of relatively disordered, i.e. porous BPSG film, has been presented in Figure 2.59(a). In this structure, additive atoms exist both in a glass matrix and in randomly distributed additive-contain clusters, which contain mainly phosphoric oxide in the form of $[nP_4O_{10}]$. The density and size of clusters were believed to be a function of the total phosphorus concentration in film and the deposition condition. The proposal about higher porosity in TEOS-ozone BPSG films as compared to other types of BPSG films has been made based on the following features:

(a) kinetic studies and the proposed mechanism of CVD SiO_2 and glass films, which assume a parallel formation of the film and clusters in the gas-phase (see Fig1.26 for more details), including a conclusion about a possibility of gas phase phosphorus oxide clusters formation discussed previously for PSG and BPSG films by Yoshimaru et al [178].
(b) accumulated results of BPSG film property studies;
(c) accumulated results of the BOSG film net charge value monitoring using the SCA technique.

The second film scheme shown in Figure 2.59(b) is the scheme of relatively ordered, i.e. non-porous film with randomly distributed additive atoms in glass matrix. This scheme was assumed to be valid for CVD processes with high k_{eff}, for instance, for silane-based CVD, PECVD, HDP-CVD. It is believed to be true for relatively high temperature deposited films as well as for thermally treated and flowed BPSG films.

Applying these schemes for the explanation of the SCA data obtained for long-stored TEOS-ozone BPSG films, the following was concluded. During the long exposure to ambient air contained moisture, the porous structure of the film with chemically active clusters was believed to be prone to form the solutions of boron- and phosphorus-contained acids inside the film pores. The processes of formation of these solutions were relatively slow and, in general, one could conclude that as-deposited TEOS-ozone BPSG films were always in a certain electrically equilibrium state. As a result, there were no significant changes during long-term SCA monitoring of as-deposited BPSG films (see Figure 2.36(a)).

However, abrupt changes of SCA values in these films were seen after vacuum pumping or water rinsing of films resulting in the fast removal of additive solutions from the pores. This was assumed to be the result of the creation of a new electrically non-equilibrium state of the film. Logically, as long as the concentration of the solution in the pores depends on the total additive concentrations, the changes of the total film charge after fast pumping should be a function of the total additive concentrations in films. This assumption was confirmed by the experiment (see Figure 2.36(b)).

The process of moisture absorption into the as-deposited non-porous (dense) film was very slow. For instance, it took about a month to see a little moisture-related peak in FTIR spectra of PECVD and, especially, HDP-CVD films. Therefore, there were no obvious changes in the state of additives inside these films. SCA values were stable even after vacuum pumping and water rinsing of as-deposited films, see Figure 2.36(a). Some small enough liquid droplets on the film surface were believed to be formed due to the interaction of additives with moisture on the film surface.

FINAL REMARKS ON PART 2

In Part 2 of the monograph, BPSG film composition, structure, properties and interrelations between glass film properties and CVD process features are reviewed and analyzed. Analysis made in Part 2 of the monograph allows the explaination of almost all CVD glass thin film features, obtained during 30 years of BPSG investigations. The basic physical and chemical properties of thin glass films are explained in terms of glass film structural disordering, which defines the oxide film density, as well as wet etch rate, shrinkage, intensity of film-moisture interaction, defect formation phenomenon, etc.

For the silicon dioxide films, the cause of structural disordering is CVD features, which were characterized using the effective constant of the deposition process k_{eff}. The low deposition rate CVD processes, such as TEOS-ozone CVD (Methods 4 and 5), reveal the worst cases of the film density. In contrast, high-speed PECVD processes provided the best film density and, therefore, the best thin film structure ordering. Observed features and structural disordering were explained by using a simplified multi-route and multi-step kinetic model of the silicon-based thin-film gas-phase deposition processes and with a classification of the studied silicon-based CVD processes.

To analyze the property features and structural disordering of phosphorus- and boron-contained silicate glass films, the total boron and phosphorus content and the boron share in the film should be taken into account along with the gas-phase deposition features. It was assumed that the latter plays a less important role because all studied CVD oxidation reactions in the presence of phosphorus compounds were found to accelerate and proceed according to a common chain-brain reaction mechanism. As a result of such acceleration, all these reactions features become quite similar. BPSG film properties were affected by the quantity of boron atoms with threefold coordination with respect to the oxygen atoms. The best BPSG structural ordering and the best film properties correspond to the boron concentrations defined by an empirical equation [B] < 0.4 [P] and this glass film structure is supposed to consist of SiO_4, PO_4, and BO_4 tetrahedra.

The properties of the borophosphosilicate glass films studied were determined by several factors. Among them were: (a) the kinetic features of the chemical vapor deposition methods, which were responsible for the porosity of the synthesized glass films; (b) the occurrence of a certain number of boron atoms in the threefold coordination with respect to oxygen in the glass structure; (c) the total oxide concentration in the glass; (d) and the concentration and sizes of phosphorus oxide clusters embedded into the glass structure.

One more feature of the phosphorus-containing glass films prepared by CVD was the possibility of forming anomalous structures of phosphorus oxide in glasses. The assumption regarding these anomalies offers an explanation for the specific features of BPSG films synthesized by Method 5. They were the following: (a) a multifold increase in the deposition rate with the suppression of the "surface sensitivity" effect and multifold decrease in the film roughness upon introduction of phosphorus compounds; (b) fluctuations of the phosphorus and boron concentrations in glass films; (c) a higher reactivity with respect to water vapors; (d) strong leaching of phosphorous and boron; (e) weak effect of the phosphorous concentration on the glass flow; (f) strong change in the charge in the glass films upon heat treatments. These features were taken into account when constructing the glass structure models proposed in the author's works. It was assumed that the sizes and concentration of clusters were proportional to the total concentration of phosphorus and boron in the glass films. The occurrence of pores in the glass films resulted in an increase in the rate of interaction with the air moisture.

The heat treatment of glasses at temperatures above their softening point leads to a decrease and/or disappearance of pores, as well as incorporation of clusters into the glass structure and/or their evaporation from the film. After heat treatment at high temperatures, BPSG films having identical compositions but prepared by different CVD methods become similar in properties.

PART 3. BPSG FILM IMPLEMENTATION IN ELECTRONICS

Part 3 of the monograph (Chapters from 15 to 20) is devoted to the description of BPSG thin film implementation aspects. As long as BPSG films were introduced in electronic technology as flow-able dielectric, the most common use of this film is to smooth sharp edges of device structures. Chapter 15 is devoted to the detailed analysis of flow aspects of BPSG films on the real IC device structures. It is important to note that data presented in Chapter 15 were obtained experimentally and explained quantitatively. We also showed that the approaches presented can be applied to explain the flow of other types of flow-able glass films, for instance, BPSG films with germanium oxide additive. Chapter 16 shows approaches for obtaining the optimized BPSG composition used in electronic technology. Chapter 17 is devoted to the aspects which are rarely discussed in a related literature, namely BPSG film durability. Some approaches to improve BPSG film post deposition thermal treatment are proposed. Finally, we generalize some data regarding BPSG film application in electronics in Chapter 18.

Chapter 15

THERMALLY INDUCED BPSG FILM VISCOUS FLOW ON STEPPED IC DEVICE STRUCTURES

BPSG films are now routinely used as a Pre-Metal Dielectric (PMD) (another term used is Inter Layer Dielectric, ILD) between the polysilicon or silicide gate and the first level of device metallization(see Figure 1.2). Once deposited, a BPSG film underwent thermal treatment at elevated temperatures that allowed implementation in IC device technology the major glass film property – its ability to soften and to flow. The main advantage of BPSG is the ability to flow at significantly lower temperatures in comparison with previously used PSG films.

As shown in Chapter 11, the effective decrease of the viscosity of PSG glass is due to introducing several percent of boron oxide into the glass in addition to the phosphorus oxide. The decreasing of PMD flow temperature was a general direction in IC technology. It was necessary for decreasing the thermal budget of VLSI and ULSI circuits.

In integrated circuit technology, the BPSG film capability to soften can be realized at both isothermal furnace anneal (FA) and rapid thermal anneal (RTA). Although the flow properties of BPSG thin films in device manufacture have been addressed by many authors, most of them treated the subject in a qualitative manner. Cross-sectional scanning electron microscope (SEM) images of processed specimens form the core of these reports. The author's research into the thermal flow of BPSG thin films started in the mid-1980s. Unlike other research, the author's group proceeded from well-known viscous properties of bulk glasses and found their quantitative correlations with thin film glass flow on stepped IC device surfaces.

Chapter 15 is a consolidation of long-term quantitative studies of BPSG film flow issues and solutions. It focuses on the detailed analysis of the main practical aspects of BPSG film flow studies. As a result of these studies, an approach for quantitative characterization of BPSG film flow was developed. In addition, the acceptance of developed approaches for films flow analysis in quaternary glass systems, such as GeO_2-SiO_2 glass enriched with B_2O_3 and P_2O_5 is shown.

15.2. A Brief Historical Review of Flow-Able PMD in IC Technology

An analysis of published data as well as the results of BPSG film implementation allowed selection of three basic directions of the flow-able film application in VLSI and ULSI technology.

Historically, the first implementation of BPSG film capability to flow at enhanced post-deposition temperatures was the so-called film "planarization" of device steps (see Figure 1.1(c,d,e,f)). The term "planarization" meant the smoothing of an insulator surface in order to prevent the breaking and short-circuiting of conducting paths that will run on the surface. The glass film anneal was used in a few hours time delay after film deposition. Different gas ambient compositions, such as oxygen, nitrogen, argon and steam, were used. This application was linked mainly to the advent of high-speed integrated multilevel device technology with short transistor channels and shallow p-n junctions. The place of PMD in the device structure is very important, because it affects a quality of a metallization layer sputtered above. For instance, non-conformal PMD film on device steps and strictly vertical side-walls of etched in PMD contact holes enhance non-conformality of metal sputtering (as shown in Figure 3.1(a,c)). In such a case, the metal thickness on a side wall is not sufficient for proper metal line conductivity. Moreover, metal bridging issues are also known for these structures because of the shortening of neighbor metal lines after etch. Thus, in order to make metallization reliable, device topography needs to be smoothening (as shown in Figure 1.1(d,f) and Figure 3.1(b,d)).

Film flow allowed the smoothening of high device steps covered with BPSG film. Numerically, it corresponded to a decrease of the initial angle A_0 (step coverage angle on device step after film deposition) to final low value A_f after anneal at a certain temperature T_a for a certain anneal time t_a (see Figure 3.2(a,b)). The main advantage of BPSG films is about 200 – 300 °C difference (depending on the additive concentration level) between flow temperatures as compared to PSG films with the same quality of planarization. The main results of the planarization properties of BPSG, which were studied mainly qualitatively, are presented in Table 3.1.

Later, the second BPSG film anneal, frequently called "reflow", was introduced in IC device technology in order to smooth sharp edges and taper contact holes. This second anneal was performed after lithography and plasma etch procedures forming openings in BPSG film – contact holes. This second anneal allowed decreasing of the side wall angle in contact holes from Θ_0 to Θ_f, as shown in Figure 1.1(e,f) and Figure 3.3(a,b,c,d).In such a case, an inert gas ambient was used in order to prevent the oxidation of the silicon surface in these holes. Some data regarding this issue can also be found in Table 3.1.

According to many researchers, it was established that practical BPSG films should contain no more than 8 – 10 wt % boron and phosphorus totally combined and that the furnace-annealing (FA) temperature should be above 900 °C. Almost complete planarization is possible under the stated conditions. However, defects were found to arise in the glass at certain concentrations of boron and phosphorus at relatively high annealing temperatures. The defects were traced to the intense interaction of the two additives with the moisture that penetrates into the glass from the air (see Chapter 13). Thus, the use of BPSG in ICs requires

the careful optimization of the film composition. Some particular quantitative investigations other than ours were presented in Table 3.1. However, there were no detailed papers and, therefore, no deep understanding on the impact of the underlying device geometry, BPSG film parameters and anneal processes on the effectiveness of BPSG film flow and device planarization.

The thermal-flow planarization of BPSG thin films was proven to be suitable for IC device technology of minimum feature sizes down to 0.35μm. Below this level, chemical mechanical polishing technology emerged as a viable alternative. Nevertheless, glass-flow techniques have found another application in ULSI circuits of feature sizes down to 0.15 – 0.18 μm. Among the technology challenges, the drive of the ULSI technology node towards the 100 nm and below resulted in the decrease of device gap spacing in conjunction with the increase of aspect ratios at transistor level. Small gaps with below 0.1 μm spacing and aspect ratios (AR) higher than 10 become a feature of sub-100 nm IC technology and these numbers are improving further [341]. The decrease of gap spacing goes along with a change of gap shape and re-entrant gap profile is a common shape in ULSI devices. These peculiarities with CVD film deposition features enhance the issue of a void-free PMD gap-fill, which is known as a "voiding" phenomenon and reflects a certain empty space between device elements (see Figure 3.4(a-d)). Thus, providing a void-free BPSG film gap-fill with film flow, e.g. elimination of key-holes formed during film deposition, became a new issue of BPSG films implementation in the mid 1990s.

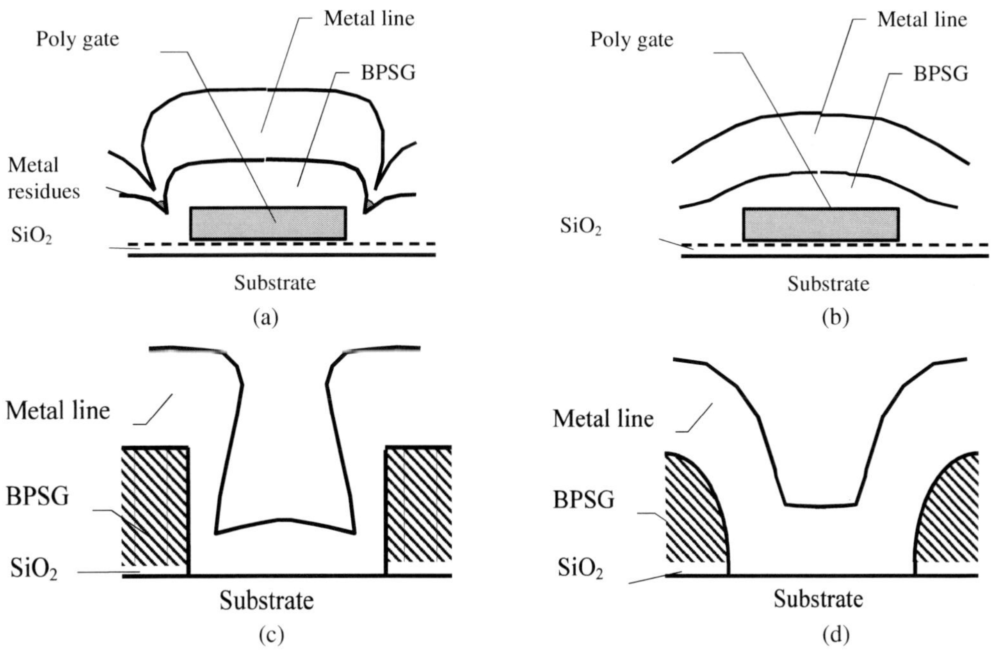

Figure 3.1. Cross-sectional schemes of stepped device structure with a polysilicon strip (it models a conventional poly gate region of a MOSFET) covered with BPSG film and plasma-etched pre-sputtered metal (a,b) and schemes of vias in BPSG films filled with sputtered metal (c,d). Left side – without BPSG film flow, right side – after BPSG film planarization using thermal flow.

Table 3.1. Summary of accumulated flow-related quantitative results for glass films used in IC technology (in a chronological order)

[Ref]	BPSG object (thermal treatment conditions)	Results, equations or correlation obtained
[52]	CVD films (FA)	After FA of BPSG films, contact hole side wall slope was found to be about 2 times more effective with boron as compared to phosphorus.
[54]	Bulk glass	Introduced glass transition temperature (T_g), that was found to correlate with glass flow temperature (T_f) as follows: $T_f=T_g+288°C$.
[339]	CVD films (FA)	At contact hole reflow, slope was found to be higher than 90° if a geometric ratio of BPSG film element thickness to this element length exceeded ~0.393.
[82]	CVD films (FA)	Proposed an empirical equation A=51.5-1.4[P]–9.5[B] for film planarization on device steps. A possibility to use a non-destructive surface profilometer technique for film flow analysis on device step was shown.
[86]	CVD films (FA/RTA)	An equation $\cot A \sim t^{1/n}$ for film planarization was proposed, where A is an angles of device step with glass film. Empirical values of "n" were defined to be ~2.1 ± 0.5 for RTA and ~2.4 ± 0.8 for FA.
[340]	CVD films (FA)	A modeling of BPSG and PSG films viscosity was done using EVOLVE-FLOW simulator, values for viscosity and surface tension were analyzed.
[175]	CVD films (RTA)	Defined activation energy of void elimination in small device gaps using BPSG films with about 5 wt. % each additive at RTA (~3 eV).
[178]	CVD films (FA)	Both boron and phosphorus affect BPSG (Method 1) planarization capability equally once used in equal oxides weight % B_2O_3 and P_2O_5 concentrations in the glass film.

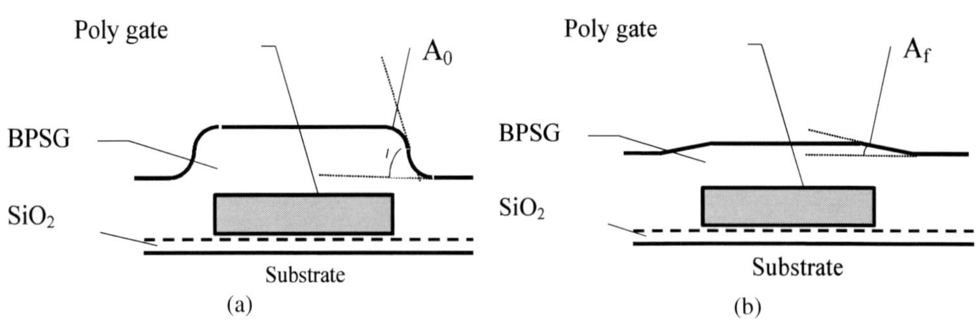

Figure 3.2. Cross-sectional schemes of SEM images: (a) in Figure 1.1(c), (b) in Figure 1.1(d) which show BPSG step angle definitions A0 and Af, accordingly.

Generally, BPSG films were successfully employed for the filling of voids that may arise in BPSG film during deposition (see an example in Figure 3.4 (a)). However, there are some issues there. Thermal-flow planarization was commonly performed by relatively high-temperature FA, which was a comparatively long process, sufficient for proper film softening and flow. In contrast, for sub-0.18 μm ULSI circuits, it was generally accepted that the thermal annealing after the formation of transistors should be carried out at temperatures below about 750 °C for FA and about 850 °C for RTA. To ensure void filling at such low temperatures, one had to increase the boron and phosphorus concentrations in the glass above a certain "commonly accepted" values. However, this measure enhanced defect formation processes in BPSG films. Thus, an optimized BPSG film flow became a major issue along with the gradual shrinking of device design rules as well as a need to decrease the thermal budget of ULSI devices.

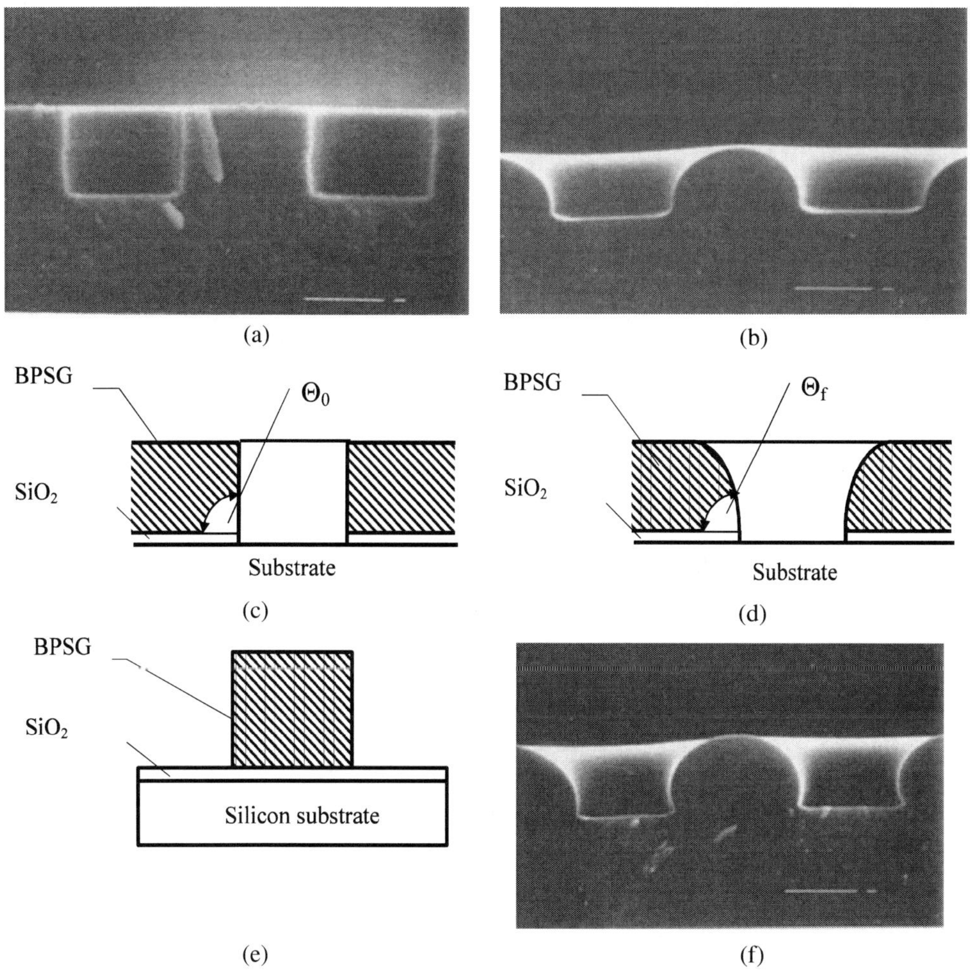

Figure 3.3. Cross-sectional (a,b) SEM images of vias in as-deposited (a) and thermally flowed (b) BPSG films [339]. Figures (c,d) are cross-sectional schemes of SEM images (a,b) and show BPSG step angle definitions Θ0 and Θf. Figures (e,f) show a cross sectional scheme of a spherical glass element formed in case of non-optimized height to length ratio of the glass element before the film anneal [339]. Reproduced by permission of ECS – The Electrochemical Society.

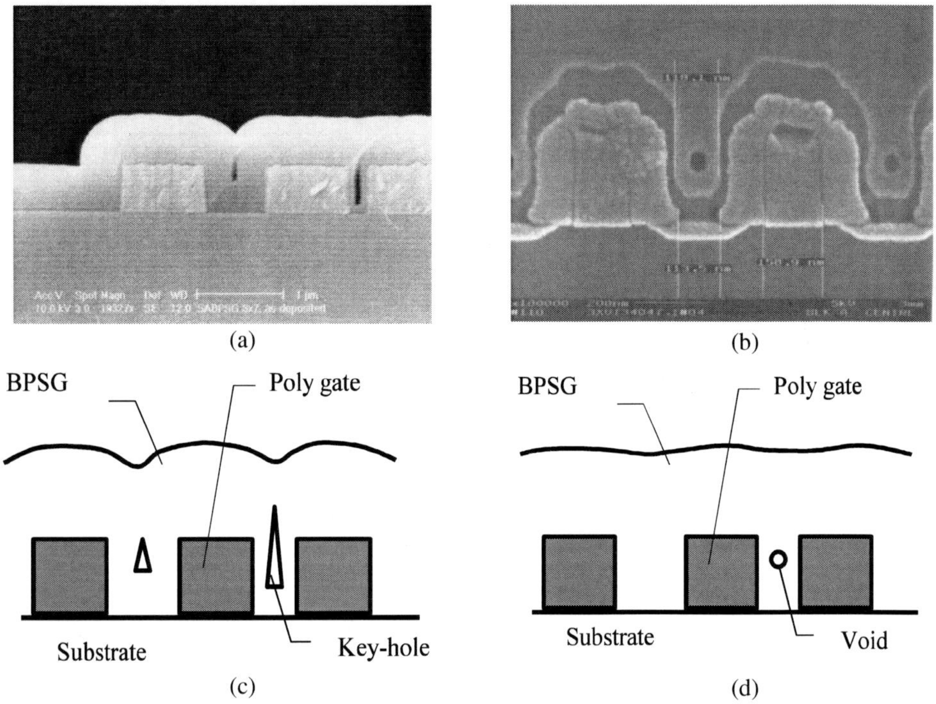

Figure 3.4. Cross-sectional -holes between device elements for as-deposited (a) and rounded voids in thermally flowed (b) BPSG films (highlighted by block arrows). Figures (c,d) are cross-sectional schemes of respective SEM images. Image (a) was presented at 4th Intern. Dielectric for ULSI Multilevel Interconnection Conf. (DUMIC), 1998 [208], image (b) was presented at 8th Intern. Symp. on IC Technology, Systems and Application (ISIC-99), 1999) [231].

The author and his co-workers based their studies [90,93,105] from a well-known relation between the viscosity (μ) and the absolute temperature T of bulk glass (3.1):

$$\mu = \mu_0 \times \exp(U / RT), \quad (3.1)$$

where U is activation energy for viscous flow. To apply this knowledge to IC technology, the following logic was assumed. A step of an as-deposited glass film on IC device structure exists in a thermodynamically non-equilibrium state. When heated causing glass flow, the step evolves to equilibrium and eventually disappears if heating is long enough (a planar surface is thus obtained). Strictly speaking, planarization occurs under the action of both surface tension and gravity forces. However, the latter is negligible for actual step heights in ICs (they are less than 1 μm). On the other hand, practical thermal-flow processes were usually performed at moderate temperatures and for limited lengths of time, with the result that perfect (to completely flat surface) planarization is rarely feasible. A deep understanding of the planarization mechanism was therefore necessary for the optimization of the flow process.

This framework allowed the quantitative analysis of the thermal-flow planarization of stepped BPSG surfaces and summarization of relevant data from the literature. It was proposed that the basic properties of BPSG films prepared by different deposition methods were similar and they could be generalized for the comprehensive characterization of BPSG material for multilevel ULSI circuit technology. Using this proposal, the approaches for quantitative characterization of BPSG planarization and optimized additive concentration range in BPSG films were developed.

It was proposed that the change of angle A_0 on device step should follow the temperature dependence for glass viscosity(see Eq.3.1). In such a proposal, similar dependence for angle was proposed to be as follows (3.2):

$$\cot A_f = \cot A_0 \times \exp(E_a \times B_T) \tag{3.2}$$

where A_0 and A_f are initial step coverage (as-deposited) and post-flowed angles of device step with glass film, and B_T is a temperature-related member. For example, we carried out an electron-microscopy study of the angle A_f as a function of the boron and phosphorus concentrations, the annealing temperature, and the angle A_0 in the case of 30-min FA. A set of empirical formulas was proposed for the flow of BPSG over a step of dimensions typical of actual chips. This model configuration will be considered below.

The proposed formulas were later confirmed by experiments on BPSG films produced by two modern CVD methods. One of them was plasma-enhanced chemical vapor deposition (PECVD) involving organic ethers in the presence of oxygen, i.e. Method 8. The other was based on the oxidation of the above-mentioned ethers with an ozone-oxygen mixture, i.e. Method 5. As a further step, the model was modified to include the dependence on annealing time. We thus arrived at a multi-factorial, quantitative picture of BPSG flow under FA. Afterwards the approach was generalized to rapid thermal annealing (RTA). Despite a much smaller amount of relevant experimental data, an empirical formula was derived for this regime as well. Later, studies of BPSG film gap-fill using FA and RTA technique were performed and an approach for quantitative BPSG film gap-fill evaluation was developed.

15.3. A Methodology for Analysis of BPSG Film Parameters on Stepped Device Surfaces

Our early studies on the thermal flow of BPSG films dealt with device structures with a minimum feature size of about 2 μm. In that case the post-etch angle between a sidewall and the substrate surface was far less than 90 ° and gaps between device components were fairly wide. These facts allowed us to essentially assume that the initial flow angle A_0 is 85° at most. Thermal flow was examined from cross-sectional SEM images of test structures fabricated on silicon single crystals by CVD, photolithography, and plasma etching. A test structure was an array of polycrystalline-silicon lines (strips) with fixed width and spacing, as shown in Figure 3.5(a). This structure with a number of lines was covered with BPSG film, as shown by schematic cross-section in Figure 3.5(b), followed by its anneal if necessary.

The glass thickness d_{BPSG} and the dimensions of the array were chosen in accord with practical chips: the strip spacing x_{poly} was set to 1 – 3 μm, d_{poly} was 0.3 – 0.5 μm for practical chips, and the ratio of d_{BPSG} to the strip thickness d_{poly} was taken as 1.5 – 2. The latter was important, otherwise the speed of BPSG flow on device step could be artificially decreased. The individual concentrations of boron and phosphorus were 3 – 6 wt %. Some wafers with specimens, spaced a fixed distance apart, were subjected to FA in a horizontal-tube or a vertical-tube quartz diffusion furnace using argon, nitrogen, or oxygen; the process lasted 12 – 60 min at atmospheric pressure and a temperature of 850 – 990 °C. RTA was performed in nitrogen at 700 – 900 °C, the flat-temperature period being 20 – 90 sec. With the latter type of annealing, the author employed a Centura-RTP system (Applied Materials, Inc.) and apparatuses from AG and Associates.

For cross-sectional SEM imaging, the specimens were broken across the poly lines (for this reason, the line array was usually a few millimeters long and wide). The same approach was also followed in the examination of contact vias in a glass film. In this case a test structure was an array of trenches or contact vias made by dry etching of a BPSG film. The vias were of different dimensions and were distributed over a large area (to facilitate breaking).

After breaking a specimen, a sample of area 1 cm^2 was split off at a site of interest and attached to a stainless-steel holder. The surface to be examined was etched for a few seconds in hydrofluoric acid with a buffer (to show the interfaces), rinsed in deionized water, and dried. Finally, a thin layer of a conductive material, such as platinum, was deposited to prevent charge buildup during SEM examination.

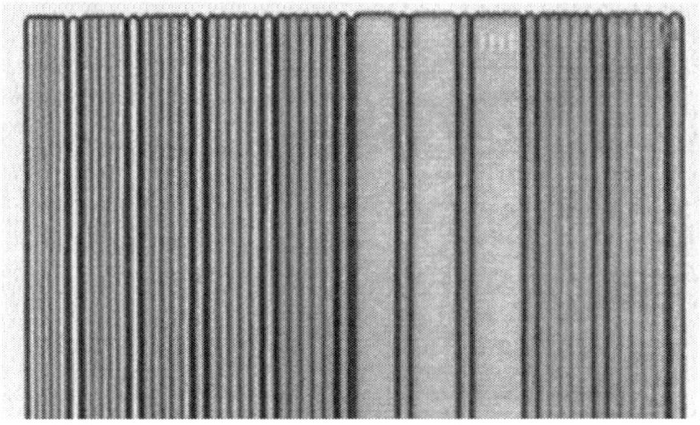

(a)

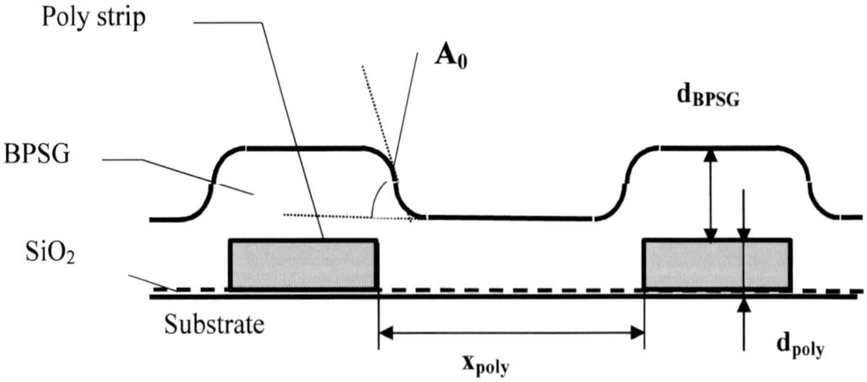

(b)

Figure 3.5. Optical microscopy top view on polysilicon strips (a) and a model of stepped BPSG surface used for SEM analysis (b).

The flow of BPSG films was characterized by the angles A_0 and A_f (Figure 3.2) on the basis of micrographs, the angles being measured with a protractor. The same SEM procedure was used for contact hole analysis to measure angles of Θ_0 and Θ_f, as shown in Figure 3.3. However, in such a case the openings were etched in BPSG film followed by anneal to make a desirable side wall slope. Perspective and cross-sectional images were analyzed. As the research progressed, we changed from one type of SEM to another. In the latest phase, we employed XL 30 FEG microscopes (Philips, Germany) with a magnification of up to 10^6 times.

The time variation of A_f was also studied nondestructively, using the B5000 NanoScope III model of large tapping mode atomic-force microscope (AFM) from Digital Instruments. Conceptually, it was possible to rapidly derive the profile of a stepped surface from a computer-generated AFM image of the surface, so that it was not necessary to fabricate special test structures in the evaluation of a glass-flow technology. Note that this approach was proposed by Williams et al. [82] as early as in 1987 (see images in Figure 3.6(a-e)), but the sensitivity of the method was insufficient. In our studies, AFM data were found to agree with SEM ones within about 5 % for angles < 80°, see Figure 3.7(a-c) [225].

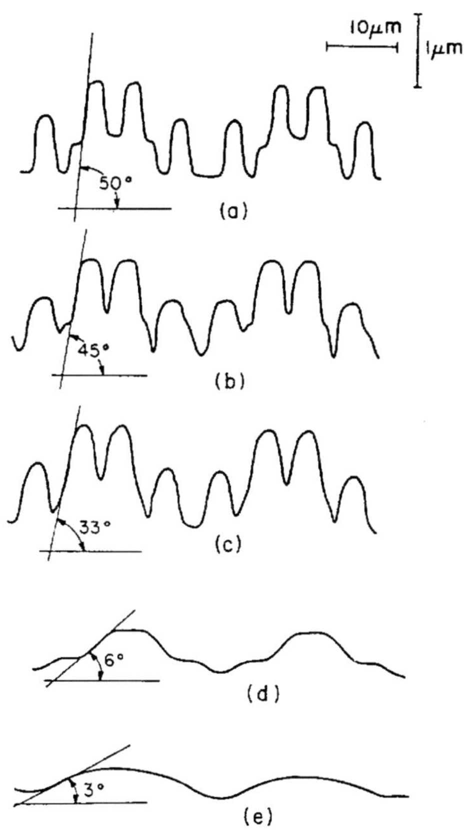

Figure 3.6. Surface profilometer trace over policide runner: a) before BPSG deposition, b) after BPSG deposition, and after 900 °C anneal for 1.9 wt % [B] – 4.7 wt % [P](c); 3.1 wt % [B] – 5.8 wt % [P] (d); 4.1 wt % [B] – 5.5 wt % [P] (e) [82]. Reproduced by permission of ECS – The Electrochemical Society.

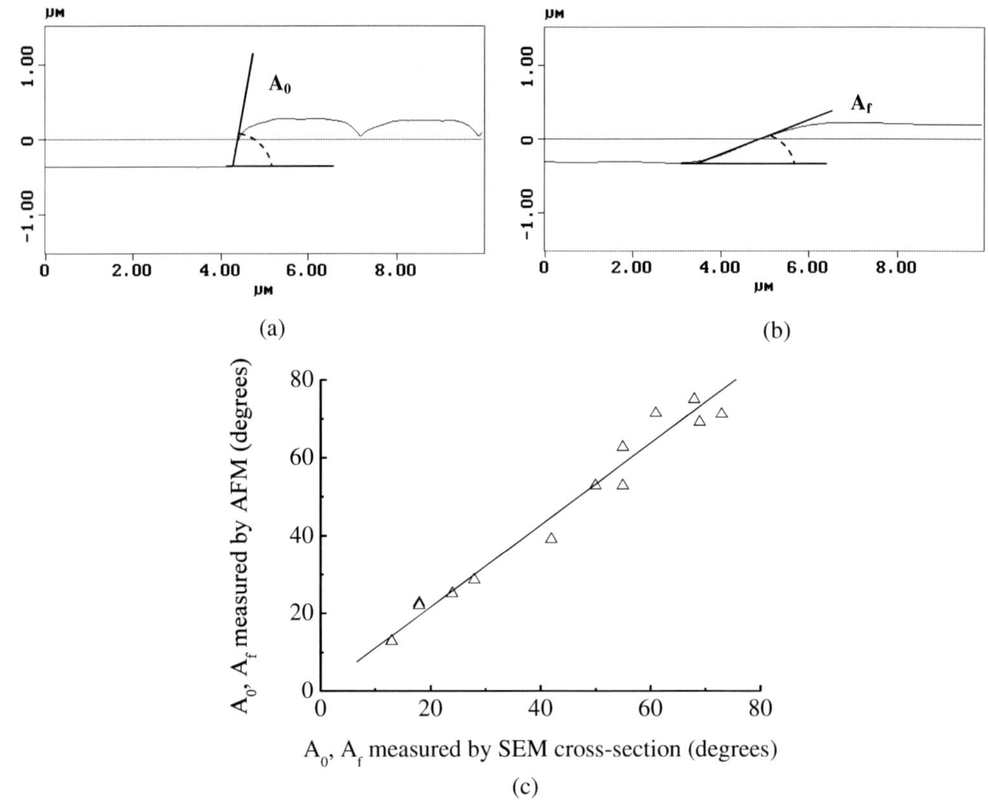

Figure 3.7. BPSG flow angles measured using AFM (a,b) and comparison of AFM and SEM measurements (c) on the same test structures.

The quality of gap-fill with flowed BPSG films was normally evaluated using similar device structures with lines, placed at different spacing one from another. After film deposition or film anneal at desired conditions, void appearance in device gaps could be observed (as shown in Figure 3.4(a)). Voids were dark rounded areas in photos from SEM and their shape and size could differ significantly. It is important to note that voids existence was not an artifact due to cross-section etch for SEM sample preparation. Their appearance was confirmed by the SEM analysis of non-etched and etched samples. However, in order to measure void size, the SEM picture must be very contrasted and this can be achieved only with proper sample etch.

15.4. Quantitative Characterization of Primary Factors Affecting BPSG Film Flow

BPSG film flow under FA conditions in dry gas. Figure 3.8(a) gives examples of A_f-T_a relations for the FA of BPSG films. More than 60 specimens were examined, differing in deposition technique, composition, and annealing conditions. As already noted, we assumed that $A_0 < 85°$, which is the case with devices of a minimum feature size over 1 µm.

It was found that the graphs can be conveniently transformed into straight lines by re-calculating them into a so-called Arrenius plot, namely into the $1/T_a$-log cot A_f graph, where T_a is stated in Kelvins, as shown in Figure 3.8(b). A number of straight lines refer to the same deposition technique met at a point whose coordinates correspond to A_0 and the temperature T_{bf} at which thermal flow begins. Indeed, T_{bf} depended on the composition of the glass, but it was difficult to calculate this dependence. It was therefore assumed that T_{bf} (beginning flow) was 700 ± 20 °C for the glasses considered.

In Figure 3.8(b), the gradients (tangents) of the lines were proportional to the E_a term of Eq.(3.2) and, therefore, depend on the concentrations of boron and phosphorus. Let us change to the concentrations of B_2O_3 and P_2O_5 stated as weight percent of elements (the respective conversion factors from [B] and [P] concentrations are about 3.2 and 2.3). The dependence of the gradient on the total additive oxides content was represented by a simple curve (as in Figure 3.8(c)). This characteristic was found to generally agree with data derived from experimental micrographs for similar surface configurations, available from the literature. (The behavior of closed circles in Figure 3.8(c) will be discussed below).

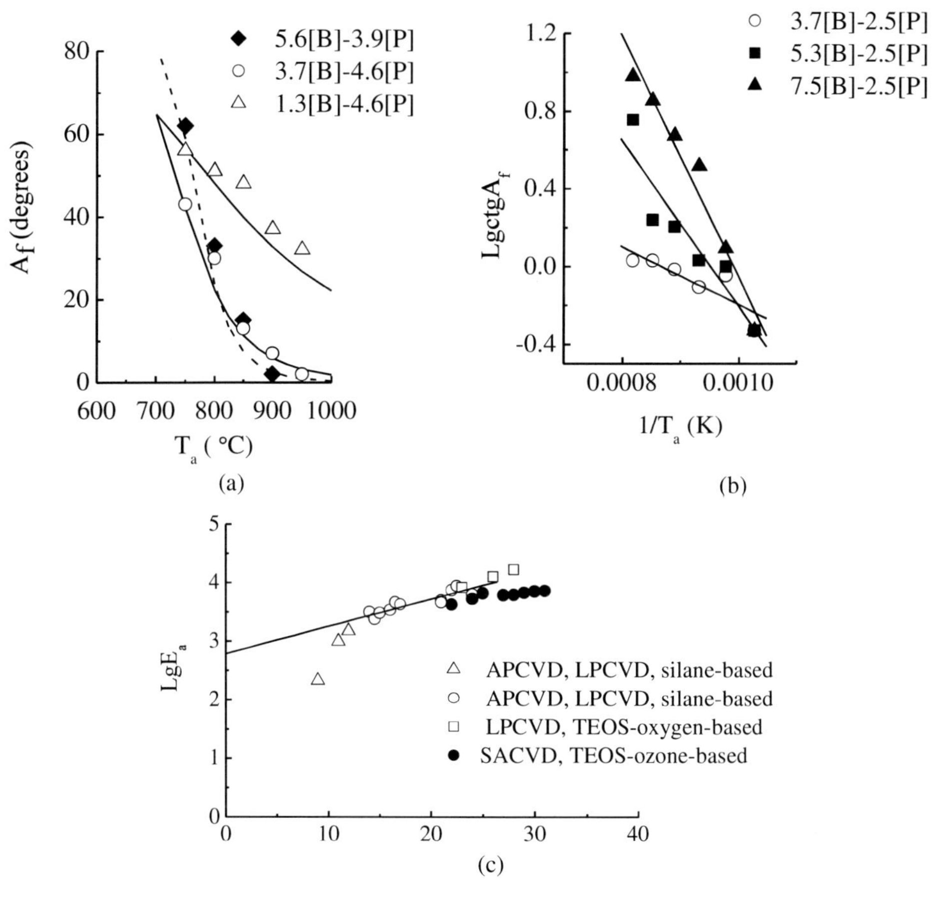

Figure 3.8. Thermal-flow BPSG film planarization by 30-min FA. Figures (a) and (b) depict the annealing-temperature dependence of the final flow angle. Figure (c) represents the activation term of Eq. (3.3) vs. the total concentration of additive oxides for annealing in dry gas. In each figure, data points and lines refer to measurements and calculations, respectively [223].

It is important to note that the concentration dependence of E_a was found to indicate that the viscous flow of BPSG over a step was equally sensitive to the concentrations of B_2O_3 and P_2O_5. By contrast, it was widely held that the effect of B_2O_3 is more than 2 times stronger than that of P_2O_5. However, changing from B_2O_3 and P_2O_5 to B and P concentrations, we can see that boron was only 1.5 times more effective than phosphorus in affecting BPSG flow. It is also important to note that there was a rapid decrease in E_a for total concentrations of B_2O_3 and P_2O_5 below 12 wt %, signifying that the viscous-flow capability of the glass was impaired. In Figure 3.8(c), the straight segment of the curve extended to the straight line:

$$\lg E_a = \lg E_{a(0)} + \chi(3.2[B] + 2.3[P]), \tag{3.3}$$

where a coefficient of proportionality $\chi = 0.05$. From here on a chemical formula or the symbol of a chemical element in brackets denotes the concentration of the compound or the element in weight percent. The last equation can be re-drawn to be equivalent to:

$$\lg E_a = \lg E_{a(0)} + \chi([B_2O_3] + [P_2O_5]) \tag{3.4}$$

In both cases of additive representation, the intercept on the vertical axis gives $E_{a(0)}$ = 562.

Eventually, the flowing properties of BPSG films at isothermal furnace conditions were described by the following empirical equation (3.5):

$$A^{FA}_{f(t_a=const)} = \arctan\left\{\tan A_0 \times 10^{-562(T_a-973)/973/T_a \times 10^{0.045(3.2[B]+2.3[P])}}\right\}, \tag{3.5}$$

where T_a is the annealing temperature in degrees Celsius. Equation (3.5) was written for a fixed annealing time (in our experiments, annealing was performed in dry gas for 30 min).

The constraints (3.6) were found to determine the range of use of Eq. (3.5) and the range of linearity for the concentration dependence of $\log E_a$:

$$\begin{aligned}
&5.2 < 1.4[B] + [P] < 13 \\
&0.35 < [B]/[P] < 5.55 \\
&A_0 < 85^\circ \\
&d_{BPSG} / d_{poly} > 1.6 \\
&t_a = 30 \text{ min} \quad (\textit{isothermal} \quad \textit{anneal}) \\
&\textit{Anneal} \quad \textit{ambient} \quad N_2, \quad O_2, \quad Ar
\end{aligned} \tag{3.6}$$

Note that constraints (3.6) were derived from experimental data for limited concentration intervals, so that Eq.(3.5) appeared to be applicable to higher additive concentrations as well. Curves calculated by Eq.(3.5), such as those shown in Figure 3.8(a) (full curves), generally agree with experimental data points. The exception was higher temperature conditions, but this case was less important in practical terms.

The above approach enables one to simulate the variation of the angle A subject to step (surface) parameters, the properties of the glass, and annealing conditions. For example, we calculated by Eq.(3.5) the concentrations of B and P for which A will evolve from A_0= 85 degrees to A_f^{FA} = 6 degrees during annealing in dry nitrogen at a given temperature. Figure 3.9 shows the results, expressed as so-called flow isotherms for the desired angle of glass planarization, together with SEM data for 850 °C annealing temperature. Notice that there is a good agreement between the calculation and the experiment.

BPSG film flow under RTA conditions in dry gas. For the RTA of BPSG in dry gas, we regarded the annealing time t_a as not being a principal difference to the thermal flow processes. This assumption was suggested by the results of Mercier [86], which are included in the Table 3.1. Looking at the $\cot A_f$ vs. $t^{1/n}$ relation for FA or RTA, one notices that the two regimes have close enough values of 1/n. Accordingly, we followed the same line of reasoning as with FA (see section above), checking the results against a limited amount of experimental data available for RTA [223]. For t_a = 40 sec, we proposed that:

$$A_{f(t_a=const)}^{RTA} = \arctan\left\{\tan A_0 \times 10^{-143(T_a-998)/998/T_a \times 10^{0.050(3.2[B]+2.3[P])}}\right\}, \quad (3.7)$$

with the same notation as in Eq.(3.5). Figure 3.10(a) shows calculated and measured variations of A_f with T_a. A deviation from the experimental data was again observed only for higher temperature conditions. Figure 3.10(b) represents the calculated concentrations of B and P for which A will evolve from A_0 = 50 ° to A_f^{RTA} = 30 ° during annealing at a given temperature. Thus, Eq.(3.7) can serve as a preliminary model of flow under RTA. Its parameters should be refined based on future experimental results.

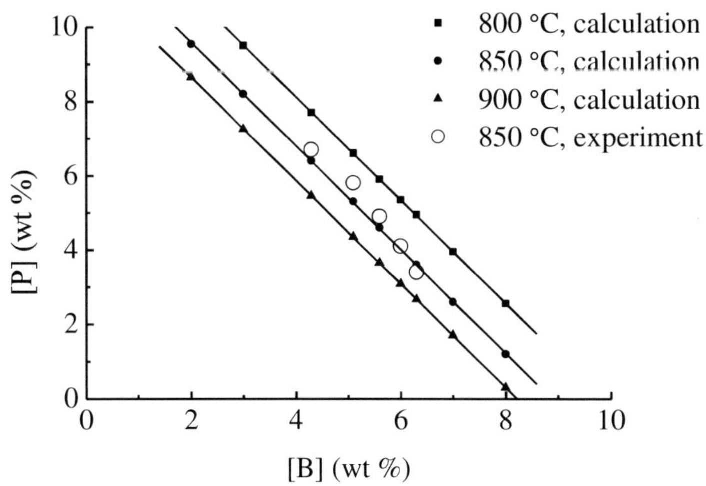

Figure 3.9. The calculated (lines) and measured (open circles) concentrations of [B] and [P] for which the flow angle A will evolve from A0 = 85 degrees to Af = 6 degrees during 30-min FA at a given temperatures [192,223].

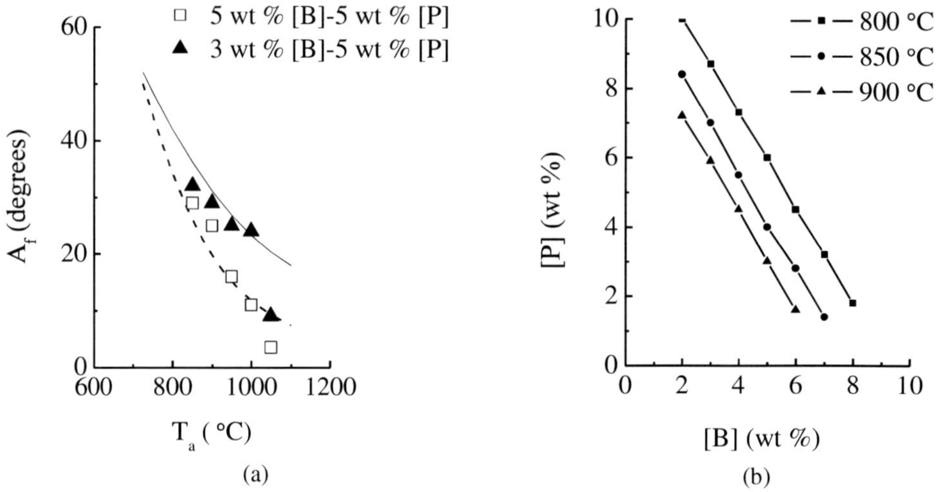

Figure 3.10. Thermal-flow planarization by 40-sec RTA [223]. Figure (a) shows measured (data points) and calculated (curves) values of the final flow angle as functions of annealing temperature. Figure (b) represents the calculated concentrations of [B] and [P] for which the flow angle A will evolve from A0 = 50 degrees to Af = 30 degrees during the annealing at a given temperature.

Annealing-time dependence. There are different ways to describe the dependence of A_f on the annealing time t_a for FA and RTA. For example, one might represent experimental data in the t_a-cot A_f or the $t_a^{1/n}$-cot A_f plane(see Table 3.1). Following the latter approach, we transformed the general equation introduced by Mercier [86] for A_f as a function of t_a to a more convenient form:

$$A_{f(t_a)} = \arctan\left\{\tan A_{t(t=const)}\left(t_{const}/t_a\right)^{\lambda}\right\}, \tag{3.8}$$

where $A_{f(ta)}$ is the value of A_f for the annealing time of interest t_a, $A_{f(t=const)}$ is an experimental value of A_f for a known annealing time, and λ is a coefficient and it is either 0.42 ± 0.10 for FA or 0.47 ± 0.16 for RTA.

Comparative analysis of primary factors affecting the film flow. The thermal flow of BPSG films is mainly governed by the initial step coverage angle A_0, the respective concentrations [B] and [P] of boron and phosphorus, the annealing temperature T_a, and the annealing time t_a. Equations (3.5) and (3.7), used in conjunction with Eq.(3.8),facilitate the evaluation of how A_f changes with these parameters. These complex equations for A_f for FA and RTA conditions at any anneal time are shown below (with the same notation as in Eqs.(3.5), (3.7) and (3.8):

$$A^{FA}_{f(t_a)} = \arctan\left\{\left[\tan A_0 \times 10^{-562(T_a-973)/973/T_a \times 10^{0.050(3.2[B]+2.3[P])}}\right] \times (t_{const}/t_a)^{0.42\pm0.10}\right\} \tag{3.9}$$

$$A_{f(t_a)}^{RTA} = \arctan\left\{\left[\tan A_0 \times 10^{-143(T_a-998)/998/T_a \times 10^{0.050(3.2[B]+2.3[P])}}\right] \times (t_{const}/t_a)^{0.47\pm0.16}\right\} \quad (3.10)$$

We investigated the dependence in the case of FA, with the parameters varied about the baseline values A_0 = 75 degrees, [B] = 4.5 wt %, [P] = 4.5 wt %, T_a = 850 °C, t_a = 30 min. The change in A_f was calculated as a function of the fractional deviation of each parameter from its baseline value. A straight line was fitted to the data points for each parameter, as presented in Figure 3.11.

The gradients of the lines in Figure 3.11 were found to be - 27, + 16, - 7, - 5, and - 1 for T_a, A_0, [B], [P], and t_a, respectively. It can be seen that:

(1) the annealing temperature T_a and the initial step coverage angle A_0 revealed the strongest influence on BPSG flow;
(2) the two major factors revealed an opposite influence on A_f.

The latter strongly confirms that the thin film BPSG deposition method strongly affects film planarization properties. For instance, PECVD BPSG films were very non-conformal and, therefore, A_0 is higher than 90 degrees. In contrast, step coverage with SACVD BPSG films were always significantly better, about 60 degrees. This simple comparison of typical step coverage values predicts the fact that PECVD BPSG film planarization is always worse. Therefore, PECVD BPSG film needs a higher temperature in order to obtain similar film planarization efficiency as compared to SACVD films. Moreover, the conclusion (2) is of crucial importance to sub-0.25 μm ULSI circuits, for which problems in gap filling combine with constraints on annealing temperature and time.

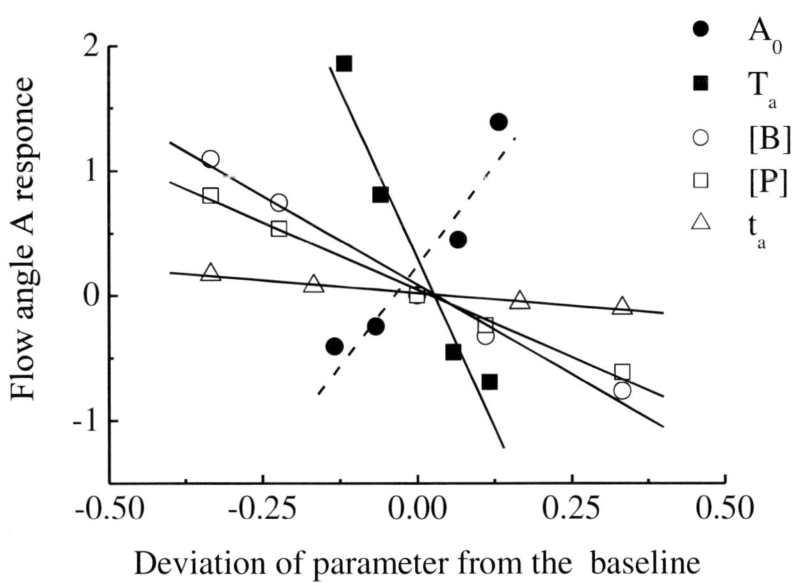

Figure 3.11. Comparison the individual effects of primary factors on BPSG flow [223].

Contact-via reflow. Little was known about the mechanism of the evolution of sidewall angles in contact vias under reflow, due to the complexities of preparing test specimens. Our SEM analysis led us to the fundamental conclusion that the sidewall angle Θ_f will change in a way similar to that for the angle A_f of a stepped glass surface (see Figs.3.2(a,b)) for the given glass composition, deposition technique, and annealing conditions. Specifically, for the glasses of the same composition, we drew an empirical relation (3.11):

$$(\Theta_0 - \Theta_f) \approx 0.5(A_0 - A_f) \qquad (3.11)$$

Equation (3.11) is supposed to be applicable to both BPSG and phosphosilicate glass (PSG) and different annealing conditions, including annealing in dry or wet gas (Figure 3.12). The coefficient 0.5 may reflect the fact that the glass in a via flows in a single direction, whereas the glass on a surface flows in two opposite directions. Thus, the reflow of contact vias might be investigated by analogy with the thermal-flow planarization of stepped glass surfaces.

15.5. Secondary Factors Affecting BPSG Flow

Equations (3.5) and (3.7) were derived under constraints (3.6). However, our experiments and data from the literature indicated the existence of other factors that may, in some situations, also affect the flow of glass under annealing, thus reducing the accuracy of calculations. This section briefly analyzes three such cases in order to encourage the reader to carefully examine all relevant factors while studying glass flow.

Method of the film deposition. Returning to Figure 3.8(c), consider the closed circles, which slightly yet consistently deviate from the straight line representing a number of deposition methods. The reason for this behavior may be the retardation of thermal flow caused by the effect of clustering in the form of pure P_2O_5 of a certain proportion of the phosphorus that is present. $P_2O_5$is meant to be linked to the $\equiv Si-O-$groups forming. This practically means that just a part of the phosphorus linked to silicon ($\equiv Si-O-PO=$) is involved in the flow process and the rest of the P_2O_5 clusters would evaporate from the film and transform into $\equiv Si-O-PO=$ groups at the elevated temperature taking a part in the glass flow process. Note that the CVD of the BPSG films involved tetraethylorthosilicate (TEOS), esters of boric acid and orthophosphoric acid, and an ozone-oxygen mixture. This is probably the most common production method of BPSG deposition in ULSI manufacturing.

Effect of a SiO_2 cap layer. A thin (from tens to hundreds angstroms) SiO_2 cap layer was frequently used to increase the stability of the BPSG film and to reduce its surface-defect density by protecting, to a degree, the film from atmospheric water. The first time that this simple solution was proposed was by Nauka et al [109]. Such a layer was produced in the same CVD process once the BPSG film was fabricated and the additive compound flows were switched off. Experiments showed that a SiO_2 cap layer could inhibit or enhance the thermal flow of the film.

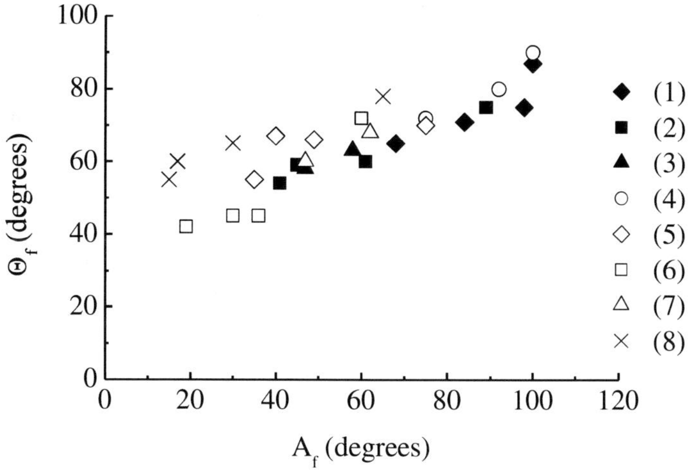

Figure 3.12. Contact-via thermal reflow: correlation between the final sidewall angle of a contact via and the final flow angle of a stepped glass surface for given glass composition and annealing conditions. Data description: (1) PSG, 6 wt %, RTA 1050 °C, steam-oxygen ambient; (2) PSG, 6 wt %, RTA 1100 °C, steam-oxygen ambient; (3) PSG, 6 wt %, RTA 1150 °C, oxygen ambient; (4) PSG, 8 wt %, RTA 1000 °C, steam-oxygen ambient; (5) PSG, 8 wt %, RTA 1050 °C, steam-oxygen ambient; (6) PSG, 6 wt %, RTA 1100 °C, steam-oxygen ambient; (7) PSG, 6 wt %, RTA 1150 °C, oxygen ambient; (8) BPSG, 4 wt % [B] – 4 wt % [P], FA, 850 °C, argon ambient. Data on PSG are derived from cross-sectional images published by Shah et al [342], data on BPSG were obtained by the author.

The former effect was discovered and explored by Bar-Ilan et al. in BPSG films produced by PECVD from esters-oxygen mixtures [143]. It was concluded that the thickness of a cap layer should not exceed 50 Å (Figure 3.13); otherwise, the layer will seriously suppress and hinder thermal-flow planarization.

The latter effect was discovered and examined by Lu et al. [212]. The BPSG films were fabricated by the TEOS-ozone based CVD Method 5 using mixtures of organic ethers at atmospheric or a subatmospheric pressure. Unlike PECVD, as-deposited films showed much higher porosity and were more liable to form defects on the BPSG film surface (see Chapter 10 and Chapter 13 in Part 2 of the monograph). A cap layer thicker than 50 Å was therefore required.

The relationship was addressed between the minimum cap-layer thickness and the total concentration of boron and phosphorus in the glass. It was approximated by a straight line, which gave a minimum cap-layer thickness of over 200 Å for B and P concentrations of 5 wt % each. As noted in the preceding paragraph, this thickness should prevent the BPSG film from being planarized. Investigation of the thermal-flow planarization of BPSG films with different compositions for cap-layer thickness ranging from 0 to 900 Å established that planarization can even be "enhanced" by a cap layer with a thickness in excess of about 300 Å (Figure 3.13). This unexpected and dramatic effect was thoroughly analyzed and traced to the elastic properties of porous silicon dioxide produced by TEOS-ozone interaction. Additionally, such a thick cap layer should impede the escape of boron and phosphorus from the BPSG layer under annealing, thus enhancing the flow [212].

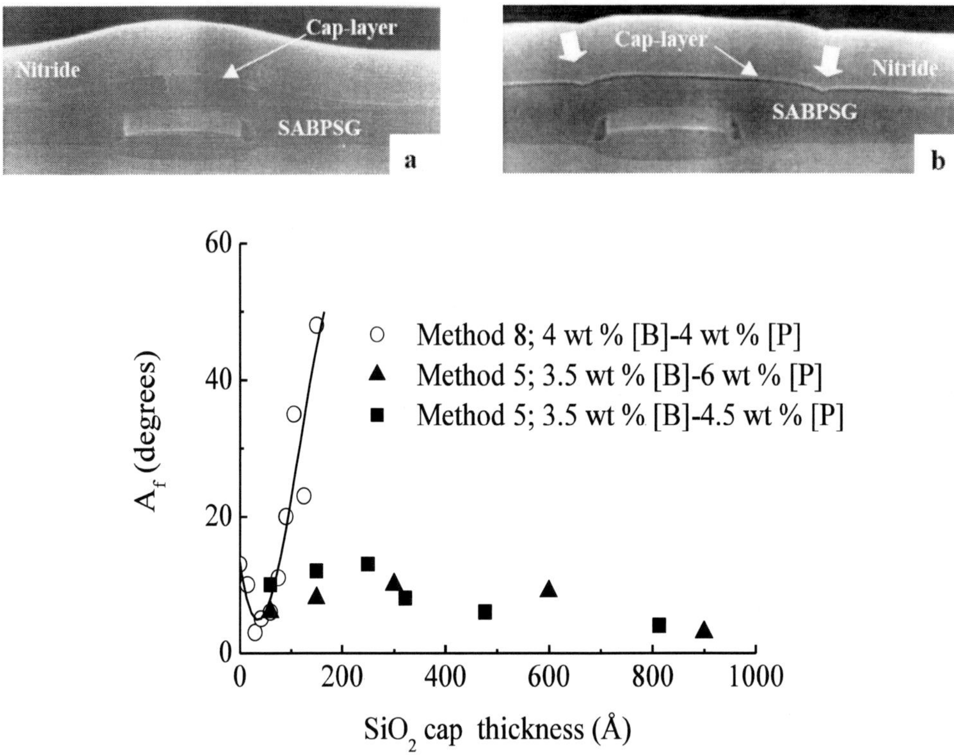

(c)

Figure 3.13. Effect of SiO2 cap layer on thermal-flow planarization: SEM cross-section of annealed SACVD BPSG film with 50 Å (a) and 800 Å (b) cap-layers (notches are highlighted by arrows) and final flow angle vs. cap-layer thickness (c) (the BPSG film is made by Method 8 and annealed at 850 °C for 30 min by Method 5 and annealed at 900 °C for 12 min). Presented at 15th Intern. VLSI Multilevel Interconnection Conf. (VMIC), 1998 [212].

Effect of ambient during BPSG film anneal. The method above could be extended to the case of wet-gas annealing. Regrettably, this area was poorly explored by experimenters. According to Tauber [7], the same standard of planarization can be achieved by wet-gas annealing at temperatures lower by about 50 °C.

Effect of glass element geometry – via reflow. It was shown by Levy et al [339] that the shape of glass element affects the glass film flow. Particularly, if the height to width ratio of etched element with vertical side walls more than 0.393, after film anneal and flow, the side wall angle can significantly exceed 90 degrees.

This means after contact via the reflow process the size of via can be less that after etch. The reason of this effect is the surface tension of the melting glass causes the rounding of the glass element or even droplet formation from the initially "vertical" glass element (see schemes and photos in Figure 3.3(e,f)).

15.6. BPSG Glass Flow Characterization by Flow Effectiveness on the Device Steps

Modern ULSI devices tend to have rectangular step profiles due to the ongoing decrease in feature size and the advent of dry etching. As a result, in most of the current cases A_0 greatly exceeds 90 degrees even when the BPSG film demonstrates perfectly conformal step coverage. In practice, deposition technologies tend to provide modest conformity. This is especially true for PECVD, despite its widespread use. As a result, this method is not suitable for most types of PMD. However, PECVD may be a preferable option, if not the only proper one, with multicomponent glasses because the complex kinetics of the reactions involved makes it difficult to maintain a desired composition of films. It follows from the above that a novel approach is required to the characterization of the flow of BPSG over submicrometer steps. This approach should be universal in the sense that it should work for BPSG thin films with any practical additive concentrations and with an arbitrary A_0.

The author for the first time introduced a dimensionless flow parameter called "glass flow capability" [244]. It was defined as $(A_0\text{-}A_f)/A_f$ or $1 - A_f/A_0$ (detail background and description can be found in cited paper). Glass flow capability varies over the range [0, 1] and it is at a maximum when the glass surface becomes perfectly planar under annealing: $A_f = 0$. The minimum glass flow capability implies the absence of planarization: $A_f = A_0$.

The value of A_f can be determined by measurement on the step surfaces. Alternatively, it can be calculated by Eq.(3.9) and Eq.(3.10), provided that the parameters meet the constraints of (3.6). It is wise to find out how flow effectiveness, $1\text{-}A_f/A_0$, varies with annealing temperature or additive concentrations, by analogy with Figure 3.8; note that these relationships can be examined for any A_0. Figure 3.14(a) shows examples of the temperature dependence for the additive concentrations of Figure 3.8(a). Figure 3.14(b) depicts the concentration dependence of the gradient k_f of straight lines fitted to experimental data points. The concentration characteristics are similar to those for the activation term E_a in Figure 3.8(c). In Figure 3.14(a), the intercepts of the lines on the T_a axis give values of T_{bf}. These essentially depend on the additive concentrations, as illustrated by Figure 3.14(c). Also notice that T_{bf} is indeed 700 ± 20 °C for the concentrations of interest [267,268]. For the additive concentrations considered and any A_0 (the process is 30-min FA in dry gas), Figure 3.14 suggests the equation:

$$1 - A_f / A_0 = 2.61\times10^{-4}([B_2O_3]+[P_2O_5])\times\{T_a - 754 + 3.21([B_2O_3]+[P_2O_5])\} \quad (3.12)$$

Here the definitions are the same as above in Eq.(3.5) and Eq.(3.7).

15.7. Comparison of Isothermal Furnace and Rapid Thermal Anneal of BPSG Films Using Glass Flow Capability

Sub-quarter micron ULSI circuits tend to be fabricated by RTA, which minimizes the overall thermal effect on the devices. On the other hand, RTA is inferior to FA in flow effectiveness and hence in the quality of planarization. There have been efforts to improve the

RTA flow of BPSG included raising the total additive concentration above 8 wt % or annealing at elevated humidity and, possibly, elevated pressure. However, most of the approaches were unsuccessful in the context of ULSI due to unacceptable defect density at higher additive levels or the oxidation of layers beneath the glass due to the wet environment. The qualitative character of experimental data published so far indicates that research was urgently required into the quantitative characterization of glass flow under arbitrary annealing conditions. Assuming that the laws of glass flow are independent of the annealing method, the author presumed [244,267,268] that the above equations provide a framework for comparing the patterns of flow in different annealing regimes. This strategy allowed one to reduce the amount of experimental work embracing different process conditions.

Studies were initiated by gathering and analyzing published experimental data in order to compare FA and RTA conditions with respect to similar BPSG films (deposition method and composition). The results are summarized in Table 3.2. Another flow parameter was introduced, termed "thermal budget", which was defined as the product of annealing temperature in Kelvin degrees and annealing time in minutes: $(T_a + 273) \times t_a$. The data were analyzed in terms of the ratio:

$$[(T_a + 273)\times t_a]_{FA} / [(T_a + 273)\times t_a]_{RTA} = m \tag{3.13}$$

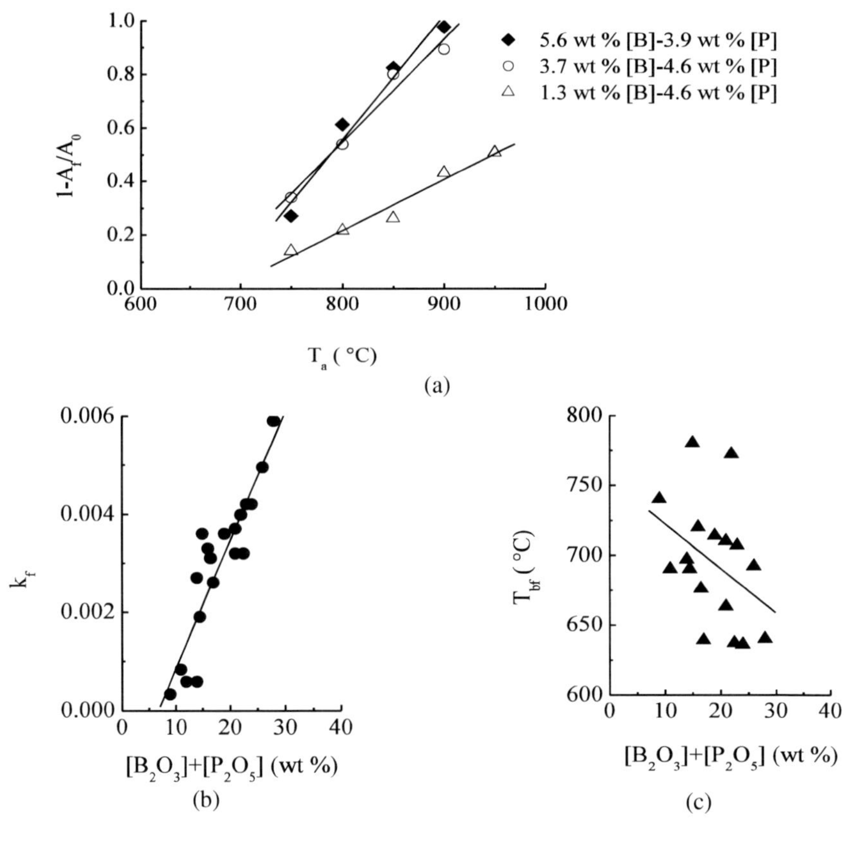

Figure 3.14. Example of calculated dependence of glass flow capability on annealing temperature or additive level [268].

Table 3.2. Comparison of FA and RTA planarization for BPSG thin films [268]

[B]-[P] (wt %)	A_0 (degrees)	A_f (degrees)	FA		RTA		m	Ref.
			T_a (°C)	t_a (min)	T_a, °C	t_a (min)		
4-4	92	65	875	30	1075	0.33	77	[72]
4-4	82	65	825	30	1025	0.33	77	[72]
3-6	52	11	907	30	1070	0.33	79	[72]
3-6	52	11	907	30	1040	0.66	41	[177]
5-5	50	3	907	30	1100	0.33	78	[177]
5-5	50	3	907	30	1050	0.66	41	[177]

The BPSG films considered were deposited under the same conditions, were equal in boron and phosphorus content, and showed the same change in A_f under FA and RTA. The values of m calculated from the published reports were found to strongly indicate a correlation between the characteristics of the two annealing regimes (see Table 3.2). At the same time, the author was not clear why the maximum and the minimum value of m differed by a factor almost as large as 2.

As a further step, the variation in $1\text{-}A_f/A_0$ was calculated with the thermal budget for the known glass composition and process conditions, using Eq.(3.9) and Eq.(3.10) for simplicity. Figure 3.15 presents example graphs for different annealing temperatures and fixed values of [B], [P], and A_0, with annealing assumed to be in dry gas. The annealing time is 15 – 60 min for FA and 20 – 120 sec for RTA. Notice the distinction in the glass flow capability between FA and RTA.

The relationship between glass flow capability and thermal budget provided an understanding of the spread in m. We observed that the glass films differed in A_0. As pointed out above, A_0 as well as T_a exert the strongest influence on BPSG flow, their respective variations having opposite effects on A_f. Figure 3.16(a) illustrates the vital importance of A_0 to glass flow capability. With this factor taken into account, we achieved good agreement between the calculation and the experiment for both FA and RTA on the basis of the tabulated experimental data (despite their scarcity), as demonstrated in Figure 3.16(b). This result supports our approach to the modeling of BPSG flow under arbitrary annealing conditions in ULSI manufacturing.

15.8. Calculating Optimum Flow Temperatures

We have seen that annealing conditions should be set so as to minimize the adverse thermal effect on the devices. More specifically, the thermal budget applied to the PMD should be at a minimum yet sufficient for adequate planarization (or gap filling).

It was generally accepted that the flow temperature T_f of thin films can be identified for an arbitrary glass composition by determining the glass-transition temperature T_g of bulk specimens and using the empirical relation $T_f = T_g + 288 \pm 16$°C, as proposed by Nassau et al [54]. However, the authors failed to give specific values of flow parameters. Furthermore, the reported values of T_g and, hence, the corresponding values of T_f, were inapplicable to RTA.

Thus, the problem of identifying an optimum annealing temperature remained unsolved for a long period of time.

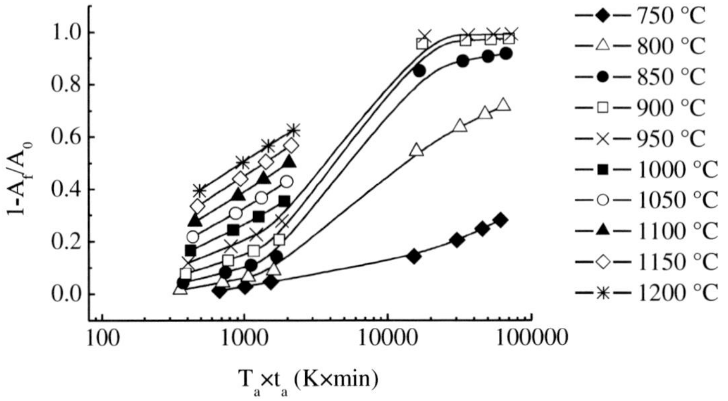

Figure 3.15. Calculated glass flow capability vs. anneal thermal budget in dry gas for 5 wt % [B], 5 wt % [P], and $A_0 = 85$ degrees [268].

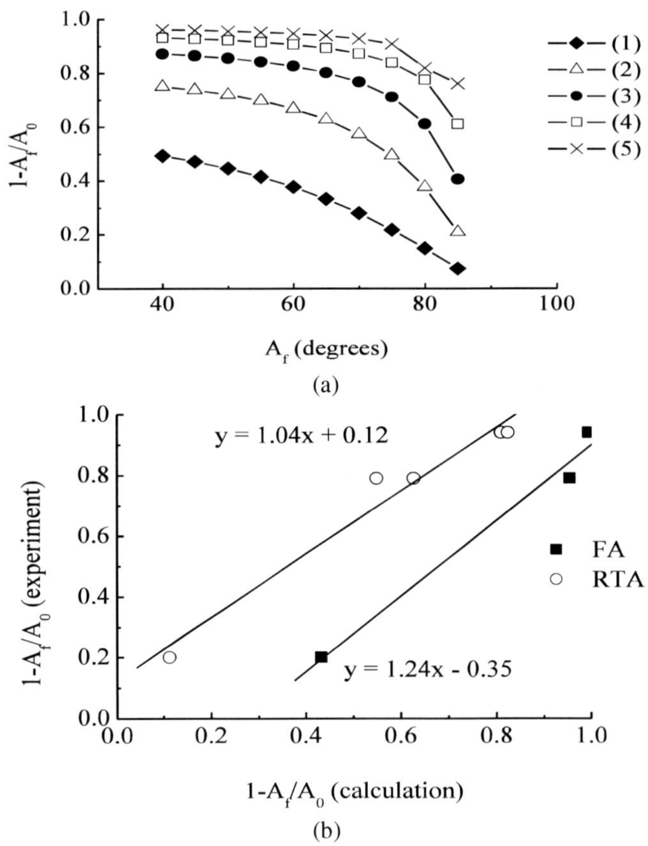

Figure 3.16. Calculated glass flow capability: (a) dependence on initial step-coverage angle and (b) comparison with experimental values (see table). Figure (a) refers to BPSG films with 4 wt % [B] and 4 wt % [P], which were subjected to FA processes with almost equal thermal budget [268].

It was believed that the optimum annealing temperature (i.e. acceptable maximized temperature) $T_{a(max)}$ is related to a maximum glass flow capability for either type of annealing. Further, $T_{a(max)}$ as a function of concentration appears to be the most informative characteristic. The optimum annealing temperature can be determined from the annealing-temperature dependence of the rate of change of glass flow capability with respect to thermal budget:

$$d(1-A_f/A_0)/d(T_a \times t_a) \tag{3.14}$$

Figures 3.17(a) and 3.17(b) give examples of the dependence for FA and RTA, respectively. They are calculated from Eq.(3.9) and Eq.(3.10) for various boron and phosphorus combinations.

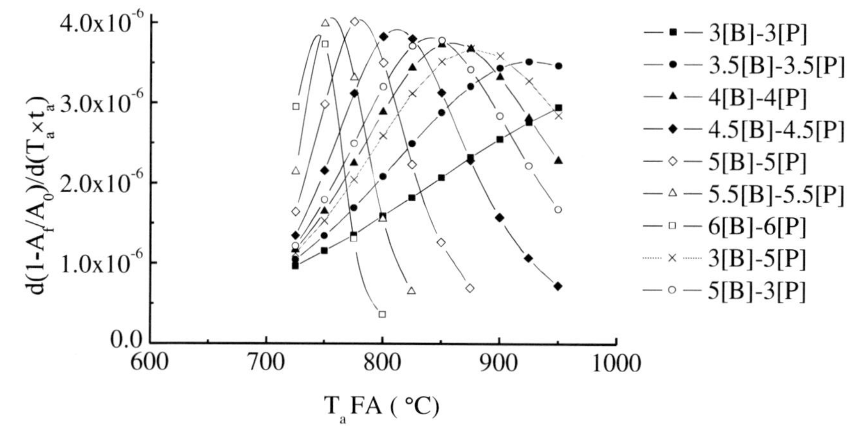

(a)

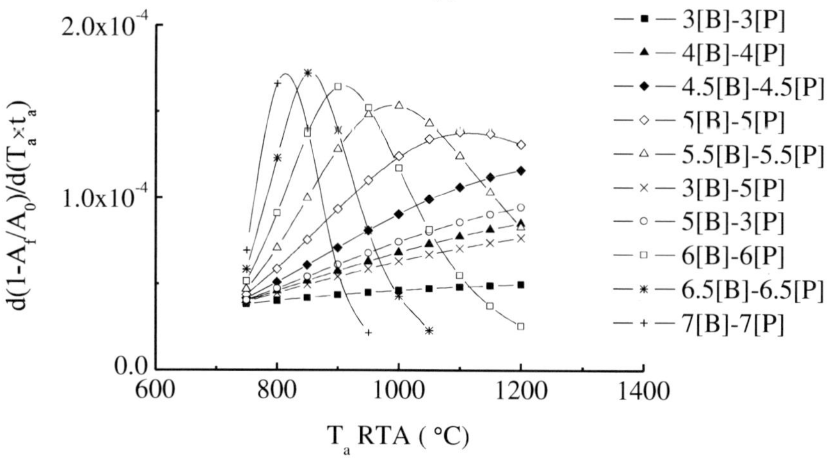

(b)

Figure 3.17. Calculated $d(1\text{-}A_f/A_0)/d(T_a t_a)$ vs. annealing temperature for the (a) FA and (b) RTA of BPSG films differing in composition and (c) experimental graphs of post-annealing step-coverage angle vs. FA annealing temperature for different glass compositions [268]. BPSG film compositions are presented in wt % [B] and wt % [P].

The comparison of calculated curves in Figure 3.17(a,b) with experimental ones obtained for the same additive concentrations and plotted in the T_a-A_f graph, as shown in Figure 3.8(a), revealed that the peaks of the former approximately correspond to the points of inflection on the latter. It appears reasonable to propose that the corresponding values of T_a are related to optimum annealing temperatures. Figure 3.18 shows values of $T_{a(max)}$ calculated in this way versus $[B_2O_3]$ + $[P_2O_5]$. The sum of the oxide concentrations is used because the flow of BPSG is equally sensitive to the oxides, as noticed above; this fact greatly simplifies the calculation. In addition, for wet gas anneal ambient these curves are believed to be shifted downward by about 50 °C.

15.9. Glass Flow Capability of GeO_2-SiO_2 Glass Films

The tendency toward lower annealing temperatures for sub-0.18 μm ULSI stimulated research into silicate glasses that could replace BPSG as a PMD. For example, GeO_2-SiO_2 glasses and multicomponent glasses that contained Si, Ge, B, P, As, Pb were evaluated as candidates [54,120,239,240]. Until recently, practical importance has only been placed on Ge-contained glasses.

The flow of Ge-contained silicate glass thin films over wafer topography has been poorly explored. Experimental results tend to be presented in qualitative form, as micrographs. Nevertheless, we believe that the flow effectiveness $1 - A_f/A_0$ could be instrumental in interpreting the published data and quantifying the flow.

The author applies his approach below to the analysis of data presented by Simpson et al [239,240]. The flow of GeO_2-SiO_2 glasses was studied over a wide range of composition, including the case of B_2O_3 and P_2O_5 additives. The glass films were produced by PECVD from hydrides. Annealing was performed in argon, steam, or forming gas. The authors were able to reduce the step-coverage angle A_0 from over 100 degrees to A_f = 0 degrees. Regrettably, planarization was characterized in general terms, on a three-point qualitative scale. We analyzed the results only for 1 hour FA in argon because the authors explored this case in greater detail.

The flow effectiveness was estimated by measuring A_0 and A_f on the SEM micrographs. Figures 3.18(a,b) show flow effectiveness versus annealing temperature and k_f versus GeO_2 concentration in mole percent, respectively. It is seen that the flow starts at about 45 mol % GeO_2 and the k_f characteristic becomes nonlinear at about 80 mol % GeO_2. Figures 3.19(a,b) allow one to derive formulas similar to Eq.(3.12). These indicate that flow effectiveness over 0.5 can be obtained at annealing temperatures below 850 °C if the GeO_2 concentration is raised above 60 wt %. Note that the corresponding value of k_f is about 0.005 (see Figure 3.19(b)). The same value was observed in BPSG film with a total additive content $[B_2O_3]$ + $[P_2O_5]$ equal to 25 wt % (see Figure 3.14(b)). This was the most common level of additives in BPSG films; it correlated with fairly low defect densities. However, we think that the corresponding level of GeO_2 was excessive for GeO_2-SiO_2 glass, as evidenced by crystallization [239].

If GeO_2-SiO_2 glass contains B_2O_3 and P_2O_5, it showed a higher flow effectiveness and a sharply reduced T_{bf}. Figure 3.19(c) illustrates this effect by means of $(1 - A_f/A_0)$ vs. T_a characteristics for GeO_2-SiO_2 (the ratio of mol % SiO_2 to mol % GeO_2 is about 1/4) and for two combinations of the B_2O_3 and P_2O_5 levels, together with estimated values of k_f and T_{bf}. Notice that the boron- and phosphorus-contained glasses exhibit an almost 200 °C decrease in T_{bf} as compared with the GeO_2-SiO_2 glass. Furthermore, boron- and phosphorus additives significantly raise k_f. Indeed, a more detailed comparison will be possible with a larger amount of experimental data on a wider variety of glass compositions.

15.10. Evaluation of BPSG Gap-Fill Process

In ULSI manufacture, glass films deposited on stepped surfaces often contain voids located in gaps between adjacent features. This is unacceptable. The shape of voids depends on the deposition technique, keyhole voids being the most common example. Other frequently encountered void types were depicted in Chapter 4. With further advances in miniaturization, voids may approach the size of the gaps themselves.

After deposition, a BPSG film is subjected to thermal annealing in order to take a planar form and make the material fill the gaps (Figure 3.4). Figure 3.20 depicts the effect of annealing on voiding for two BPSG annealing processes commonly used in ULSI manufacture: long high-temperature isothermal annealing (FA) completely eliminating the voids, and pulse annealing (RTA) that remained with some small voids. It can be seen that the thermal annealing of BPSG film must be performed at a fairly high temperature and during a fairly long period. Otherwise, keyhole voids become spherical ones (see Figure 3.4(b)) instead of disappearing, which is also unacceptable.

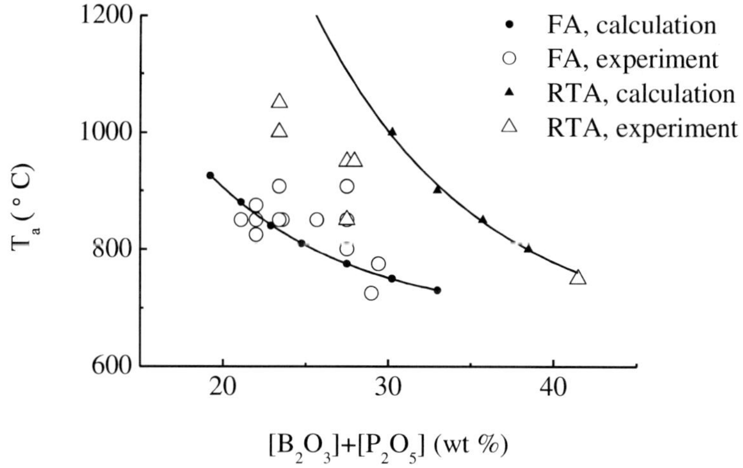

Figure 3.18. Calculated optimum annealing temperature and BPSG film gap-fill capability for almost vertical gaps vs. total additive concentration for FA and RTA. The data are obtained by measurement (open marks) or computation (closed marks) [268].

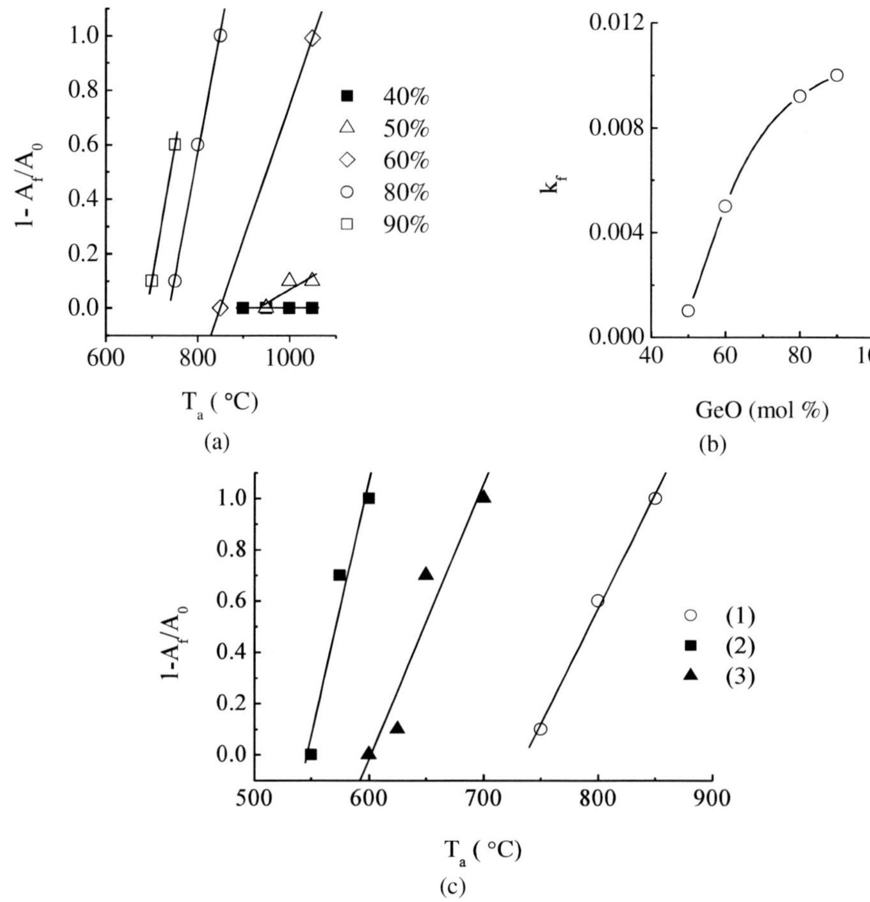

Figure 3.19. Using glass flow capability in the evaluation of thermal-flow planarization for GeO2-SiO2 glass films with A0> 90 degrees: (a) glass flow capability vs. annealing temperature subject to GeO2 concentration, (b) kf vs. GeO2 concentration, and (c) glass flow capability vs. annealing temperature for GeO2-SiO2 glasses with and without additives. The experimental data treated by Vasilev in [268] were taken from Simpson et al. [239]. Details for graph (c):

(1) 20 mol % SiO2, 80 mol % GeO2; kf = 0.009, Tbf = 737 °C;
(2) 14 mol % SiO2, 73 mol % GeO2, 12 mol % P2O5, 1 mol % B2O3; kf = 0.020, Tbf = 546 °C;
(3) 16 mol % SiO2, 77 mol % GeO2, 6 mol % P2O5, 0 mol % B2O3; kf = 0.011, Tbf = 604 °C.

The author proposed a new, less laborious and expensive approach to the evaluation of deposition--planarization processes in terms of gap-fill capability [244,257]. His approach was based on the relationship, in graphic form, between the flow effectiveness and the parameters of gaps that can be adequately filled by glass flow. Based on background analysis above, he concluded that flow characteristics of flow-able glass on the device step were strongly affected by three major parameters:

- the shape of the structure incoming to the flow-able film deposition;
- the shape of growing on this structure film (or CVD features);
- the glass flow capability.

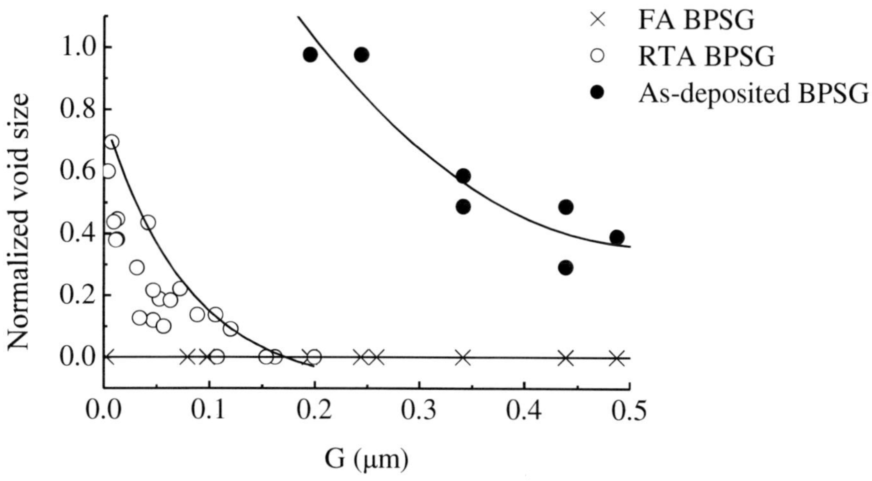

Figure 3.20. Planarization used for improving gap-fill capability in the case of a BPSG film. The RTA was performed at 850 °C for 30 sec, the gap depth being 0.25 μm. The FA was performed at 900 °C for 12 min. The data on as-deposited film were obtained for a gap depth of 0.5 μm. The vertical void size was normalized to the gap depth [257].

Taking the information above as a starting point for developing an approach of quantitative evaluation of CVD film gap-fill capability, the following simple proposal was done: gap-fill improvement at thermal treatment of flow-able film and the characteristics of structure, which is covered with this flow-able film, can be linked, as shown schematically in Figure 3.21(a). In this drawing the "x" axis was proposed to represent a certain complex parameter accumulating a set of parameters, like glass properties (boron and phosphorus concentrations) and anneal conditions (T_a, t_a, anneal ambient, etc.). It reflects a capability of flow-able film to flow. The author named this parameter "Glass Flow Capability" (GFC). Axis "y" represents a certain complex structure parameter, which was defined in Chapter 4 as a ratio of AR to the gap size G, or a ratio of structure height (H) to gap size in power two (G^2), e.g. H/G^2, and named as "Structure Complexity" (SC). The SC is supposed to include a set of parameters, such as gap characteristics, aspect ratio, etc. It is clear from general considerations that the SC of ULSI structures is enhanced along with a decrease of IC device technology nodes.

Following the proposed logic, the tighter structures should have a higher H/G^2 value. In fact, the decrease of device technology nodes rules means that gaps become a few times smaller, but the thickness of structures (e.g. gap height) become about a twice thinner. For instance, for selected gap height ~0.2 μm of a "vertical" structure with gap values of G_v=0.1 μm and G_v=0.05 μm, the structure complexity can be calculated to be $SC_{0.1}$=20 and $SC_{0.05}$=80, respectively.

The GFC parameter is characterized by the dimensionless value ($1 - A_t/A_0$), defined above. It varies from zero to unity; it is higher for a material with a better flow capability. This flow parameter can be evaluated by SEM analysis or calculated using Eq.(3.9) and Eq.(3.10). The former requires no knowledge of the composition of the material and flow conditions as long as A_t has been found. The latter strategy is applicable to BPSG only and requires that A_0 and anneal conditions be specified.

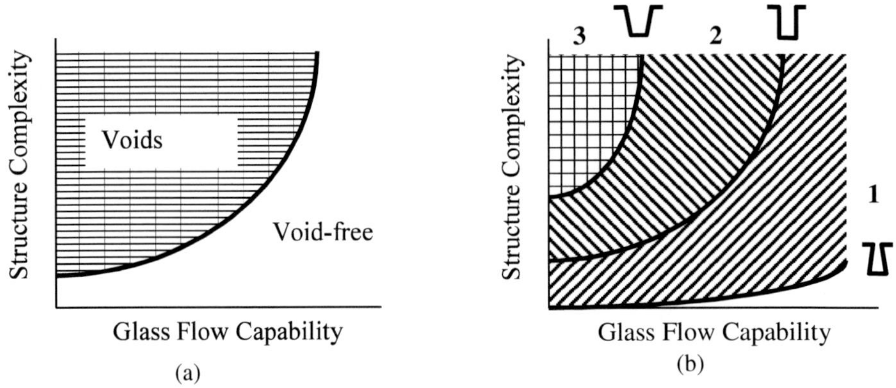

Figure 3.21. Graphic representation of deposition-planarization processes in terms of deposition gap-fill capability (vertical axis) and glass flow capability (horizontal axis). Figures (a) and (b) are an elementary and a sophisticated form of comparative process evaluation, respectively. In Figure (b), the numerals designate (1) plasma deposition processes, (2) low-pressure and atmospheric pressure processes, and (3) High-Density-Plasma CVD processes and the cross-sectional schemes reflect the conventional shape of gap: tapered, vertical or re-entrant.

The scheme in Figure 3.21(a) can help us to do a simple qualitative analysis. Obviously, an enhancement of glass flow capability will lead to the possibility of better gap-fill even for very complicated structures. This means that we can achieve gap-fill for very high values of SC. Theoretically, at extremely high thermal anneal conditions, where GFC can be defined to be infinity, any device gaps can be filled with very liquid glass, producing void-free structures even with extremely low gaps. Following the logic, the zero GFC value on the graph, which reflects "zero" glass flow capability, corresponds to the SC of as-deposited CVD films. At the same time, if GFC is getting worse due to the changes of any factor responsible for film flow parameter (either glass composition, of thermal anneal conditions), an improvement of gap-fill with flow-able glass film becomes less possible. Eventually, a curve in Figure 3.21(a) should reflect a certain boundary between voiding conditions and a void-free area. Thus, any coordinates above the drawn line should reflect void existence in structures, and at the same time, any coordinates below the drawn curve should give us void-free structures after proper film flow. It is clear that an improvement of GFC achieved with wet anneal ambient, or high-pressure anneal, will lead to the shift of a particular point from the void area towards the void-free area. At the same time, a decrease of structure complexity will also lead to the void-free gap-fill.

All considerations above assume a structure with vertical side-wall slope. However, additional structure definitions, such as "vertical", "tapered" and "re-entrant" are also needed despite the fact that they cannot be defined quantitatively in most cases. Taking into account our classification of structures made in Chapter 4, and simple general considerations as well, the graph in Figure 3.21(a) can be re-drawn in order to analyze the gap shapes, as well as CVD process effects on glass flow gap-fill capability. These graphs are presented in Figure 3.21(b). It is clear that the structure complexity, which is able to be filled with flow-able glass film will follow the sequence: $SC_{\text{"re-entrant"}}<SC_{\text{"vertica"}}<SC_{\text{"tapered"}}$. In fact, significantly tighter structures can be filled with flow-able films if structure tapering is used. This figure also reflects that as-deposited film gap-fill is supposed to be much better for "tapered" structures as-compared to the "vertical" or "re-entrant" structures, which is shown as much higher

structure complexity value of "tapered" structures at GFC=0. From the deposition process point of view, an improvement of growing film conformality should give us an improvement of as-deposited gap-fill, e.g. smaller key-holes. Obviously, the trend of smaller key-hole formation during film deposition will follow showed above sequence: LPCVD≥APCVD>PECVD. Eventually, the better as-deposited film gap-fill will lead to the better post-flowed gap-fill. For instance, use of fluorinated subatmospheric pressure BPSG films with better as-deposited step coverage as compared to non-fluorinated one, helped improve as-deposited gap-fill and, therefore, post-anneal gap-fill [243]. Therefore, a best HDP CVD BPSG method will allow achievement of the best gap-fill at relatively low thermal budget conditions.

Based on the general analysis performed, it was concluded that the whole vision of gap-fill issue presented in Figure 3.21(a,b) linked qualitatively very well with the device structure types, CVD process features, film features and anneal characteristics, revealing no logical discrepancies. It allows the use of this logic as a basis for further quantitative characterization of the gap-fill issue with flow-able glass films.

An analysis of three selected groups of structures with different shapes of gaps was performed by the author in [244]. He used summarized experimental results of our gap-fill studies for "re-entrant", "vertical", and "tapered" structure types and data from some other references. Most of the data presented were found to be for TEOS-ozone-based BPSG films, deposited either at APCVD, or at SACVD conditions. The author also put on these graphs a few additional data points defined as "critical", which were found to correspond to the exact void-free gap-fill achieved or reported in the literature. This can be seen in Figure 3.22 for the "vertical" structure type.

There was a very high spread of data especially for the "tapered" gap shape. This confirms that the definition for tapered structures was not clear and it was very difficult to compare one structure to another. At the same time, the expected trend of voiding at not sufficient enough film flow was obvious. Comparing all of the experimental graphs, the author concluded that it was very difficult to find a functional correlation of structure complexity and glass flow capability. That was due to the wide spread of data obtained for different types of structures. Nevertheless, it was believed that the summarized data presented confirm the idea of the approach.

To date, there has not been sufficient published experimental data for BPSG films flow, showing numerically what conditions are sufficient for proper gap-fill. In section 15.8, the author introduced the optimum annealing temperature (i.e. acceptable maximized temperature) $T_{a(max)}$, which is related to a maximum flow effectiveness for either type of annealing. Assuming that this temperature is also optimized for gap-fill purposes, one can plot experimental data points, as shown in Figure 3.18 for both FA and RTA conditions. These data points reflect reported good gap-fill data achieved at dry inert ambient and for mostly slightly tapered structures. It can be seen that for FA the correlation can be evaluated to be good. At the same time, for RTA conditions experimental data points are placed at significantly lower anneal temperatures as compared to the calculations. This comparison indicates that anneal temperature for proper gap-fill can be chosen slightly lower as compared to calculated maximums. The author found that temperatures corresponded to the ~0.8 from obtained maximums can meet the experimental data point for both RTA and furnace anneal as well. In this case, almost all published data points are placed above calculated curves reflecting certain void-free areas above these curves.

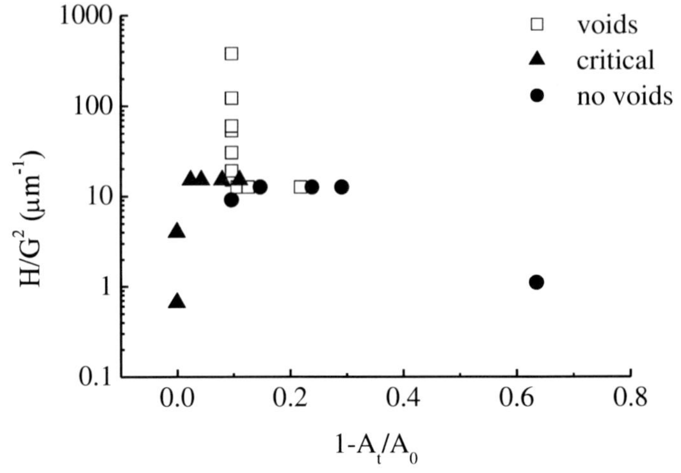

Figure 3.22. Summarized experimental and published gap-fill data for "vertical" gap shapes re-calculated into the basic complex parameters: structure complexity H/G2 and glass flow capability (1-Af/A0) [244].

Another important feature of the approach presented is a possibility of simple calibration of gap-fill data with film planarization after anneal. Scanning microscope analysis for gap-fill can be performed together with a measurement of flow angle A_t on the single step. In this case, knowing the value A_0 as well, an experimental value of glass flow capability can be easily calculated. Flow angles can be measured using the atomic-force microscope technique [225] (see Figure 3.7). We believe, this kind of experimental calibration for gap-fill together with calculation using the presented approach can be effectively used in the case of more complex glass film use, such as GeO_2 contained BPSG films.

Conclusion

In this chapter, a detailed overview of a 30-year experimental investigation into the thermal flow of BPSG thin films on stepped device surfaces is presented. The viscous flow over a stepped surface under FA or RTA was examined by SEM for different BPSG compositions. A few numerical equations were considered for the calculation of changes in flow angle during thermal-flow planarization, subject to the primary factors affecting glass flow: annealing temperature, initial flow angle, the boron and phosphorus concentrations of the glass, and annealing time. The impact of each factor was quantified and the factors were compared numerically. Consideration was also given to secondary factors which may have a significant influence on the process in some situations. The empirical formulas proposed can be directly employed in VLSI and ULSI device technology and manufacturing, reducing the amount of costly measurement.

Two important parameters were considered for the characterization of the flow of BPSG over ULSI steps in details: glass flow capability and thermal budget. They are related to annealing temperature, annealing time, topographical parameters, and glass additive concentrations. The approach to using these parameters reduces the cost of measurement, facilitates the interpretation of data, works for both FA and RTA, and is applicable to ULSI gap filling as well as glass planarization. It also enables the optimization of the annealing temperature for thin films of multicomponent glasses, such as GeO_2-SiO_2 glass contained B_2O_3 and P_2O_5.

Chapter 16

BPSG FILM COMPOSITION OPTIMIZATION

In order to use BPSG films in IC technology successfully, these films must be carefully optimized. This means the film composition needs to be defined based on both device technology and film property requirements. BPSG films can be deposited by several techniques and, therefore, the properties of these films differ. To optimize BPSG film composition, the author selected three major BPSG film properties, which must be considered for VLSI and ULSI device technology: (1) a film capability to flow at thermal anneal; (2) an ability to interact with ambient moisture, and (3) an ability to form the BPSG film defects. In order to optimize the film composition properly, all these major properties must be comprehensively characterized and quantitatively described. In Chapters 12, 13 and 15, quantitative characteristics of these basic BPSG film properties have been presented. To define an optimized additive concentration range in any type of BPSG film, an approach for quantitative description of BPSG film was developed by the author and published or presented in [192,213,224].

16.1. A QUANTITATIVE APPROACH FOR BPSG FILMS COMPOSITION OPTIMIZATION

An example of the optimized range of BPSG additive concentrations is shown in Figure 3.23. In this example, the optimized additive composition area drops in between the curve for FA condition at a temperature of 850 °C (this anneal provides sufficient enough angle of film on the device step of about $A_f = 15$ degrees) and the curve for 24-hours defect free-area in as-deposited BPSG films. The first curve was calculated using Eq.(3.5) introduced in Chapter 15. Concentration limits of this equation [B]/[P], previously summarized in Eq.(3.6), are also taken into account and shown in Figure 3.23.

The second curve for defect-free BPSG boundary was defined by equation (12.1) in Chapter 13. The addition limitation for boron content in BPSG film $[B] \leq [P]$ should also be taken into account to reduce the moisture absorption ability of BPSG film and, therefore, to make BPSG films highly reliable. This limitation was discussed in Chapter 12 and it restricts the film's ability to absorb moisture from ambient into the film depth (see Figure 2.61). Quantitatively all these ranges were characterized by a set of the following empirical equations (3.1)-(13.3):

$$0.35 < [B]/[P] < 5.55 \quad (3.15)$$

$$30.4 - 1.39[B] - 0.023T_a < [P] < -0.17[B]^2 - 0.19[B] + 9.6 \quad (3.16)$$

$$[B] \leq \sim 0.4[P] \quad (3.17)$$

where the additive concentrations are presented in elemental weight percent and T_a is the temperature of film anneal in Celsius degrees.

Comparison of data for 20 types of empirically optimized BPSG films presented in [207] showed that almost all of them dropped into the above defined area presented in Figure 3.24.This coincidence confirms the validity of the approach. The summarized BPSG film defect data showed that the total additive content in BPSG films cannot exceed a certain limit, which can be chosen based on device and technology needs. Eqs.(3.15)-(3.17) allow the definition of the optimized range of additive concentrations, which predict good BPSG film flow for device planarization with reduced moisture absorption and defect problems. Any other optimized additive concentration ranges for different initial conditions of film step coverage or film anneal can be easily obtained using related equations, as shown in detail in [192,213,224]. As an example of other initial conditions, an agreement between calculations and experimental data for 900 °C thermal anneal can be seen in Figure 3.24. It should be noted that for TEOS-ozone BPSG films, which revealed more pronounced defect formation effects, a similar calculation can be used. However, the time for the defect-free area was estimated to be about 18-hours after film deposition and the beginning of film exposure to air.

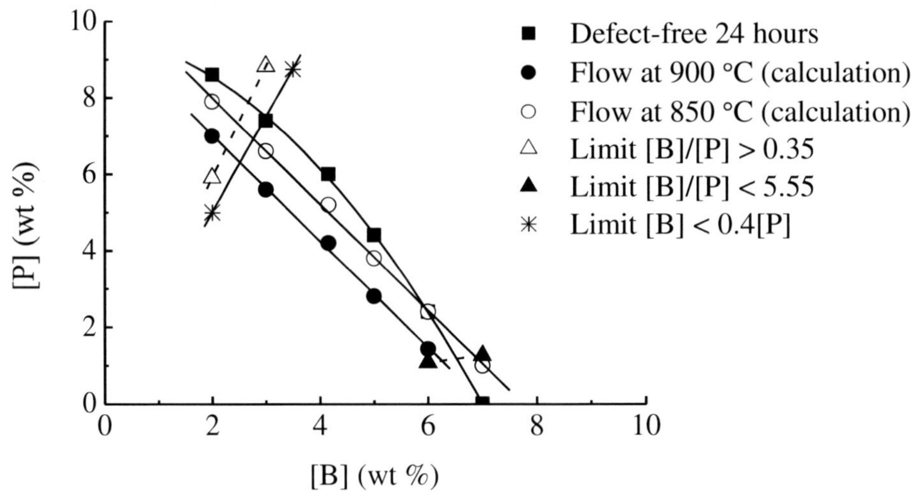

Figure 3.23. Example of optimized range of additive concentration calculation in BPSG films [213]. Limits for boron and phosphorus content are presented in elemental wt %.

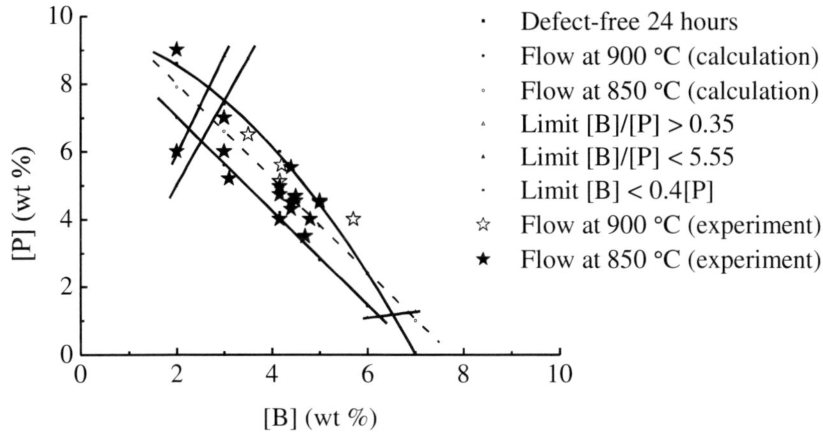

Figure 3.24. Correlation of experimental data and calculations for different BPSG film compositions and FA anneal conditions [207]. Limits for boron and phosphorus content are presented in elemental wt %.

16.2. Optimization of the Additive Content in Multicomponent Glass Films

It is strongly believed that a similar logic and approach for additive range optimization can be used for multicomponent films, which can flow at significantly lower temperatures as compared to BPSG films. Among these films are germanium-doped BPSG films [239,240], which were considered a few times as a good option for advanced ULSI technology. In this case, for three separate additives a three-dimensional diagram need to be used instead of two-dimensional presented in Figure 3.23. The lines for film flow will transform into planes and curves for defect-free boundaries will transform into a curved surfaces. Thus, an optimized GeBPSG film composition would be expected to be somewhere in the space between these surfaces. Unfortunately, at the moment, multicomponent glass films have not been studied sufficiently and there is no data to draw such diagrams quantitatively. Nevertheless, we believed that relatively simple approaches presented in this chapter will allow obtaining optimized additive concentration ranges in multi-components films in a future.

Chapter 17

BPSG FILM DURABILITY

BPSG film durability (another term in use is stability) is a very rarely studied topic. Some stability data regarding the film deposited using Method 1 were presented in the mid-1980s by Kern and co-workers [56,57]. The film durability was linked to the additives in the film, particularly to their concentrations and their activity with respect to the ambient. Being exposed to corrosive or aqueous compound solutions, or treated at enhanced temperatures, the boron and phosphorus oxides can leach out of the glass film changing the film properties. The authors came to the conclusion that as-deposited BPSG films should never be exposed to corrosive or aqueous compound solutions before anneal. As for additives loss during high-temperature film treatment in dry ambient, the authors found it to be about 30 % in the 200 – 300 Å region near the film surface and concluded that for practical application these small changes were not really important.

Intensive studies and development of BPSG film technology in the last 15 years were inspired by the fast sub-half micron ULSI technology development. This led to the conclusion that existent knowledge of BPSG film stability was insufficient. For advanced IC devices and new methods of BPSG film deposition, the film durability issues should be studied using new approaches, methods, and tools.

BPSG film post-deposition thermal treatment at isothermal conditions provided a good film densification, i.e. an improvement of film moisture resistance and sufficient BPSG film softening for film planarization and gap-fill. Normally, this range of relatively high thermal budget conditions corresponded to the temperature 850 – 950 °C and time for 10 – 30 minutes. Arranging these numbers as a thermal budget, i.e. as T(K)×t(min), these FA conditions can be defined to be ~ 3.7×10^4 K×min. Similar changes in BPSG film softening can also be achieved with Rapid Thermal Anneal (RTA) conditions at temperatures up to 1100 °C and anneal time less than 1 minute, which corresponds to the range of thermal budget of about $(0.4\text{-}1.4)\times10^3$ K×min.

Such high thermal budget conditions are not applicable for all types of sub-quarter micron ULSI devices. Particularly, modern microelectronics necessitate a considerable decrease in the thermal budget of devices in the course of their manufacturing. For instance, logic devices with silicide gates require lower thermal budget RTA conditions, which correspond to a temperature lower than 850 °C and anneal time less than 2 minutes (this corresponds to thermal budget as low as $< 0.4\times10^3$ K×min).

As we show in Figure 2.23, this range of BPSG film thermal treatment conditions drops in between so-called glass transition temperature T_g and glass flow temperature T_f, and is closer to T_g. Based on common sense, certain stepped changes of BPSG film properties or severe instability of BPSG films, deposited at temperatures less than T_g, and annealed at conditions close enough or just a little bit higher than that of T_g, might be expected. Among expected changes, for instance, might be an enhanced glass ability of moisture absorption and defect formation, long-term instability of film properties and boron and phosphorus states, etc. This expected to be due to the differences in the glass matrix stability at low anneal temperatures and additive species instability at temperatures higher than their evaporation temperature. Relatively low temperatures of heat treatment of BPSG films are insufficient for the viscous flow of glasses but they are sufficient for the activation of phosphorus-containing clusters in the porous glass matrix. For instance, according to the classical chemistry, the sublimation temperature of pure phosphorus pentoxide is equal to 359 °C.

In this respect, there arises the problem of an insufficient durability of the glass films under consideration, which can affect a long-term normal operation of devices. In this chapter, we summarize studies of low thermal budget annealed films investigated in detail previously [214,242,269,270].

17.2. BPSG Film Durability

Chosen to be a standard, BPSG films contained 4.5 wt % P and 4 wt % B were prepared by Method 5 (i.e. with the highest evaluated film porosity). Experiments were conducted in accordance with the following procedure: the first set of measurements was done after film deposition. After that, it was repeated after anneal (0.3 – 2 min at temperatures of 700 – 950 °C using a RTA. For comparison purposes, the film samples were also isothermally treated at 750 °C. All samples were exposed for 5 days to clean-room air with everyday measurements of film thickness, refractive index, defect appearance (control under optical microscope), surface charge, film stress, FTIR spectra registration. After that, films were rinsed with deionized water and monitored another 4 days followed by an autoclave testing at 121 °C, 1.2 atm pressure and 100 % humidity for 168 hours. After testing, a full set of parameters was measured again. In addition, in order to obtain a difference between samples due to the assumed difference in absorbed moisture, films were conducted to a hot pumping at 440 °C and the pressure of 10^{-2} torr followed by a nitrogen purge at 2 torr at the same temperature for 2 minutes. After this treatment, the last set of measurements was performed.

Contrary to expectations, no stepwise changes in the BPSG film properties were observed under the above annealing and storage conditions therefore indicating the film stability. However, the electrical properties of the glass films changed after annealing. In particular, the use of the SCA technique (see the results and thorough discussion in [214,242]), made it possible to evaluate the changes in terms of the net averaged charge and the charge distribution (standard deviation within wafer, StD SCA) over the wafer surface BPSG films after heat treatments. Despite the fact that SCA data cannot be used directly for the explanation of charge phenomena in the films, it was believed that this technique was an effective tool to detect the beginning of changes in electrical film properties with temperature. After film deposition, average SCA values (measured with MDC-1 tool by Material Development Corporation) in studied BPSG films were found to be 5.5×10^{10} q/cm^2 with StD SCA within wafer 0.06. Changes of surface charge value and surface charge uniformity on the wafer (StD SCA) with the thermal budget anneal conditions are shown in Figure 3.25(a,b).

Both changes in average charge and charge non-uniformity within wafers in Figure 3.25(a,b) were found to increase with the thermal budget value. Therefore, these changes were not observed upon low-temperatures heat treatments, but upon high-temperature heat treatments, which lead to the softening and viscous flow of the glasses. Highest changes were observed for temperature 900 °C and temperature 850 °C was second. Despite huge differences in the thermal budget values between RTA and FA conditions, changes are higher in the case of higher anneal temperature, as can be clear seen in Figure 3.25(a,b). This confirms that the temperature is the major parameter affecting charge changes in the films.

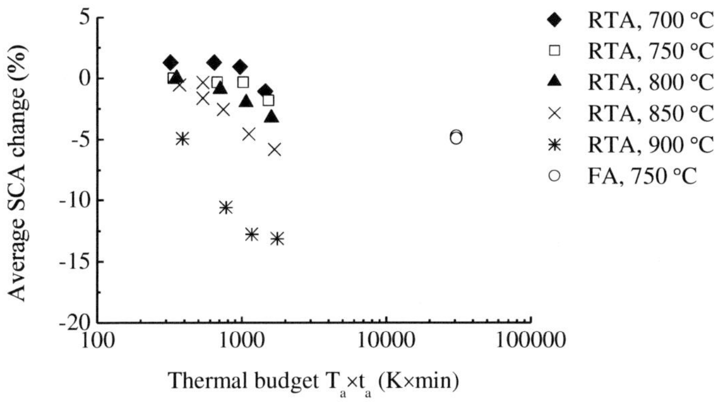

(a)

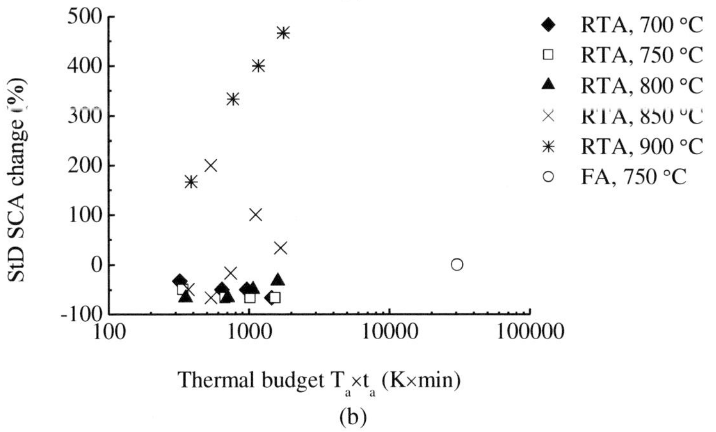

(b)

Figure 3.25. Changes (in percent from obtained as-deposited values) of average surface charge value (a) and charge deviation within wafer (b) with anneal thermal budget [242].

This result was analyzed in the light of previously published data regarding a correlation of change in average charge (ΔSCA) and charge non-uniformity within a wafer for different

BPSG film compositions (from 3 to 6 wt % of each boron and phosphorus) with FA at 850 – 900 °C and RTA at 925 °C (both in nitrogen ambient) [214,222] (see Figure 3.26(a)). Two features can be seen from this comparison. The first one is that FA provides similar changes of average total charge, but lower and uniform charge deviation within the wafer. The second one is that RTA reveals significantly higher charge non-uniformity within wafers.

The nature of the change in the electrical properties of the glasses upon heat treatment still remains unclear. However, these changes can be explained by the change in the phosphorus and boron states in the glass film. The additive out-diffusion during a short anneal time are different within the wafer. At the same time, long anneal time provides uniform changes of additive status and additive out-diffusion from the film within wafer. The results obtained can be conveniently analyzed by dividing the thermal budget range into three sub-ranges.

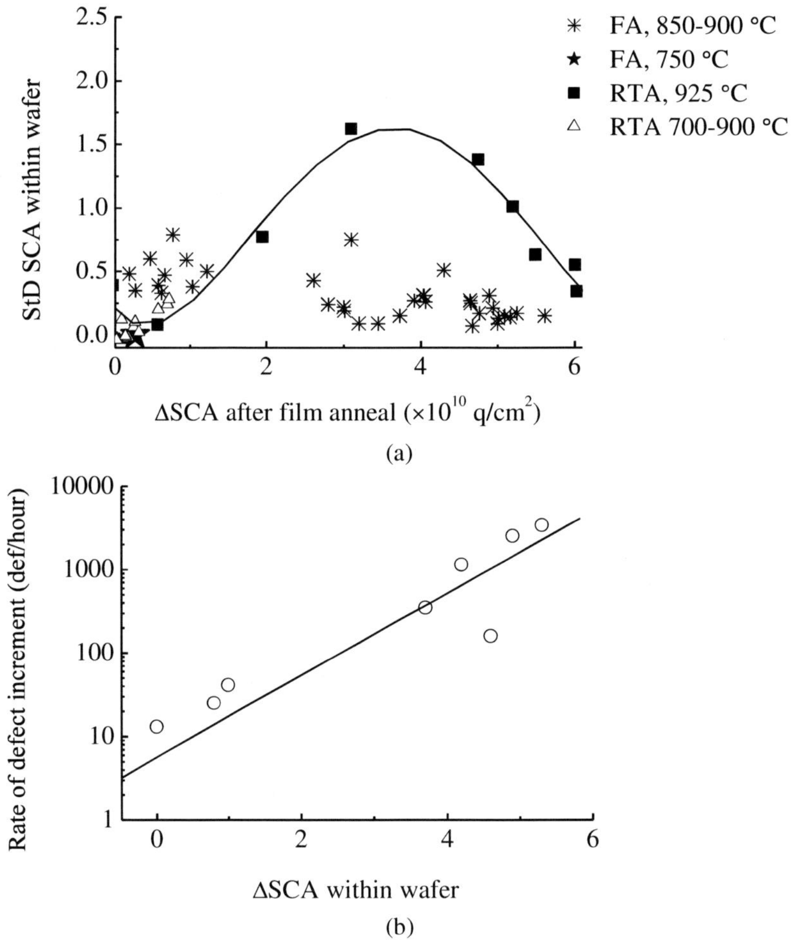

Figure 3.26. Correlation of total charge changes after the film anneal and charge non-uniformity within wafers (a) and correlation of charge changes and rate of BPSG surface defect increment at heavy thermal budget FA conditions (b).

Low thermal budget. At a low thermal budget, LTB (when the softening and viscous flow of glasses is absent or insignificant), the properties and the effective density of the films do not change noticeably. The net charge in these films is uniform over the wafer surface and is stable in time. This can be the clear confirmation of the stable status of boron and phosphorus oxides in the film at such a low thermal budget conditions. Particularly, no evaporation of additives from the film occurs at such conditions. As a result, an effect of surface liquid defect formation is significantly lower as compared to high temperature anneal, which causes out-diffusion of additives followed by their condensation on the surface with defect nucleation. Thus, low temperature RTA is better for additive stability, but low heat treatment temperatures are not efficient in microelectronics due to a very weak glass flow.

High thermal budget. At high thermal budget (HTB) conditions the film densification is better and the viscous flow of glasses occurs at high rates. At the same time, these temperatures were found to cause the highest changes of obtained net surface charge values and charge uniformity on the wafer. In fact, SCA allowed obtaining differences between BPSG film properties, namely at 900 °C and, partly, at 850 °C as compared to other studied anneal conditions. The nature of these charge changes is still not clear, however, these anneal conditions were found to correlate with the beginning of film flow, as well as with the typical post-densification defect appearance on the BPSG film surface. Such correlation, for instance, is shown in Figure 3.26(b) as a dependence of a rate of defect increment on the wafer surface, measured with laser-scanning methods(calculated based on data presented in Ref.187) on the changes of average total charge in SACVD BPSG films with total concentration from 6 to 11 wt % (calculated based on data presented in Ref.214). The obvious correlation of these two parameters can be clearly seen.

It is believed that the average net charge and charge non-uniformity changes indicate a beginning of additive out-diffusion from the surface and subsurface film areas at such thermal budget conditions. The short anneal is not sufficient to achieve a new stable status of additives, as well as their uniform evaporation from the film within wafer. In fact, despite of much higher thermal budget value for isothermal anneal at 750 °C, the change of SCA values are much smaller as compared to short-time anneal at temperatures higher than 850 °C. We believe obtained data reflects a difference in additive status in BPSG film within the substrate area, which is enhanced by RTA. In contrast, the heat treatment at high temperatures for long periods is responsible for the formation of a new structure with a stable state of phosphorus and boron in the glass and, hence, leads to new stable electrical properties of the film. In this case, new steady-state charge distribution after heat treatment is characterized by a minimum non-uniformity over the wafer surface. However, as was noted above, high-temperature heat treatments can be unacceptable in modern technology of integrated circuits.

Intermediate thermal budget. The case of intermediate thermal budget (ITB) is the most interesting and important for microelectronics. Under these conditions (for example, at T > 850° C), the heat treatment results in a certain densification of glass, its viscous flow, and the most non-uniform charge distribution over the wafer surface. In this case, the processes of glass softening, transformation of additive clusters into the glass matrix, and diffusion of phosphorus and boron from the deposited film proceed at a rather low rate. As a result, rapid annealing under intermediate conditions leads to the formation of an imperfect film structure that is intermediate between the structures corresponding to heat treatments at low and high temperatures.

The structural inhomogeneity of the glass over the wafer surface differs substantially, and these films are unstable in their electrical properties. It should be emphasized that no changes in the electrical properties were observed for the considerably denser BPSG films produced by Method 8. This suggests a higher stability of the physicochemical and electrical properties of these films.

A few conclusive proposals and suggestions were made by the author.

First, studies generally did not highlight any step changes or significant changes of BPSG film properties (or any negative trends at low thermal budget film anneal), which might be considered as constraints for BPSG film to serve as Pre-Metal Dielectric in ULSI devices.

Secondly, based on common sense, BPSG films deposited using Method 5 after low-thermal budget RTA can be defined to be less stable as compared to the film after FA at 750 °C. This result was expected due to the significantly shorter RTA time which was insufficient for proper film densification. Here the author would agree with Dyer [217] that the film density (or indirect parameter - shrinkage value) could reflect such generalized parameters as film stability (or instability). However, currently, there is no clear definition of particular numbers to evaluate BPSG film stability, especially taking into account a variety of methods of film deposition.

17.3. Functional Dependences of BPSG Film Durability

The generalization of the results obtained for the aforementioned heat treatment temperature ranges allowed the author to describe the durability of glasses by the parameter "BPSG film durability" as follows [270]:

$$\text{BPSG film durability}_{(\text{as-deposited})} \sim \gamma_d, [B]^{-1}, [P]^{-1}, d_{cap}, t_e^{-1}, \psi_d \tag{3.18}$$

$$\text{BPSG film durability}_{(\text{LTB})} \sim \gamma_T, [B]^{-1}, [P]^{-1}, d_{cap}, t_e^{-1}, \psi_d \tag{3.19}$$

$$\text{BPSG film durability}_{(\text{ITB})} \sim \gamma_T, [B]^{-1}, [P]^{-1}, d_{cap}, t_e^{-1}, \psi_{T,\tau} \tag{3.20}$$

$$\text{BPSG film durability}_{(\text{HTB})} \sim \gamma_T, [B]^{-1}, [P]^{-1}, d_{cap}, t_e^{-1}, \tag{3.21}$$

where γ_d and γ_T are as-deposited and thermally annealed film density, [B] and [P] are additive concentrations, d_{cap} is thickness of capping silicon dioxide layer on the top of BPSG film, t_e is ambient exposure time, ψ_d and $\psi_{T,\tau}$ are the dimensionless parameters, characterized the general state of boron and phosphorus oxides in the glass structure in as-deposited and annealed state. Their values are assumed to be in the range from zero to unity for the BPSG films after deposition and heat treatment. For instance, for the films after deposition and low-temperature heat treatment, we assume $\psi_d < 1$. For the films heat treated at high temperatures, when the transformation of the initial glass structure is completed, the parameter $\psi_{T,\tau}$ is determined by unity. Finally, at intermediate thermal budgets that closely correspond to conditions of rapid annealing and depend on its temperature and time, we propose $\psi_{T,\tau} << 1$.

Chapter 18

BPSG FILMS IN ELECTRONIC TECHNOLOGIES

BPSG pioneer introduction as an IC production process dates back to 1980 and to RCA (USA). The moving force of the film implementation was the substantially lower temperature of the film flow as compared to PSG films used in IC manufacturing as a PMD material. This was required to produce specific contours in PMD films in advanced devices with short-channel lengths and shallow junctions to minimize the lateral and vertical diffusion of dopants. The fabrication of ICs for use in environments that may contain high levels of radiation also required that processing be done at temperatures considerably lower than those used for the production of commercial integrated circuits. Low process temperatures were also desirable to minimize process-induced crystallographic damage, unwanted impurity diffusion, and distortion effects. With BPSG as a PMD, it was found possible to define processes with a maximum wafer temperature during processing as low as 850 °C. This reduction allowed using BPSG technology for applications that ranged from commercial CMOS circuits in plastic packages to high-reliability, radiation-hardened, short-channel bulk and silicon-on-sapphire (SOS) circuits [50].

BPSG film has now become a routine PMD material. As shown in the previous part of the monograph, BPSG thin films appears to have all of the appropriate physical, chemical, mechanical, and electrical properties that make it suitable for use as a dielectric material in IC technology. BPSG film deposited by all studied CVD techniques exhibited undoubtedly good dielectric properties being studied with "classical" techniques (see Chapter 7) that made them to be an attractive isolation material for device applications.

For a number of applications, BPSG films exhibited properties that made them more attractive than films of the PSG material. From the beginning of their appearance in 1982, BPSG films were considered for use in VLSI and ULSI applications based on a few types of physical and chemical properties:

- lower flow (fusion) temperature PMD material;
- gettering properties;
- moisture absorption effect;
- lower chemical etch rate in ammonium fluoride-buffered hydrofluoric acid solutions;
- lower as-deposited tensile stress;
- lower-temperature doped-oxide diffusion sources.

During a long period of the film studies and employment in the IC technology, the only PMD BPSG application as a lower fusion temperature glass material undoubtedly got widespread use. All of the selected directions of BPSG film application reflect some specific film features, which can be realized in other precision technologies. In this regard, these features are briefly discussed in this Chapter.

18.1. BPSG Films as Pre-Metal Dielectric in VLSI and ULSI Circuits

General considerations of BPSG film applications. The primary application of BPSG is as a low-temperature flow-able PMD (see Figure 1.2). It is used to ensure adequate metal coverage over steps and into contact vias so that metal discontinuities or local regions with thin metal over steps do not occur. In many applications, two flow (fusion) processes are used: one to smooth the surface topography over steps such as those of a polysilicon gate (called "flow"); and a second, often called "reflow", to taper contact openings made through the fused ULSG layer. For these applications, it is important to control the top contour of the flowed glass film. In general, the intent should be not to planarize the surface completely, but to taper and soften sharp step sand edges. In circuits of relatively low density, some tendency toward planarization during the first BPSG film flow can be tolerated, but in high-density circuits more precise control of BPSG film contours is required. In recent years, BPSG planarization needs became less necessary due to the implementation of sub-half micron device technology as a Chemical Mechanical Polishing technique. Nevertheless, the BPSG thermal flow process is still in use because it is necessary to fill the narrow gaps between neighbor device elements to eliminate the voiding issues.

The choice of BPSG film composition is determined primarily by temperature limitations of the IC device technology. For example, in the case of processes used for the fabrication of radiation-hardened CMOS circuits, wafer temperatures after growth of the gate oxide were typically kept below the growth temperature of the oxide. For example, a glass composition was selected to permit significant film flow ability below 900 °C in the RCA hardened CMOS/SOS devices [50]. In general, practical considerations suggested the use of the composition with the highest flow temperature that was compatible with the process. As the composition of BPSG was adjusted to lower the flow temperature, a decrease of the film stability or an increase in defect density tended to occur. To decrease this inherent BPSG film property, so-called SiO_2 "cap-layer" with very thin (30 – 100 Å) film thickness deposited in-situ immediately after BPSG film deposition is used, especially for highly additive-enriched BPSG films.

For a typical application of BPSG as a PMD layer, a common approach to the selection of a glass composition was based upon the achievement of an adequate flow in 30 min at the maximum usable temperature. The first flow was normally done in a dry or steam ambient, and the second flow was performed either in nitrogen or in a sequential process, that involves both dry oxygen and nitrogen.

Normally, BPSG films were used in IC device technology as a part of PMD stacked dielectric. To prevent boron and phosphorus diffusion from BPSG films into the gate region, the film was usually used with so-called "barrier" or "liner" dielectrics. In most cases, this dielectric was 500 – 1000 Å silicon dioxide film. It was deposited using any known CVD technology, for instance, TEOS pyrolysis, PECVD TEOS oxidation, TEOS oxidation with ozone-oxygen mixture and so on. For convenience, these barrier oxides were deposited in the same reactor, with the same chemical compounds, and at the same temperature mostly using PECVD processes to obtain the best liner film density.

However, PECVD methods, giving the best film quality in terms of film density, provide very non-conformal film deposition on the device steps (see Chapter 4 of the monograph). In such a case, in the sub-quarter micron device elements all shapes of the narrow gaps between gates transform into the re-entrant shapes. BPSG film deposition on the re-entrant structures covered with liner oxides enhances the voiding issue. Therefore, more efforts to optimize the film composition and film anneal conditions are required to eliminate this issue. Another issue with PECVD liner films is plasma-induced damage that can be seen, particularly with the use of SCA techniques. To optimize net charge value within the wafer, careful optimization of the PECVD liner process was necessary [343].

In addition to silicon dioxide under- and cap-layers used in combinations with BPSG films, IC technology development required the use of a thin dielectric layer of low etch rate beneath PSG films. Such layers, called "etch-stop layers" (ESLs), were made best with silicon nitride. This material was produced mostly by reaction between silane and ammonia in plasma at about 400 °C. A major drawback with this process was that the hydrogen content in silicon nitride films was about two orders of magnitude higher than in BPSG (see, for instance, data in Figure 2.11). Hydrogen diffusion under thermal annealing is thought to degrade the performance of metal–oxide–semiconductor transistors, as it should enhance boron migration from the gate through the gate oxide [344]. For this reason, interest was developed in high-density-plasma processes, which did not use ammonia [345].

Finally, there were a few attempts to use stacks of BPSG films, in which additive content and, therefore, flow temperature were different in the range of 50 – 100 °C [346]. The top film was considered to be less enriched with additives as compared to the internal layer. However, two- or three layer BPSG stacks are more overcomplicated than useful. In fact, it is very difficult to control repeatability of the process, especially in terms of additive content and their distribution within the wafer. As a result, practically BPSG film stacks did not find any reasonable application in device technology.

Furnace anneal vs. RTA. In addition to the selection of the appropriate BPSG composition, the conditions used for the film flow can be adjusted to optimize the final contours of a patterned PMD film or to achieve proper gap-fill. As described earlier, parameters such as temperature, time, film thickness and step coverage, as well as anneal gas ambient and pressure, can be used to influence the flow of a glass. Anneal temperature and time can to some extent be traded-off. In this regard, either furnace isothermal, or rapid thermal anneal, can be selected to be sufficient to meet the device and BPSG film requirements. It is necessary to notice, that excess flow during the first film flow can make the etching of contact holes difficult and uncontrollable since the thickness of the glass layer can vary in the region of the contact to be opened. In the case of a second flow (reflow), which follows the etching of contact openings, very little transport of glass is desired. The top edge of the contact opening should be simply rounded. Details of the film flow at both used film anneal conditions, cross-sectional images and calculations of BPSG film flow on device steps can be found in detail in Chapter 15.

The other major parameter that could be conveniently varied is the ambient used for the flow processes. It was found that for a given glass composition, the lowest flow temperature (down to about 50°C) occurs with a steam ambient. However, film flow under oxidizing conditions may not always be possible or desirable, especially during reflow of the glass after contacts are opened. It should be clear that flow and reflow characteristics of the glass could be affected also by the presence of the SiO_2 cap-layer, which can either suppress or enhance the flow ability of the glass depending on its history (see Chapter 15).

All of the above was linked to the flow characteristics of the glass films. There was no information regarding the anneal effects on device characteristics. The only paper [141] was devoted to an anomalous threshold voltage shift of MOSFET under RTA BPSG flow and reflow. The observed shift was attributed to the induction of a great amount of donor-like states in the SiO_2/Si interface and explained by the thermal strain induced as a result of a very high RTA ramp rate. Post RTA furnace anneal helped to eliminate this effect. Unfortunately, there was no indication about the particular BPSG film deposition process and film properties.

Application of CMP technique for PMD BPSG films. For IC devices at feature size of 0.35 um and below, BPSG film global planarization across the whole chip is needed to improve the limited lithography depth-of-focus and increase the yield of metal interconnect. Chemical Mechanical Polishing technology was recognized as one of the most important technologies to achieve global planarization [347]. There were a few papers published regarding BPSG film CMP and their basic results were similar [147,200]. An example of the CMP technique application for BPSG films deposited using Method 5 is briefly described below. The effect of BPSG additive concentrations on the CMP polishing rate was studied using identical process conditions. BPSG films were thermally annealed at 750 °C for 45 minutes.

The experiment was done on an Ebara's EPO213 dual-head chemical mechanical polisher with integrated cleaning stations. Eight inch silicon wafers were used. Rodel's IClOOO/Suba IV stacked pads and ILD1300 ammoniated silica slurry were used for the polishing. The pad conditioning method and stable polishing process for TEOS oxide was described elsewhere [200]. It was found that the polishing rate and uniformity of the BPSG film were much higher than those of TEOS oxide when the optimized polishing parameters for TEOS oxide were used. The higher CMP rate resulted in higher wafer-to-wafer variability, which agreed with data reported by Stamper et al [175]. With optimized CMP process parameters, a polishing rate of ~2600 A/min with less than 5 % uniformity was achieved for BPSG films with 4 wt % of boron and 4 wt % of phosphorus.

The polishing rate increases with both boron and phosphorus concentration. The phosphorus was found to have a slightly stronger effect on the polishing rate than boron. The polishing rate of BPSG as a function of total additive concentration (in wt %) is shown in Figure 3.27. The polishing rate increases almost linearly with total boron and phosphorus concentration in the range of concentration studied. A possible mechanism for the increasing of the polishing rate with additive concentration assumes the difference in additive structures with respect to the silica structure. Additives created dangling bonds in the oxide and, thus, weaken the structure of the oxide network with the increase of additive content. Weakening the structure of oxide films may allow the additive oxide to dissolute faster in the high pH silica slurry and therefore result in a higher polishing rate.

18.2. BPSG Film Gettering Capability

The gettering capability of BPSG films is due to a fairly high concentration of phosphorus oxide, a well-known getter for alkali-metal ions. It was demonstrated in experiments using the ^{22}Na isotope [50,56]. The element was introduced by applying $^{22}NaCl$ to the film surface or by ion implantation, followed by thermal processing at 430 – 500 °C. The distribution of sodium indicated that its ions were retained by a gettering mechanism rather than a barrier one. A similar mechanism was identified in PSG films, but these showed a smaller amount of gettering, as evidenced by a larger proportion of sodium ions escaping retention.

18.3. BPSG Moisture Absorption Effect

Chapter 12 summarizes a large amount of data devoted to the inherent BPSG film property – moisture absorption from ambient. This effect happens at the beginning of the as-deposited film exposure to the air followed by defect formation on the film surface. In Chapter 16, the author considered an approach for quantitative optimization of BPSG film composition. This approach allows obtaining a defect-free film composition during a convenient time delay of 24 hours after film deposition. According to Kern et al [50], as-deposited BPSG films should never be exposed to corrosive or aqueous compound solutions at elevated temperatures. This constraint is due to the boron and phosphorus oxides leaching from the film, similarly to the other CVD films of silicate glasses, such as PSG. BPSG films were recommended to be stored in a dry ambient until they can be flowed.

Hot-carrier-induced MOSFET degradation depends strongly on the quality of the gate oxide. This quality is affected by both the fabrication process and the materials covering the gate oxide. In general, insulating materials easily inject water molecules into the gate oxide because their hygroscopicity is high enough to absorb moisture from ambient. Thus, based on common sense, the moisture content in the PMD material is supposed to be as low as possible.

The authors [176] studied water-related hot-carrier degradation effects caused by BPSG films (Method 1, APCVD) using thermal desorption spectroscopy (TDS) to study quantities and configurations of water molecules in the films. In addition, they used a BPSG film containing deuterium-substituted water to evaluate the behavior of water in interlayer dielectrics. This BPSG film was deposited using deuterium-substituted alcohol in addition to the compounds used for CVD. Deuterium was used to distinguish the incorporated water molecules from those absorbed after the film growth.

Analyzing TDS spectra, BPSG film was found to store a larger amount of water after firm deposition as compared to silicon dioxide films, used as PMD material in parallel. It was expected that the water-related hot-carrier degradation occurs more easily in BPSG samples than in SiO_2 samples because the amount of water reaching the gate oxide should be greater in the BPSG samples. Contrary to expectations, there were no differences between BPSG and SiO_2 samples found. Moreover, hot-carrier-induced degradation was found to be smaller in BPSG than in SiO_2 PMD. These results suggest that the degree of hot-carrier degradation did not necessarily depend on the amount of water contained in the PMD. Although BPSG films contained more water than SiO_2 films, water-related hot-carrier degradation in BPSG devices was superior to that in SiO_2 devices. Eventually, the authors came to the conclusion that the hygroscopic BPSG film during film deposition caused the diffusion of water molecules into the underlying oxides. However, after deposition water re-absorption from the underlying oxides takes place and BPSG film acts as a desiccating agent that pumps water molecules out of the top gate and field oxides, thus reducing undesirable effects.

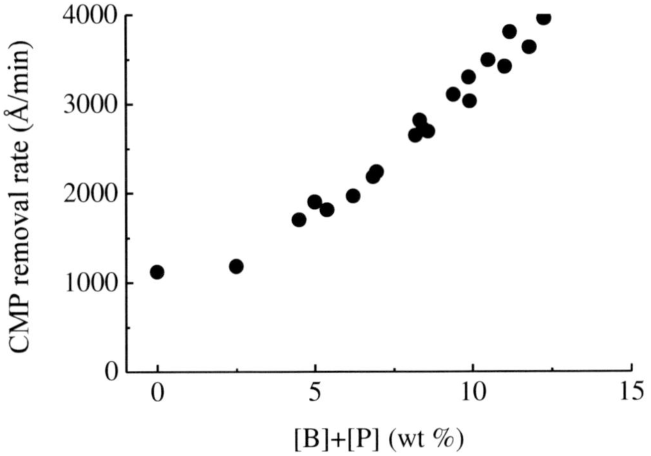

Figure 3.27. Chemical Mechanical Polishing removal rate as a function of total additive content (in elemental wt %) in thermally annealed BPSG films [200].

18.4. BPSG Film Implementation in Low-Etch Rate Applications

The effect of the decrease of BPSG film etch rate in buffered hydrofluoric acid solutions was used for express-control of the boron content in the film. In such a case, it is necessary to calibrate the etch rate values, performed preferably under fixed isothermal conditions (for instance, 25 °C), using a standard analytical technique. The same approach, but the film dissolution in so-called P-etch solution, can be used for express-control of the phosphorus concentration in BPSG films. It is known that this solution is very sensitive to the phosphorus concentration in glass films [18,50]. Using both solutions and calibration curves, which are shown, for instance, in Figure 2.17(a,b), it was possible to realize the immediate evaluation of both additives in BPSG films.

The etch rate of BPSG films in buffered hydrofluoric acid solutions can be comparable to that of SiO_2 films, and this feature can be desirable for certain applications. Typical etch rates for as-deposited and thermally annealed (i.e. densified) BPSG films deposited using Method 1 are listed in Table 2.6, together with those for SiO_2, BSG, and PSG films for comparison. More details of BPSG film wet etch-related issues can be found elsewhere [159].

The etching of glass films and SiO_2 films in both gas- and liquid-phase mixtures containing HF can be catalyzed by water [157,158]. Insufficient water to initiate the etching reaction produces an induction period which can be used to selectively etch glass film relative to SiO_2. It was found that additives in glass films also enhance the etching reaction by producing low volatility etching products which trap water and may directly catalyze the reaction.

Normally, dielectric films used in device technology are etched using plasma-activated, or so-called "dry methods". In this case, the dry etch rate was found to be practically equal for SiO_2 and glass thin film. For instance, we found a similar dry etch rate value of about 3600 Å/min for annealed BPSG films deposited using Methods 5 and 8, and as-deposited HDP-CVD SiO_2 and PSG films. This means that PECVD etch rate will be similar through the any SiO_2-based film stacks no matter it includes BPSG films or not.

18.5. BPSG Use as Low-Stress Film

Low-temperature CVD films of SiO_2 tend to be in tensile stress as deposited. The addition of phosphorus oxide or boron oxide to CVD SiO_2 films is known to decrease tensile stress, thus permitting use of relatively thick films on Si without cracking. However, since CVD films with higher phosphorus concentrations can result in Al metal corrosion in plastic-encapsulated devices, the maximum phosphorus content that can be used in low-temperature-deposited (400 °C) CVD phosphosilicate glass films over Al metal is approximately 4 wt % [P] [18]. Thus, it was expected that the lowering of the phosphorus content can be achieved by the use of the boron additive in the film. As-deposited at 430 °C CVD BPSG films are in very low tensile stress in the range of 9×10^8 dyne×cm^{-2}. Thus it may be considered for applications in which low film stress is desirable. Examples of such applications can be as follows: IC device structures with thick (> 5 μm) dielectric films; passivation layers in device structures with thick metallization layers (> 2 μm); passivation layers over Al metallization in device structures subjected to subsequent high-temperature steps and so on. However, over the years there was no detailed information about low-stress BPSG film applications in production processes and devices.

Considering BPSG film PMD applications, one could realize that as-deposited film stress value is not that important due to the film flow processing used in the main device technologies. For instance, for two popular production processes designated above as Method 5 and Method 8, typical stress values for as-deposited BPSG films are in the range of $(4.3\div6.5)\times10^8$ dyne×cm^{-2} (tensile), but $(4.0\div6.0)\times10^8$ dyne×cm^{-2} (compressive), respectively. Post-deposition BPSG film anneal allowed eliminating stress differences due to the film flow.

18.6. BPSG Film Application as Low-Temperature Diffusion Source

CVD BPSG films can also serve as lower-temperature doped-oxide solid-to-solid diffusion sources [50]. For example, BPSG film with 4 wt % [B]–4 wt % [P] deposited on a 70 ohm cm p-type wafer and subjected to a 60-min heat treatment at 900 °C, produced a 5-ohm/square n-type surface layer. Although the BPSG films are a source of both B and P atoms, the principal dopant in the silicon is P (because of relative differences in the distribution coefficients of the two dopants at the Si-SiO_2 interface). The principal effect of the boron oxide in the glass is to lower the viscosity, thus permitting more phosphorus atoms to reach the Si/SiO_2 interface. A bare silicon wafer adjacent to BPSG in furnace will be boron-doped because of the relatively high volatility of boron oxide.

18.7. BPSG Film Application Peculiarities

As a final point of BPSG film application, it is necessary to mention that the gamma-ray hardness of BPSG films deposited by Method 1 at atmospheric pressure was reported to be much higher than that of PECVD from a hydride–nitrous oxide mixture, while the film compositions were the same [164]. The high radiation hardness of the former glass was explained due to a balance between electron and hole traps in the glass. The radiation susceptibility of the latter was attributed to phosphorus only, which was considered responsible for hole centers.

As noted in Part 1, weak oxidizing agents such as nitrous oxide are inadequate in terms of deposition kinetics. In addition to that, it is necesary to note that there were a few publications devoted to some issues with PECVD BPSG films deposited with the use of silane and nitrous oxide [256,288,290] according to Method 7 (see details in Chapter 6). Summarizing these observations, one can conclude that the most reasonable cause of differences of the properties of the BPSG films is non-complete oxidation of phosphorus and the formation of phosphorus species in P^{3+} and P^0 state instead of P^{5+} state as per normal to form completely oxidized P_2O_5. However, it can be noted that Method 7 is used quite rarely.

18.8. Other BPSG Film Application Data

In [265], dielectric properties of BPSG films (named borophosphosilicate glass-ceramics) were evaluated. The films were synthesized by a sol-gel process using boric acid, phosphoric acid and tetraethoxyorthosilicate (TEOS) as starting compounds. The borophosphosilicate glass-ceramics was sintered below 950 °C in air. The structure of these glass-ceramics was studied by X-ray diffraction and SEM. It was found that the SiO_2 glass phase wraps up BPO_4 crystallites. $Si_3(PO_4)_4$ crystallites exist when the SiO_2 content is decreased. The authors found that between 400 and 950 °C, the liquid faction of glass and the crystallization of crystal concurred to keep the chip in shape. The size of the BPO_4 crystalline grains was about 100 nm. The measurements for the dielectric properties of the sintered glass-ceramics show a dielectric constant of less than 5 and a dielectric loss factor of less than 3×10^{-3}. It was concluded that this glass-ceramics showed a good potential for its use as dielectric materials in super high frequency multi-layer chip inductors.

The authors of [252,263] studied relaxation mechanisms of strained films bonded to oxide and low viscosity glass films. Relaxation of compressively strained heteroepitaxial $Si_{0.7}Ge_{0.3}$ films bonded to high and low viscosity glass compliant layers was investigated in [252]. These structures were formed by transferring $Si_{0.7}Ge_{0.3}$ films to Si substrates covered with thermal SiO_2 and BPSG films. Relaxation was studied through thermal annealing experiments. For the low viscosity BPSG, relaxation was observed near 800 °C and was accompanied by buckling of the $Si_{0.7}Ge_{0.3}$ film. At this temperature, no change in the

$Si_{0.7}Ge_{0.3}$ film bonded to thermal SiO_2 was observed, and through this comparison relaxation on BPSG is interpreted as the result of viscous flow of the glass. Finally, film buckling was successfully avoided by patterning the strained films into small areas prior to annealing, and is an indication of film expansion. In [263], a process was recently developed to fabricate a structure comprising, from top to bottom, a SiGe thin film, a glass film, and a Si wafer. The SiGe film was patterned into islands. On annealing, the glass flowed and the islands relaxed. The resulting strain-free islands were used as substrates to grow epitaxial optoelectronic devices. The authors determined the conditions under which the islands relax by expansion without significant wrinkling, and demonstrate that a cap layer suppresses wrinkles, relaxing a large island crack-free.

The authors of [282] studied the stability of luminescent trivalent Cerium in silica host glasses modified by boron and phosphorus. Ce-doped BSG, phosphosilicate PSG, and BPSG films with different compositions were prepared by the sol-gel method, as silica films were used for comparison. The homogeneous rare-earth dispersion in the host matrix and the stabilization of the Ce^{3+} oxidation state enhanced the intensity of the photoluminescence emission in PSG and BPSG with respect to BSG and silica. The energy of the Ce^{3+} emission band in PSG and BPSG matrixes agreed with the phosphate environment of the rare earth.

FINAL REMARKS ON PART 3

Part 3 of the monograph describes different aspects of BPSG thin film implementation in electronic technologies. The main application of BPSG thin films in IC technologies is based on the film flow capability. This is necessary to smooth sharp edges of device structures and to eliminate voids in narrow IC device gaps covered with BPSG films. Analysis of BPSG film flow capability data has been performed quantitatively for traditional furnace thermal anneal, and Rapid Thermal Anneal. Detailed analysis of flow aspects of BPSG films has been presented using examples of the real IC device structures with steps and gaps. It is important to note that BPSG flow data were obtained experimentally using the scanning electron microscopy technique. It has been shown that the approaches presented can be applied to explain the flow of other types of flow-able glass films, for instance, BPSG films with a germanium oxide additive. Eventually, multi-factorial dependencies of the film flow capability have been presented. This summarized knowledge can be effectively used for time-consuming evaluation of the film flow on real IC device structures.

The quantitative BPSG flow description together with the quantitative moisture absorption and defect formation description made in Part 2 of the monograph, allowed the author to state simple quantitative approaches for obtaining the optimized BPSG composition used in electronic technology. These approaches and drawn formulas were proven using a number of experimental data, which showed good correlation.

Some rarely discussed aspects regarding the film durability have also been analyzed. Deep understanding on the causes of potential film instability and approaches to improve BPSG film post deposition thermal treatment have been proposed by the author for the first time.

Generalized data and equations presented in this chapter provide sufficient information for researchers and engineers regarding directions and details of BPSG film application in electronic-related technologies.

CONCLUSION

This monograph summarizes the 30years of studies of borophosphosilicate glass (BPSG) thin film developed for the use in electronic technologies. The author would like to highlight that all data presented and analyzed in this book were obtained experimentally by many research groups from different countries and research institutions and electronic companies. This makes the monograph content and the author's conclusions valid and general.

The first part of the monograph describes important thin film synthesis aspects, such as CVD methodology, BPSG thin film CVD processes, silicon dioxide and glass film growth kinetics, CVD step coverage and gap-fill features, and some details of BPSG production processes. Based on the analysis performed, commonalities and differences of CVD processes have been demonstrated. All CVD-related topics together allow a complete view on CVD thin film processes and reaction schemes.

The second part of the monograph contains a description of BPSG film properties, film structure, glass flow capability and gap-fill analysis, BPSG film-moisture interaction and defect formation phenomenon. Special attention is given to the interrelations between BPSG film structures and properties with definitions of the classical bulk glass chemistry that allows an examination of the relationship between the glass film structure and thin film peculiarities. The described differences of glass film properties have been linked to thin film CVD process features and reaction schemes.

Part 3 covers different aspects of BPSG film application in electronic technologies. The film implementation in device technology is analyzed based on the film properties. Accumulated knowledge of BPSG film technology allows a quantitative characterization of the film flow property on IC device structures. An approach to optimize the film composition in order to get durable BPSG film is presented.

The author has eventually realized his goal. The interrelation of chemical vapor deposition kinetic features, thin film material properties, and electronic technology aspects have been linked together, providing a comprehensive understanding of BPSG thin films.

ACKNOWLEDGMENTS

The author would like to thank the management of Research and Development Division of Novosibirsk Semiconductor Device Manufacturing (Russian Federation) and personally Ms. N.A. Epova for her initiation and long-term support of the research works on BPSG thin films, the management of R&D Department of Chartered Semiconductor Manufacturing (Singapore), personally Dr. C.K. Lau and Dr. Lap Chan, for the invitation to work at Chartered and for providing excellent operating conditions for research and development of thin film technology and BPSG thin film research.

The author would like to express his appreciation to major co-authors from Russia and Singapore: Ms. T.G. Dukhanova, Mr. S. Khee, Dr. J. Z. Zhen, Dr. J. L. Sudijono, Dr. A. Cuthbertson, Mr. W. Lu, and to collegues and subordinates of research groups, who helped him to perform experimental works and to discover BPSG film features: Mrs. S.M. Maroshina, Mrs. A.N. Kovaleva, Mrs. G.I. Bureeva, Mrs. I. Nikolaeva, Mr. J. Hua, Ms. H.R. Wang, and, especially, Dr. K. V. Loiko and Dr. I. Peidous for unforgettable research time spent in Singapore.

Special thanks the author would like to say to Prof. S.M. Repinski (Russia) for long-term support and helpful discussion of thin film topics and Prof. T.E. Wade (USA) for invitations to present thin film research materials at DUMIC and VMIC conferences in Santa Clara.

The author would like to thank SibIS LLC for assistance in meeting the publication costs of this monograph.

REFERENCES

[1] Vassiliev, VY. *Candidate thesis in Chemical Sciences*, 1990, Novosibirsk, Russa (In Russian).

[2] Vassiliev, VY. *Doctoral thesis in Chemical Sciences*, 2002, Novosibirsk, Russia (In Russian).

[3] Vasilev, VYu; Repinskiy, SM. *Russ. Chem. Rev*., (Engl. Transl.), 2005, vol.74, 413-441.

[4] Vasilyev, VY. In *Ruthenium Properties, Production and Applications (Chemical Engineering Methods and Technology)*;Watson DB. (Ed.), Nova Science Publishers, 2011, 1-84.

[5] Powell, CF., Oxley, JH., Blocher, JM., Jr. (Eds.). *Vapor Deposition*; Wiley, New York, NY, 1966

[6] Kern, W., Ban, VS. In *Thin Film Processes*; Vossen JL., Kern W. (Eds.), Academic Press, New York, NY, 1978, 257-331.

[7] Tauber, R. In *Silicon Processing for VLSI Era*; Wolf, S., Tauber, R. (Eds), vol.1, Lattice Press, Sunset Beach, CA, 1986, 161-197.

[8] Adams, AC. In *VLSI Technology*, 2nd edition, Sze SM. (Ed.), McGrow-Hill Int. Ed., 1988, 233-271.

[9] Ting, CH. In *Handbook of Multilevel Metallization for Integrated Circuits*. Wilson SR., Tracy CJ., Freeman JL. (Eds), Noyes Publications, Park Bridge, NJ, 1993, 202-273.

[10] Cheng, HC. In *ULSI Technology*. Chang CY., Sze SM. (Eds.), McGrow-Hill Int. Ed, 1996, 205-269.

[11] Rosler, RS. *Solid State Technol*., 1977, vol.20, no.4, 63-70.

[12] Kern, W.; Rosler, RS. *J. Vac. Sci. Technol*., 1977, vol.14, no.5, 1082-1099.

[13] Hollahan, JR.; Rosler, RS. In *Thin Film Processes*. Vossen JL., Kern W. Eds.; Academic Press, New York, 1978, 257-331.

[14] Kern, W.; Heim, RC. *J. Electrochem. Soc.,* 1970, vol.122, no.4, 562-568.

[15] Kern, W.; Heim, RC. *J. Electrochem. Soc.,* 1970, vol.117, no.4, 568-573.

[16] Nagasima, N.; Suzuki, H.; Tanaka, K.; Nishida, S. *J. Electrochem. Soc.,* 1974, vol.121, no.3, 434-438.

[17] Schnable, GL.; Kern, W.; Comizzoli, RB. *J. Electrochem. Soc.,* 1975, vol.122, no.8, 1092-1103.

[18] Kern, W.; Schnable, G.L., Fisher, AW. *RCA Review*, 1976, vol.37, no.1, 3-54.

[19] Kern, W.; Schnable, GL. *IEEE Trans. Electron. Devices*, 1979, vol.26, no.4, 647-657.

[20] Levy, RA.; Vincent, SM.; McGahan, TE. *J. Electrochem. Soc.,* 1984, vol.132. no.6, 1462-1480.
[21] Brown, WA.; Kamins TJ. *Solid State Technol.*, 1979, vol.22, no.7, 51-57.
[22] Reinberg, A.R. *J. of Electronic Materials*, 1979, vol.8, no.3, 345-375.
[23] Hammond, M. *Solid State Technol.*, 1979, vol.22, no.12, 61-66.
[24] Rosler, RS.; Engle, GM. *Solid State Technol.*, 1979, vol.22, no.12, 88-92.
[25] Rosler, RS.; Engle, GM. *Solid State Technol.*, 1981, vol.24, no.4, 172-177.
[26] Winkle, LW.; Nelson, CW. *Solid State Technol.*, 1981, vol.24, no.10, 123-128.
[27] Kern, W. *Semiconductor International,* 1982, vol.4, no.3, 89-103.
[28] Ojha, SM. In *Physics of Thin Films.Vol.12.*Hass G., Francombe MH, Vossen JL, (Eds), Academic Press, 1982, New York, 237-296.
[29] Adams, AK. *Solid State Technol.*, 1983, vol.26, no.4, 135-139.
[30] Hess, DW. *J. Vac. Sci. Technol. A*, 1984, vol.2, no.2, 244-252.
[31] Sherman, A. *Thin Solid Films*, 1984, vol.113, no.2, 135-149.
[32] Tolliver, DL. In *VLSI Microelectronics: Microstructure Science.* Vol.8. Einspruch N.G. (Ed.). Academic Press, New York, NY, 1985, 2-24.
[33] Chin, BL., Ven, EP. *Solid State Technol.*, 1988, vol.31, no.4, 119-127.
[34] Singer, P. *Semiconductor International*, 1984, vol.7, no.5, 72-77.
[35] Vasileva, LL.; Vassiliev, VY.; Popov, VN.; Sukhov, MS. *Obz. Elektron. Tekhn. (Reviews on Electron. Technology)*, Series 2, no.8, TsNII “Elektronika”, Moscow, 1984, 1-27 (In Russian).
[36] Vassiliev, VYu.: Sukhov, MS. *Obz. Elektron. Tekhn. (Reviews on Electron. Technology)*, Series 7, no.3, TsNII “Elektronika”, Moscow, 1985, 1-46 (In Rissian).
[37] Vassiliev, VYu.; Sukhov, MS. *Obz. Elektron. Tekhn. (Reviews on Electron. Technology)*, Series 7, no.4, TsNII “Elektronika”, Moscow, 1985, pp 1-52 (In Rissian).
[38] Shepelev, SN.; Vassiliev, VYu.; Popov, VN. *Obz. Elektron. Tekhn. (Reviews on Electron. Technology)*, Series 2, no.7, TsNII “Elektronika”, Moscow, 1986, 1-53 (In Russian).
[39] Sherman, A. *Chemical Vapor Deposition for Microelectronics: Principles, Technology, and Applications.* Noyes Publications, Park Bridge, NJ, 1987, 1-217.
[40] Rossi, RC. In *Handbook of Thin-Film Deposition Processes and Techniques*. Shuegraf KK. (Ed.), Noyes Publications, Park Bridge, NJ, 1988, 80-111.
[41] Shepelev, S.N.; Vassiliev, VYu.; Popov, VN. *Obz. Elektron. Tekhn. (Reviews on Electron. Technology)*, Series 2, no.2, TsNII “Elektronika”, Moscow, 1988, 1-62 (In Russian).
[42] Repinsky, SM. In *Modern Problems of Physical Chemistry of Semiconductor Surface*, Rshanov AV (Ed), Nauka, Novosibirsk, 1988, 90-152 (in Russian).
[43] Vasileva, LL. In *Modern Problems of Physical Chemistry of Semiconductor Surface*, Rshanov AV (Ed), Nauka, Novosibirsk, 1988, 90- (in Russian) 152-176 (In Russian)
[44] Vassiliev, VYu. *Obz. Elektron. Tekhn. (Reviews on Electron. Technology)*, Series 2, no.3, Institute “Electronics”, Moscow, USSR, 1990, 1-56 (In Russian).
[45] Repinsky, SM. Khimicheskaya Fizika (*Chem. Physics),* 1992, vol.11, no.10, 1383-1390 (In Russian).
[46] Vassiliev, VYu. Khimicheskaya Fizika (*Chem. Physics),* 1992, vol.11, no.10, 1406-1413 (In Russian).
[47] Singer, P. *Semiconductor International*, 1995, vol.8, no.2, 55-62.

[48] Vassiliev, VYu. *Russian Microelectronics*, 1999, vol.28, no.3, 175-182.
[49] Ramiller, C.; Yau, L. *SEMCON/WEST* '82, Technical Program Proc., 1982, 29-37.
[50] Kern, W.; Schnable GL. *RCA Review*, 1982, vol. 43, no.3, 423-457.
[51] Avigal, I. *Solid State Technol.*, 1983, vol.26, no.10, 217-224.
[52] Tong, JE.; Schertenleib, K.; Carpio, RA. *Solid State Technol.*, 1984, vol.27, no.1, 161-170.
[53] Shanfield, S.; Bay, S. *J. Electrochem. Soc.*, 1984, vol.131, no.9, 2202-2203.
[54] Nassau, K.; Levy RA.; Chadwick, DL. *J. Electrochem. Soc.*, 1985, vol.132, no. 2, 409-415.
[55] Levy, RA.; Nassau, K. *J. Electrochem. Soc.*, 1985, vol.132, no.3, 409-415.
[56] Kern, W.; Kurylo, WA.; Tino, CJ. *RCA Review*, 1985, vol.46, no.2, 117-153.
[57] Kern, W.; Smeltzer, RK. *Solid State Technol.*, 1985, vol.28, no.6, 171-179.
[58] Foster, T.; Hoeye, G., Goldman, J. *J. Electrochem. Soc.*, 1985, vol.132, no.2, 505-507.
[59] Learn, AJ. *J. Electrochem. Soc.*, 1985, vol.132, no.2, 390-393.
[60] Learn, A. J. *J. Electrochem. Soc.*, 1985, vol.132, no.2, 405-409.
[61] Learn, A.: Baerg, B. *Thin Solid Films*, 1985, vol.130, no.1/2, 103-111.
[62] Susa, M.: Hiroshima, Y.; Senda, K.; Kurijama, T.; Matsumoto, S.; Terakawa, S.; Takamura, T. *J.Appl. Phys.*, 1985, vol.58, no.10, 3880-3883.
[63] Sharp, AC., Patel, J. *Vacuum*, 1985, vol.35, no.10-11, 441-443.
[64] Chu, PK.; Grube, SL. *Anal. Chem.*, 1985, vol.57, 1071-1074.
[65] Fu, CY. *Int. Electron. Dev. Meeting (IEDM)*, 1985, Technical digest, 602-605.
[66] Tela, V.; Ha, C.; Chowaniec, C. *Proc. 2nd Int. IEEE VLSI Multilevel Interconnection Conf. (VIMIC)*, 1985, 273-279.
[67] Vassiliev, VYu.; Dukhanova, TG. *Elektronnaya Promischlennost (Electronic Industry)*, 1986, no.1, 23-26 (In Russian).
[68] Levy, RA.; Nassau, K. *Solid State Technol.*, 1986, vol.29, no.10, 123-129.
[69] Watkins-Johnson Company information material. *Solid State Technol.*, 1986, vol.29, no.5, 100-101.
[70] Baker, RL.; Jennings, SR.; *Proc. V-MIC Conf.*, 1986, 484-490.
[71] Mitchener, JC.; Mahawili, I. *Proc. V-MIC Conf.*, 1986, 298-307.
[72] Mercier, JS.; Madsen, LD.; Calder, ID. *Proc. Mat. Res. Soc. Symp.*, 1986, vol.52, 251 258.
[73] Susa, M.; Hiroshima, Y.; Senda, K.; Takamura, T. *J. Electrochem. Soc.*, 1986, vol.134, no.7, 1517-1518.
[74] Wong CCD.; Brown, WA. *Proc. Int. Conf. Semicond. Integr. Circuit Technol.*, 1986, 295-296.
[75] Yoshimaru, M.; Matsuhashi, H.; Ajioka, T.; Matsu, H. *Proc. Int. Conf. Semicond. Integr. Circuit Technol.*, 1986, 292-294.
[76] Vassiliev, VYu. In *Semiconductor tensometry. Physical and Technological Problems.* Shadrin VS (Ed.), Novosdibirsk Electrotechnical Institute, Novosibirsk, 1986, 70-77 (In Russian).
[77] Elferink, JBO.; Dunselman, K.; Evers, EJ.; Habraken, FHPM.; Weg, W.F.; Holsbrink, J. *Surface and Interface Anal.*, 1986, vol.9, no.1 6, 293 296.
[78] Becker, FS.; Pawlik, D.; Schafer, H.; Staudigl, G. J. *Vac. Sci. Technol.* B., 1986, vol.4, no.3, 732-744.

[79] Levy, RA.; Gallagher, PK.; Schrey, F. *J.Electrochem. Soc.,* 1987, vol.134, no.2, 430-437.
[80] Levy, RA.; Gallagher, PK.; Schrey, F. *J. Electrochem. Soc.,* 1987, vol.134, no.7, 1744-1749.
[81] Levy, RA.; Kametani, TY. *J. Electrochem. Soc.*, 1987, vol.134, no.6, 1565-1567.
[82] Williams, DS.; Dean, EA. *J. Electrochem. Soc.*, 1987, vol.134, no.3, 657-664.
[83] Hughes, MC.; Wonsidler, DR. *J. Electrochem. Soc.*, 1987, vol.134, no.6, 1488-1494.
[84] Hurley, KH.; Bartolomew, DB.; Bordonaro, DT. *Semiconductor International*, 1987, vol.10, no.10, 7-12.
[85] Partlow, WD.; Samuels, BC. *J. Electrohem. Soc.,* 1987, vol.134, no.7, 1740-1743.
[86] Mercier, JS. *Solid State Technol.,* 1987, vol.30, no.7, 85-91.
[87] Becker, FS.; Röhl S. *J. Electrochem. Soc.*, 1987, vol.134, no.7, 1744-1749.
[88] Nishimoto, Y.; Tokumasu, N.; Fukuyama, T.; Maeda, K. *Proc. 19th Conf. Solid State Devices and Materials,* 1987, 447-450.
[89] Vassiliev, VYu.; Dukhanova, TG. *Elektronnaya Promischlennost (Electronic Industry),* 1988, no.1,31.
[90] Vassiliev, VYu.; Duhanova, TG.; Kosyakov, VI.; Shestakov, AA. *Elektronnaya Promischlennost* (*Electronic Industry)* 1988, no.5, 40-41 (In Russian).
[91] Haaland, DM. *Anal. Chem.*, 1988, vol.60, 1208-1217.
[92] Raley, NF.; Losee, DL. *J. Electrochem. Soc.,* 1988, vol.132, no.10, 2640-2643.
[93] Dukhanova, TG.; Vassiliev, VYu.; Veretenin, YI. *Obz. Elektron. Tekhn. (Reviews on Electron. Technology),* Series 3, no.4, TsNII "Elektronika", Moscow, 1988, no 4, 1–72 (In Russian).
[94] Cargo, JT.; Hughes, MC. *J. Electrochem. Soc.,* 1989, vol.136, no.4, 1239-1241.
[95] Law, K.; Wong, J.; Leung, C.; Olsen, J.; Wang, D. *Solid State Technol.*, 1989, vol.32, no.4, 60-62.
[96] Nishimoto, Y. et al. *V-MIC Conf. Proc.*, 1989, 382 (Cited form Ref. 9).
[97] Katsumata, Y., et al. *ECS Ext. Abst.*, vol. 89-2, 1989, 309 (Cited from Ref. 9).
[98] Chou, KS.*J. Non-Crist. Solids.*, 1989, vol.110, no.1, 122-124.
[99] Ikeda, Y.; Numasawa, Y.; Sakamoto, M. *J. Electronic Mat.*, 1990, vol.19, no.1, 45-49.
[100] Zeitzoff, PM.; Hossain, TZ.; Boisvert, DM.; Downing, RG. *J. Electrochem. Soc.,* 1990, vol.137, no.12, 3917 – 3922.
[101] Nishimoto, Y. et al. *CVD XI ECS Proc.*, PV 90-12, 1990, 410 (cited from Ref. 9).
[102] Schnable, GL.; Fisher, AW.; Shaw, JM. *J. Electrochem. Soc.*, 1990, vol.137, no.12, 3973-3974.
[103] Lee, P.; Galiano, M.; Keswick, P.; Wong, J.; Shin, B.; Wang, D. *Proc. 7th Int. VLSI Multilevel Interconnection Conf. (VMIC)*, 1990, 396-398.
[104] Tsai, YH.; Hsu, SL.; Tseng, FC.; Yoo, CS. *Thin Solid Films*, 1990, vol.185, 363-371.
[105] Vassiliev, VYu.; Dukhanova, TG. Elektronnaya Tehnika (*J. Electronic Technology),* Series 3, 1991, no.4, 1723-1728 (In Russian).
[106] Fujino, K.; Nishimoto, Y.; Tokumasu, N.; Maeda, K. *J. Electrochem. Soc.*, 1991, vol.138, no10, 3019-3024.
[107] Pignatel, GU.: Sisson, JS.; Weiner, WP. *J. Electrochem. Soc.*, 1991, vol.138, no.6, 1723-1728.
[108] Chittipeddi, S.; Velaga, AN.; Nanda, AK.; Cochran, WT.; Graver, RN. *Proc. 8th Int. VLSI Multilevel Interconnection Conf. (VMIC)*, 1991, 68-74.

[109] Nauka, K.; Liu, C. *J. Electrochem. Soc.*, 1991, vol.138, no.8, 2367-2370.
[110] Vassiliev, VYu, Dukhanova, TG. Khimicheskaya Fizika (*Chem. Physics),* 1992, vol.11, no.12, 1699-1710 (In Russian).
[111] Turzevich, AC.; Zaitev DA.; Kabakov MM.; Shilo, OA.; Misyuchenko, VM.; Tokhonov VI. Elektronnaya Tehnika (*J. Electronic Technology)*, Series 3, 1992, no.1, 24-27 (In Russian).
[112] Ahmed, K.; Geisert, C. *J. Vac. Sci. Technol. A.,* 1992, vol.10, no.2, 313-315.
[113] Imai, S.; Yabuuchi, Y.; Terai, Y.; Yasui, T.; Kudo, C.; Nacao, I.; Fukumoto, M. *Appl. Phys. Lett.*, 1992, vol.60, no.22, 2761-2763.
[114] Yieh, E., et. al. *ECS Ext. Abst.* 92-1, 1992, 248 (Cited from Ref. 9).
[115] Dobkin, DM. *J. Electrochem. Soc.,* 1992, vol.139, no.9, 2573-2579.
[116] Tolles, R.; Bath, HM.; Doris, B.; Jairath, R.; Leggett, R.; Sivararn, S. Proc. *SPIE Conf.,* 1992, vol. 1805 Submicrometer Metallization, 42-51.
[117] White, LK.; Miszkowski, NA.; Kurylo, WA.; Shaw, JM. *J. Electrochem. Soc.,* 1992, vol.139, no.3, 822-826.
[118] Rojas,S.; Gomarasca, R.; Zanotti, L.; Borghesi, A.; Sassella, A.; Ottaviani, G.; Moro, L.; Lazzeri, P. *J. Vac. Sci. Techn. B*, 1992, vol. 10, no.2, 633-642.
[119] Matsuda, A.; Tohge, N.; Minami, T. *J. Mater. Sci.*, 1992, vol. 27, 4189-4194.
[120] Fisher, SM.; Chino, H.; Maeda, K.; Nishimoto, Y. *Solid State Technol.*, 1993, vol.36, no.9, 55-60.
[121] Fujino, K.; Nishimoto, Y.; Tokumasu, N.; Maeda, K. *J. Electrochem. Soc.*, 1993, vol.140, no.10, 2922-2927.
[122] Maeda, K.; Fisher, SM. *Solid State Technol.*, 1993, vol.36, no6, 83-87.
[123] Bouchard, H.; Azelmad, A.; Currie, J.F.; Meunier, M.; Blain, S.; Darwall, T. *Proc. of MRS Symposium*, 1993, vol.308, 63-68.
[124] Lo, GQ.; Kwong, DL.; Lee, S. *IEEE Trans. Electron Dev.*, 1993, vol.40, no.5, 1032-1035.
[125] Osorio, SPA.; Montero, I.; Perriere, J.; Martinez-Duart, J.M. *Appl. Surf. Sci.*, 1993, vol. 70/71, 772-776.
[126] Shenasa, M.; Coniff, J.; Kudla, P. *Electrochem. Soc. Proc. Volume PV 93-25*, 1993, 91-99.
[127] Carpio, R.; Lamaze, P.; Downing, R.G. *Electrochem. Soc. Proc.Volume PV 93-25*, 1993, 40-47.
[128] Coniff, J.; Shenasa, M.; Krott, L.; Woessner, S. *Electrochem. Soc. Proc. Volume* PV 93-25, 1993, 84-90.
[129] Ibok, EE.; Garg S. *Electrochem. Soc. Proc. Volume PV 93-25*, 1993, 52-63.
[130] Wilson, J.; Stepens, T. *Proc. of MRS Symposium*, 1993, vol.284, 211-216.
[131] Rojhantalab, H.; Moinpour, M.; Peter, N.; Dass, MLA.; Hough, W.; Moghadam, F. *Proc. of MRS Symposium*, 1993, vol.280, 147-151.
[132] Thakur, RPS.; Gonzalez, F.; Hawthorne, R.; Ward, V.; Jeng, J. *Proc. of MRS Symposium*, 1993, vol.303, 283-288.
[133] Hung, L.; Thallikar, G.; Myers, FR.; Cale, TS. *Proc. of 1st Int. Dielectric for ULSI Multilevel Interconnection Conf. (DUMIC)*, 1995, 204 210.
[134] Steiner, KG.; Velaga, AN.; Osenbach, JW. *Proc. of the 10th Int. VLSI Multilevel Interconnection Conf. (VMIC),* 1993, 125-127.

[135] Wang, CK.; Liu, LM.; Wei, CS. *Proc. of the 10th Int. VLSI Multilevel Interconnection Conf. (VMIC)*, 1993, 119-121.

[136] Yamashita, K.; Seto, M.; Kameda, M. *Electrochem. Soc. Proc. Volume*,PV 93-25, 1993, 64-71.

[137] French, PJ.; Wolffenbuttel, RF. *J. Micromech. Microeng.*, 1993, vol.3, 135-137.

[138] Hossain, SD.; Pas, MF. *J. Electrochem. Soc.*, 1993, vol.140, no.12, 3604-3606.

[139] Olsen, J.; Moghadam, F. In *Handbook of Multilevel Metallization for Integrated Circuits*. Wilson SR., Tracy CJ., Freeman JL. (Eds), Noyes Publications, 1993, 346-460.

[140] Ossikovski, R.; Blayo, N.; Drevillon, B.; Firon, M.; Delahaye, B.; Mayeux, A. *Appl. Phys. Lett.*, 1994, vol.65, no.10, 1236-1238.

[141] Hsieh, JC.; Fang, YK.; Chen, CW.; Tsai, NS.; Lin, MS.; Tseng, FC. *IEEE Trans. Electron Dev.*, 1994, vol.41, no.3, 458-460.

[142] Kobayashi, K.; Murota, M. *Glass Technology*, 1994, vol.35, no.4, 186-188.

[143] Bar-Ilan, H.; Rabinovich, I. *Proc. 12th Int. VLSI Multilevel Interconnection Conf. (VMIC)*, 1995, 76-82.

[144] Pyun, DS.; Kim, SB. *Proc. 12th Int. VLSI Multilevel Interconnection Conf. (VMIC)*, 1995, 151.

[145] Robles, S.; Russel, K.; Galiano, M.; Kithcart, V.; Nguen, BC. *Proc. 12th Int. Multilevel Interconnection Conf. (VMIC)*, 1995, 122-125.

[146] Li, J.; Fardi, B.; Thach, MD.; Moinpour, M.; Mogdaham, F. *Proc. 1st Int. Dielectric for ULSI Multilevel Interconnection Conf. (DUMIC)*, 1995, 124-130.

[147] Liu, CW.; Dai, BT.; Yeh, CF. *Thin Solid Films*, 1995, vol.270, no.1/2, 607-611.

[148] Noritake, C.; Kondo, I.; Kondo, K.; Takenaka, O.; Kinbara, A. *Thin Solid Films*, 1995, vol.270, no.1/2, 445-449.

[149] Dasari, RP.; Krishnan, K. Semiconductor Fabtech, 1995, no.2, 191-197.

[150] Chen, LJ. *Proc. 12th Int. VLSI Multilevel Interconnection Conf. (VMIC)*, 1995, 439.

[151] French, PJ.; Wolffenbuttel, RF. *J. Micromech. Microeng.*, 1995, vol.5, 125-127.

[152] Liao, H.; Thallikar, G.; Myers, FR.; Cale, TS. *Proc. 1st Int. Dielectrics for Multilevel Interconnection Conf. (DUMIC)*, 1995, 204-209.

[153] Noren, C.; DePinto, G.; Hiatt, M.; Kent, R. *Semiconductor International*, 1995, vol.18, no.6, 93-102.

[154] Noren, C.; DePinto, G.; Hiatt, M.; Kent, R. *Semiconductor International*, 1995, vol.18, no.7, 175-180.

[155] Hsu, R.; Kumta, PN.; Feist, TP. *J. Mater. Sci.*, 1995, vol. 30, 3123-3129.

[156] Syms, RRA.; Schneider, W.; Huang, W.; Holmes, AS. *Electron. Lett.*, 1995, vol.31, no.21, 1833-1834.

[157] Watanabe, H.; Ohnishi, S.; Honma, I.; Kitajima, H.; Ono, H.; Wilhelm, RJ.; Sophie, AJL. *J. Electrochem. Soc.*,1995, vol.142, no.1, 237-243.

[158] Watanabe, H.; Kitajima, H.; Honma, I.; Ono, H.; Wilhelm, RJ.; Sophie, AJL. *J. Electrochem. Soc.*,1995, vol.142, no.4, 1332-1340.

[159] Kunii, Y.; Nakayama, S.; Maeda, M. *J. Electrochem. Soc.*,1995, vol.142, no.10, 3510-3513.

[160] Gray, DC.; Butterbaugh, JW.; Hiatt, CF.; Lowing, AS.; Sawin, HH. *J. Electrochem. Soc.*,1995, vol.142, no.11, 3859-3863.

[161] Robles, S.; Russel, K.; Galiano, M.; Kithcart, V.; Nguen, BC. *J. Electrochem. Soc.*, 1996, vol.143, no.4, 1414-1421.

[162] Ahmed, W.; Ahmed, E.; Dost, AA. *J. Mat. Sci.: Materials in Electronics*, 1996, vol.7, 127-131.

[163] Yoshimaru, M.; Wakamatsu, H. *J. Electrochem. Soc.*,1996, vol.143, no.2, 666-671.

[164] Fuller, R.; Evans, H.; Gamlen, C.; Czagas, B.; Morrison, M.; Decrostar, D.; Lowry, R.; Lenaha, P.; Frye, C. *IEEE Transactions on Nuclear Science*, 1996, vol.43, no.6, 2565-2571.

[165] Kirchhoff, M.; Ilg, M.; Cote, D. *Berichte der Bunsengesellschaft für physikalische Chemie*, 1996, vol.100, no.9, 1434-1437.

[166] Somashekhar, A.; O'Brien, S. *J. Electrochem. Soc.*, 1996, vol. 143, no.9, 2885-2891.

[167] Hsu, R.; Kim, JY.; Kumta, PN.; Feist, TP. *Chem. Mater.*, 1996, vol. 8, 107-113.

[168] Tomoko, S.; Koji, Kimoto.; Takashi, A.; Yasuhiro, M. *Jap. J. Appl. Phys. Pt.2. Lett.*, 1996, vol. 35, no.12B, L1717-1719.

[169] Mayer, B. *Semiconductor International,* 1996, vol. 9, no.3, 344-365.

[170] Chu, JCS.; Tu,T.; Chen, KC.; Su, W.; Chang, T. *Proc. 14th Int. VLSI Multilevel Interconnection Conf. (VMIC)*, 1997, 499-501.

[171] Dyer, TW.; Rafftesaeth, KP.; Sa, JHY.; Nguyen, CH.; Kuei, JC.; Young, DTW.; Brondo, MR. *Proc. 3rd Int. Dielectric for ULSI Multilevel Interconnection Conf. (DUMIC)*, 1997, 197-204.

[172] Yoo, KS.; Ahn, KC.; Sohn, YS.; Kim, CT. *Proc. 2nd Int. Dielectric for ULSI Multilevel Interconnection Conf. (DUMIC)*. 1996, 247-252.

[173] Nagahara, R.; Lee, SK.; You, HM. *Proc. 13th Int. VLSI Multilevel Interconnection Conf. (VMIC)*, 1996, 433.

[174] Parikh, PJ.; Lee, JD.; Schaffer, WJ.; Westphal, JW.; Fry, HW. *Proc. 13th Int. VLSI Multilevel Interconnection Conf. (VMIC)*, 1996, 434.

[175] Stamper, AK.; Armbrust, DS.; Hartswick, TJ.; Chamgerlin, TS.; Smith, LA.; Yieh, EY. *Proc. 13th Int. VLSI Multilevel Interconnection Conf. (VMIC)*, 1996, 104-106.

[176] Ikeda, K.; Okazaki, Y.; Nakayama, S.; Rsuchiya, T. *Digest of Technical Papers of Symposium on VLSI Technology*, 1996, 116-117.

[177] Iyer, R.; Thakur, RPS.; Rhodes, H.; Liao, R.; Rosler, R.; Yieh, E. *J. Electrochem. Soc.*, 1996, vol.143, no.10, 3366-3372.

[178] Yoshimaru, M.; Matsuhashi, H. *J. Electrochem. Soc.*, 1996, vol.143, no.9, 3032-3037.

[179] Matsumoto, S.; Hattori, T.; Hata, Y.; Ogawa, H. *Proc. 13th Int. VLSI Multilevel Interconnection Conf. (VMIC)*, 1996, 122-124.

[180] Jaso, MA.; Gambino, JP.; Dobuzinsky, DM.; Armacost, M.; Ohiwa, T. *Proc. 13th Int. VLSI Multilevel Interconnection Conf. (VMIC)*, 1996, 407-412.

[181] Kondo, M.; Oda, K.; Ohue, E.; Shimamoto, H.; Tanabe, M.; Onai, T.; Washio, K. *Proc. IDEM Conf.*, 1996, 245-248.

[182] Fuller, R.; Evans, H.; Gamlen, C.; Czagas, B.; Morrison, M.; Lowry, R. Proc. *13th Int. VLSI Multilevel Interconnection Conf. (VMIC)*, 1996, 339-341.

[183] Ilg, M.; Kirchhoff, M.; Nguyen, S. *Electrochem. Soc. Meeting. Volume* PV 96-1, 1996, Abstract no.760.

[184] Xia, LQ.; Yieh, E.; Gee, P.; Campana, F.; Nguyen, BC. *J. Electrochem. Soc.*, 1997, vol.144, no.9, 3208-3213.

[185] Steiner, KG.; Olmer, LJ. *Proc. 3rd Int. Dielectric for ULSI Multilevel Interconnection Conf.*, 1997, 265-272.
[186] Lu, W.; Vassiliev, VY.; Zheng, JZ.; Chan, L. *Proc .3rd Intern. Dielectric for ULSI Multilevel Interconnection Conf. (DUMIC)*, 1997, 219-222.
[187] Vassiliev, VY.; Lu, W.; Zheng, JZ.; Wang, HR., Chan, L. *Proc. 3rd Intern. Dielectric for ULSI Multilevel Interconnection Conf. (DUMIC)*, 1997, 171-174.
[188] Sudijono, JL.; Vassiliev, VY.; Zheng, JZ. A *Proc. 7th Intern. Symp. on IC Technology, Systems and Application (ISIC-97)*, 1997, 597-600.
[189] Vassiliev, VY.; Zheng, JZ. *Proc. 7th Intern. Symp. on IC Technology, Systems and Application (ISIC-97)*, 1997, 522-525.
[190] Tang, SK.; Mridha, S.; Vassiliev, VY.; Zheng, JZ.; Chan, L. *Proc. 7th Intern. Symp. on IC Technology, Systems and Application (ISIC-97)*, 1997, 601-604.
[191] Loiko, KV.; Vassiliev, VY.; Chua, CT.; Lim, DHY. *Proc. of Intern. Conference on Electrical Insulation and Dielectric Phenomena (CEIDP)*, 1997, 150-153.
[192] Vassiliev, VY.; Zheng, JZ. *Electrochem. Soc. Proc. Volume* PV 97-25, 1997, 1199-1206.
[193] Gambino, JP.; Stengl, R.; Jaso, MA.; Yoshikawa, S.; Tsunashima, Y.; Asao, Y.; Levine, E. *Proc. 14th Int. VLSI Multilevel Interconnection Conf. (VMIC)*, 1997, 609-611.
[194] Kronninger, W.; Redl, W.; Hockele, U.; Frank, M.; Despande, M.; Yau, WF.; Kronger ,W.; Rausch, W. *Proc. 14th Int. VLSI Multilevel Interconnection Conf. (VMIC)*, 1997, 606-608.
[195] Schaffer, W.; Schey, D.; Bowman, M.; McGahan, W.; Boisen, J.; Bran, M.; Strausser, Y. *Solid State Technol.*, 1997, vol.40, no.9, 117-126.
[196] Dutta, AK. *J. Electrochem. Soc.,* 1997, vol.144, no.3, 1073-1081.
[197] Zhang, S.; Franke, JE.; Niemczyk, TM.; Haaland, DM.; Cox, JN.; Banerjee, I. *J. Vac. Sci. Techn. B.*, 1997, vol.15, no.4, 955-960.
[198] Mizushima, I.; Arai, N.; Kai, N.; Wakamiya, M.; Tsunashima, Y.; Mori, S. *Proc. 14th Int. VLSI Multilevel Interconnection Conf. (VMIC)*, 1997, 533.
[199] Lee, C.; Lee, SC. *Solid State Electronics*, 1997, vol.41, no.6, 921-923.
[200] Zheng, JZ.; Huang, VSK.; Lu, W.; Vassiliev, VY.; Feng, C.; Zhou, BB.; Tay CWS. *Proc. 5th An. Singapore Technical Conference on Assembly, Packaging, Test and IC Technology (SEMICON 97)*, 1997, 269-273.
[201] Yoshida, T.; Hashimoto, S.; Hosokawa, H.; Ohwaki, T.; Mitsushima, Y.; Taga, Y. *J. Appl. Phys.*, 1997, vol. 81, no.10, 7031-7038.
[202] Naimpally, A. *Dev. Chem. Eng. Mineral Process.,* 1997, vol. 5, no.3/4, 281-294.
[203] Masakazu, F.; Hiroshi, K.; Ikuo, O.; Katsuji, H. *Jap. J. Appl. Phys. Pt. 1.*, 1997, vol. 36, no.9A, 5475-5480.
[204] Lie, DYC.; Xia, W.; Yota, J.; Joshi, AB.; Zwingman, R.; Williams, R.; Kerametlian, V.; Cerney, D.*Proc. SPIE Conf.*, 1997, vol. 3212, 258-267.
[205] Goyal, KO.; Westphal, J.W. 1997 JCPDS-International Centre for Diffraction data, 1-8.
[206] Applied Materials Inc. *Solid State Technology*, 1997, vol.40, no.5, 117-126.
[207] Vassiliev, VY. (Invited lecture) In *Visual Booklet of Short Courses at IV International Dielectric for ULSI Multilevel Interconnection Conference,* 1998, Santa Clara, USA, 111-192.
[208] Vassiliev, VY.; Zheng, JZ.; Liao, M.; Lin, YS. *Proc. 4th Intern. Dielectric for ULSI Multilevel Interconnection Conf. (DUMIC),* 1998, 90-97.

[209] Vassiliev, VY.; Dukhanova, TG. *Electrochem. Soc. Proc. Volume,* PV 98-23, 1998, 258-263.

[210] Brichko, Y. *Proc. 15th Int. VLSI Multilevel Interconnection Conf. (VMIC)*, 1998, 633.

[211] Tang, SK.; Vassiliev, VY.; Mridha, S.; Chan, L. *Proc. 4th Intern. Dielectric for ULSI Multilevel Interconnection Conf. (DUMIC)*, 1998, 287-290.

[212] Lu, W.; Vassiliev, VY.; Zheng, JZ. *Proc. 15th Intern. VLSI Multilevel Interconnection Conf. (VMIC)*, 1998, 580-585.

[213] Vassiliev, VY.; Zheng, JZ.; Lin, C. *Electrochem. Soc. Proc. Volume*, PV 98-6, 1998, 244-255.

[214] Vassiliev, VY.; Tang, SK.; Zheng, JZ.; Liao, M. *Proc. 4th Intern. Dielectric for ULSI Multilevel Interconnection Conf. (DUMIC)*, 1998, 253-260.

[215] Chang, HH.; Ni, CF.; Lin, YH.; Kao, KY. *Proc. 15th Int. VLSI Multilevel Interconnection Conference (VMIC)*, 1998, 607-609.

[216] Guu, F.; Ku, RY.; Chang, S.; Tang, RA. *Proc. 15th Int. VLSI Multilevel Interconnection Conf. (VMIC*), 1998, 622-624.

[217] Dyer, TW. *J. Electrochem. Soc.*, 1998, vol.145, no.11, 3950-3956.

[218] Rao, NP.; Nijhawan, S.; Kim, T.; Wu, Z.; Campbell, S.; Kittelson, D.; McMurry, P.; Chung, CC.; Mastromattee, E. *J. Electrochem. Soc.*, 1998, vol.145, no.6, 2051-2057.

[219] Saxena, A. *Intern. J. of Thermophysics,* 1998, vol.19, no.3, 961-967.

[220] Kim, JY.; Kumta, PN. *J. Phys. Chem. B*,1998, vol. 102, no.30, 5744-5753.

[221] Niemczyk, TM.; Zhang, L.; Haaland, DM.; Radigan KJ. *J. Vac. Sci. Technol. A*, 1998, vol. 16, no.6, 3490-3494.

[222] Vassiliev, VY.; Huang, L.; Sudijono, J.; Cuthbertson, A. *Proc. 16th Intern. VLSI Multilevel Interconnection Conf. (VMIC)*, 1999, 480-482.

[223] Vassiliev, VY. *Electrochem. Soc. Proc. Volume*, PV 99-31, 1999, 322-333.

[224] Vassiliev, VY.; Zheng, JZ.; Tang, SK.; Lu, W.; Hua, J.; Lin, YS. *J. Electrochem. Soc.,* 1999, vol.146, no.8, 3039-3051.

[225] Tang, SK.; Vassiliev, VY.; Mridha, S.; Chan, L. Thin Solid Films, 1999, vol.352, no.1-2, 77-84.

[226] Vassiliev, VY. *J. on Dielectrics for ULSI Multilevel Interconnection*, 1999, vol.1, no.1, 15-28.

[227] Xia, LQ.; Conti, R.; Galiano, M.; Campana, F.; Chandran, S.; Cote, D.; Restano, D.; Yieh, E. *J. Electrochem. Soc.,* 1999, vol.146, no.5, 1884-1888.

[228] Kapkin, K.; Sisson, J.; Al-Lami, S.; Pye, JT.; Evans, B. *Proc. 5th Dielectric for ULSI Multilevel Interconnection Conf. (DUMIC)*, 1999, 245-248.

[229] Kapkin, K.; Sisson, J.; Al-Lami, S.; Pye, JT.; Brichko, Y. *Proc. 16th Int. VLSI Multilevel Interconnection Conf. (VMIC)*, 1999, 446-448.

[230] Vasilev, VYu. Zhurnal Fizicheskoi Khimii (*J. Phys. Chem.),* 1999, vol.73, no.7, 1191-1198 (In Russian).

[231] Vassiliev, VY.; Lin, C.; Gn, FH.; Cuthbertson, A. *Proc. of 8th Intern. Symp. on IC Technology, Systems and Application (ISIC-99)*, 1999, 186-189.

[232] Vasilev, VYu. Zhurnal Fizicheskoi Khimii (*J. Phys. Chem.*) 1999, vol..73, no.8, 1351-1356 (In Russian).

[233] Vasilev, VYu. *Russian Microelectronics*, 1999, vol.28, no.3, 183-192.

[234] Choi, CG,;Jeong, MY.; Choy, TG. *J. Mater. Sci.*, 1999, vol.34, 6035-6040.

[235] Qiao, J.; Jin, B.; Phatak, P.; Yu, J.; Geha S.*J. Vac. Sci. Technol. B*, 1999, vol.17, no.5, 2373-2377.

[236] Niemczyk, TM.; Zhang, S.; Franke, JE.; Haaland, DM. *Applied Spectroscopy*, 1999, vol. 53, no. 7, 822-828.

[237] Hoffmann, T.; LeDuc, P.; Senez, V. *J. Vac. Sci. Techn. B*, 1999, vol.17, no.6, 2603-2609.

[238] Yoshida, T.; Aoki, K., Mitsushima, Y. *J. Vac. Sci. Technol. B*, 1999, vol.17, no.4, 1585-1588.

[239] Simpson, DL.; Crosswell, RT.; Reisman, A.; Temple, D.; Williams, CK. *J. Electrochem. Soc*., 1999, vol.146, no.10, 3860-3871.

[240] Simpson, DL.; Crosswell, RT.; Reisman, A.; Temple, D.; Williams, CK. *J. Electrochem. Soc*., 1999, vol.146, no.10, 3872-3885.

[241] Crosswell, RT.; Reisman, A.; Simpson, DL.; Temple, D.; Williams, CK. *J. Electrochem. Soc*., 1999, vol.146, no.12, 4569-4579.

[242] Wang, CY.; Vassiliev, VY.; Tan, KW.; Sudijono, JL. *Proc. 6th Intern. Dielectric for ULSI Multilevel Interconnection Conf. (DUMIC)*, 2000, 244-252.

[243] Vassiliev; VY., Wang, CY.; Sudijono, JL.; Cuthbertson, A.; Chu, T.,;Tan, P.C.; Heng, Y.S.; Gang, Z.; Bhatnagar, A.; Rong, P.; Gee, P. *Proc. of 6th Intern. Dielectric for ULSI Multilevel Interconnection Conf. (DUMIC)*, 2000, 181-184.

[244] Vassiliev, VY. *Proc. 6th Intern. Dielectric for ULSI Multilevel Interconnection Conference (DUMIC)*, 2000, 121-132.

[245] Vasilev, VYu. *Fiz. Khim. Stekla*, 2000, vol.26, no.1, 116-129 (In Russian).

[246] Vasilev, VYu. *Fiz. Khim. Stekla,* 2000, vol.26, no.1, 130-145 (In Russian).

[247] Belyansky, M., Conti, R., Upham, A., Liucci, F., Parks, C., Lee, K.H., Strane, J. *Electrochem. Soc. Proc. Volume*, PV 2002-11, 47-52.

[248] Crosswell, RT.; Reisman, A.; Simpson, DL.; Temple, D.; Williams, C.K. *J. Electrochem. Soc*., 2000, vol.147, no.4, 1513-1524.

[249] Simpson, DL.; Crosswell, RT.; Reisman, A.; Williams, CK.; Temple, D. *J. Electrochem. Soc*., 2000, vol.147, no.4, 1560-1567.

[250] Berney, H.; Hill, M.; Kelleher, AM.; Hynes, E.; O'Neill, M.; Lane, WA. *J. Micromech. Microeng*., 2000, vol.10, 223–234.

[251] Ermolaeva, AI.; Koshelev, NI.; Dvornikov, SA. *Glass Physics and Chemistry (Russian)*, 2000, vo.26, no.3, 300-302.

[252] Hobart, KD.; Kub, FJ.; Fatemi, M.; Twigg, ME.; Thomson, PE.; Kuan, TS.; Inoki, CK. *J. Electron. Mater*., 2000, vol.29, no.7, 897-900.

[253] Gartner, M.; Modreanu, M.; Bosch, S.; Stoica, T. *Microelectronic Reliability*, 2000, vol.40. no.4-5, 617-620.

[254] Walsh, DS.; Doyle, BL. *Nuclear Instruments and Methods in Physics Research Section B: Beam Interactions with Materials and Atoms*, 2000, vol.161-163, 629-634.

[255] Lee, JW.; Chung, HK.; Kim, BH.; Jang, HM.; Moon, JH. *J. Kor. Cer. Soc.,* 2000, vol.37, no.1, 77-81.

[256] Albreht, G.; King, B. *Electrochem. Soc. Proc. Volume*,PV 2000-6, 13-33.

[257] Vassiliev, VY.; Sudijono, JL.; Cuthbertson A. *Solid State Technol.,* 2001, vol.44, no.3, 129-136.

[258] Muscat, AJ.;Thorsness, AG.;Montaño-Miranda,G. *J. Vac. Sci. Technol. A*, 2001, vol.19, no.4, 1854-1861.

[259] Kimura, S.; Mitsui, Y. *Applied Spectroscopy*, 2001, vol.55, no.3, 106A-128A.

[260] Park, B.; Conti, R.;Economikos, L.; Chakravarti, A.; Ellenberger, J. *J. Vac. Sci. Technol. B*, 2001, vol.19, no.5, 1788-1795.

[261] Vasilev, VY. *Russian Microelectronics*, 2002, vol.31, no.4, 263-273.

[262] Garcia, M.; Montero, I.; Ripalda, JM.; Galan, L. *J. Appl. Phys.*, 2002, vol.91, no.6, 3626-3631.

[263] Huang, R.; Yin, H.; Liang, J.; Sturm, JC.; Hobart, KD.; Suo, Z. *Acta Mechanica Sinics* (English Series), 2002, vol.18, no.5, 441-456.

[264] Lee, J.; Choi, JH.; Cheong, WS. J. Kor. Phys. Soc., 2002, vol.40, no.4, 615-618.

[265] Li, B.; Yue, Z.; Zhou, J.; Gui, Z.; Li, L. *Materials Letters*, 2002, vol.54, no.1, 25-29.

[266] Oberemok, O.; Lytvyn, P. *Semicond. Physics, Quantum Electronics & Optoelectronics*, 2002, vol.5, no.1, 101-105.

[267] Vasilev, V.Y., *Russian Microelectronics,* 2003, vol.32, no.3, 125-135.

[268] Vasilev, VY. *Russian Microelectronics,* 2003, vol.32, no.4, 190-199.

[269] Vassiliev, VYu. *J. Electrochem. Soc.*, 2003, vol.150, no.12, F211-F218.

[270] Vasil'ev, VYu. *Fiz. Khim. Stekla (Glass Physics and Chemistry)*, 2003, vol.29, no.5, 461-470.

[271] Thorsness, AG.; Muscat, AJ. *Electrochem. Soc.*, 2003, vol.150, no.12, F219-F228.

[272] Armelao, L.; Barreca, D.; Bottaro, G.; Canevali, C.; Morazzoni, F.; Scotti, R.; Tondello, E. *Surface Science Spectra*, 2003, vol.10, no.1, 40-46.

[273] Kobayashi, K. *Materials Science and Engineering B*, 2003, vol. 98, 181-183.

[274] Cheong, WS.;Lee, SK.; Roh, JS.*J. Vac. Sci. Technol. B*, 2003, vol.21, no.3, 975-980.

[275] Carboni, R.; Pacchioni, G.; Fanciulli, M.; Giglia, A.; Mahne, N.; Pedio, M.; Nannarone, S.; Boscherini, F. *Appl. Phys. Lett.,* 2003, vol.83, no.21, 4312-4314.

[276] Vasilev, VYu. *Russian Microelectronics*, 2004, vol.33, no.5, 271-284.

[277] Canevali, C.; Scotti, R.; Vedda, A.; Mattoni, M.; Morazzoni, F.; Armelao, L.; Barreca, D.; Bottaro, G. *Chem. Mater.* 2004, vol.16*,* 315-320.

[278] Mekkakia, MN.; Boudghene, SA. *Thin Solid Films*, 2004, vol.458, no.1/2, 309-313.

[279] Wu, YH.; Hsiao, SC.; Lin, PH.; Ku, CL. *Electrochem. Solid-State Lett.*, 2004, Vol.7, no.1, G1-G4.

[280] Yamakawa, K.; Hori, M.; Goto, T.; Den, S.; Katagiri, T.; Kano, H. *Appl. Phys. Lett.*, 2004, vol.85, no.4, 549-551.

[281] Vasilev, VYu. *Russian Microelectronics*, 2005, vol.34, no.2, 67-77.

[282] Canevali, C.; Mattoni, M.; Morazzoni, F.; Scotti, R.; Casu, M.; Musinu, A.; Krsmanovic, R.; Polizzi, S.; Speghini, A.; Bettinelli, M. *J. Am. Chem. Soc.*, 2005, vol.127, 14681-14691.

[283] Kordas, G. *J. Non-Crystalline Solids*, 2005, vol.351, no.27-29, 2358-2360.

[284] Aitken, BG.; Youngman, RE. *Eur. J. Glass Sci. Technol. B*, 2006, vol.47, no.4, 381–387.

[285] Turtsevich, AS., Nalivaiko, OYu., Anufriev, LP. *Russian Microelectronics*, 2007, vol.36, no.4, 251-260.

[286] Hashim, U.; Hamat, HN.; Ayub, RM. *Chiang Mai J. Sci.*, 2007, vol.34, no.1, 35-46.

[287] Bhatnagar, P., Panda, S., Edleman, NL, Allen, SD., Wise, R., Mahorowala, A. *J. Appl. Phys.,* 2007, vol. 101, no.7, article 076102.

[288] Sokolov, YV.; Wang, Q.; Daggubati, M. *J. Electrochem. Soc.,* 2008, vol.155, no.3, H188-H195.

[289] Jagadeesha, T.; Kim, L.; Gonsalvis, J.; Gowda, T. *Ind. J. Eng. Mater. Sci.*, 2009, vol.16, no.10, 341-346.
[290] Sokolov, YV.; Wang, Q. *J. Electrochem. Soc.*, 2010, vol.157, no. 2, G39-G43.
[291] Pande, AA.; Levitin, G.; Hess, DW. *J. Electrochem. Soc.*, 2010, vol.157, no. 6, G147-G153.
[292] Nishimoto, Y.; Tokumasu, N.; Maeda, K. *Proc. 1st Intern. Dielectric for ULSI Multilevel Interconnection Conf. (DUMIC)*, 1995, 15-21.
[293] Lassig, SE.; Li, J.; McVittie, JP.; Apblet, C. *Proc. 1st Intern. Dielectric for ULSI Multilevel Interconnection Conf. (DUMIC)*, 1995, 190-196.
[294] Korczynski, E. *Solid State Technol* , 1996, vol.39, no.4, 63-73.
[295] Gobil, Y.; Fugier, P.; Morand, Y.; M'Saad, H.; Veiron, T. *Proc. 2nd Int. Dielectric for ULSI Multilevel Interconnection Conf. (DUMIC)*, 1996, 108-115.
[296] Nag, S.; Ramamurthy, R.; Lei, WJ.; Monteil, C.; Hickey, M. *Solid State Technol.*, 1998, vol.41, no.9, 69-78.
[297] Vassiliev, VY.; Lin, C.; Fung, D.; Hsieh, J.; Hui, J. *Proc. 5th Intern. Dielectric for ULSI Multilevel Interconnection Conf. (DUMIC)*, 1999, 235-244.
[298] Vassiliev, VY.; Lin, C.; Tan, Z.; Chu, T. *Proc. 8th Intern. Symp. on IC Technology, Systems and Application (ISIC-99)*, 1999, 561-564.
[299] Armburst, DS.; Chapple-Socol, J.; Serriani, L.; Chamberlin, T.; Carlsen, K. *Proc. 5th Int. Dielectric for ULSI Multilevel Interconnection Conf. (DUMIC)*, 1999, 106-115.
[300] Vassiliev, VY.; Sudijono, J.; Cuthbertson A. *Electrochem. Soc. Proc. Volume*, PV 99-31, 1999, 310-321.
[301] Vassiliev, VY.; Lin, C.; Fung, D.; Hsieh, J.; Sudijono JL. *Electrochemical and Solid State Letters*, 2000, vol.3, no.2, p.p.80-83.
[302] Sato, M.; Yoshinobu, A. *J. Appl. Phys.,* 1986, vol.256, no.9, L764-766.
[303] Vassiliev, VYu.; Sudijono, J.; Cuthbertson, A. US Patent № 6500771, issued 31.12.2002.
[304] Kalninya, RP.; Feltin IA.; Freiberga, LA. *(Izvestiya Akademii nauk SSSR. Seriya Neorganicheskiye materiali (USSR Academy of Sciences transactions. Inorganic materials series),* 1969, vol.5, no.9, 1540-1546.
[305] Semenov, NN. *Chain Reactions,* Science, Moscow, 1986, in Russian.
[306] Shintani, A. *J. Electrochem. Soc.*, 1977, vol.124, no.11, 1771-1776.
[307] Shintani A. *J. Electrochem. Soc.*, 1980, vol.127, no.2, 426-429.
[308] Whitby, E.; Hashino, M. *J. Electrochem. Soc.*, 1996, vol.143, no.10, 3397-3404.
[309] Vassiliev, VYu.; Kurchenko, AN. In *Semiconductor tensometry*, Shadrin VS (Ed.), Novosdibirsk Electrotechnical Institute, Novosibirsk, 1986, 11-22 (In Russian).
[310] Watkins-Johnson Company advertizers WJ-999, WJ-1000, WJ-1500, WJ-2000.
[311] Quester Company advertisers APT-4300, APT-5850.
[312] Vassiliev, V.Y. US Patent № 5876798, issued 02.03.1999.
[313] Dobkin, MD.; Makhtari, S.; Schmidt, M.; Pant, A.; Robinson, L. *J. Electrochem. Soc.,* 1995, vol.142, no.7, 2332-2340.
[314] Balazs Analytical Laboratory (now Balazs NanoAnalysis). www.balazs.com
[315] Charles Evans and Associates (now Evans Analytical Group). www.eaglabs.com
[316] Kolesova, VA. *Fiz. Khim. Stekla*,1983, vol.9, no.3, 257-266 (In Russian).

[317] Wang, CY.; Lim, EH.;. Vassiliev, VY.; Ang, TC.; Sudijono, JL.; Zheng, JZ.; Cuthbertson, A.*Proc. 17th Int. VLSI Multilevel Interconnection Conf. (VMIC)*, 2000, 446-448.

[318] Kamienieski, E. *J. Appl. Phys.,* 1983, vol.54, no.11, 6481-6487.

[319] Ajuria, SA.; Fitch, JT.; Workman, D.; Mele, TC. *J. Electrochem. Soc.*, 1993, vol.140, no.7, L113-L115.

[320] Lipkin, LA. *J. Electrochem. Soc.*, 1993, vol.140, no8, 2328-2332.

[321] Keithley Instruments, Inc., information material on Quantox tool, 1997.

[322] Saxena, AN. (Invited lecture) In *Visual Booklet of Short Courses at IV International Dielectric for ULSI Multilevel Interconnection Conference,* 1998, 7-57.

[323] Stamper, AK.; Lasky JB.; Adkisson, JW. *J. Vac. Sci. Technol. A.*, 1995, vol.13, no.3, 905-911.

[324] Mena, RA.; Baumert, DC.; Ullo, CA.; Lee, H.; Lewis, JM.; Larson, SB.; McMullan, R.; Alhburn, BT. *Proc. 4th Intern. Dielectric for ULSI Multilevel Interconnection Conf. (DUMIC)*, 1998, 271-274.

[325] Hwang, GS.; Giapis, KP. *Proc. 3rd Intern. Symp. on Plasma Process-Induced Damage (P^2ID)*, 1998, 164-167.

[326] Lin, YM.; Jang, SM.; Yu, CH.; Lei, TF.; Chen, JY. *J. Electrochem. Soc.*, 1997, vol.144, no.7, 2525-2530.

[327] Schravendijk, BJ.; Carl; D.; Nagy, F.; Hoek, W. *Proc. of 1st Intern. Dielectric for ULSI Multilevel Interconnection Conf. (DUMIC)*, 1995, 141-146.

[328] Nekrasov, V.A., *Osnovy obshchei khimii* (The Fundamentals of General Chemistry), Moscow: Khimiya, 1973, vol. 2, 9–12.

[329] Boric Oxide, Boric Acid, and Borates, in *Ullman's Encyclopedia of Industrial Chemistry*, VCH, 1991, vol. A4, 5th ed., 263–280.

[330] Boron Compounds, *Encyclopedia of Chemical Technology*, New York: Wiley, 1992, vol. 4, 360–418.

[331] Phosphoric Acids and Phosphates, in *Ullman's Encyclopedia of Industrial Chemistry,* Weinheim: VCH, 1991, A19, V ed., 465-500.

[332] Zahariasen, WH. *J. Amer. Chem. Soc.*, 1932, vol.54, no.10, 3841-3851.

[333] Horn, WF.; Hummel, FA. *J. Soc. Glass. Technol.*, 1955, vol. 39, no. 117, 113T.

[334] Englert, WJ.; Hummel, FA. *J. Soc. Glass. Technol.,* 1955, vol. 39, no.187, 121T.

[335] Horn WF.; Hummel, FA.*J. Brit. Ceram. Soc.*, 1979, vol. 78, 77.

[336] Myuller, RL., Borisova, ZU. In *Solid State Chemistry*. New York: Consultants Bureau, 1966.

[337] Kolesova, VA. *Fiz. Khim. Stekla*,1981, vol.7, no.6, 641-646 (in Russian).

[338] Nemilov, SV. Thermodynamic and Kinetic Aspects of the Vitreous State, CRC Press, 1995.

[339] Levy, RA.; Nassau, K. *J. Electrochem. Soc.*, 1986, vol.133, no.7, 1417-1424.

[340] Hung, L.; Thallikar, G.; Myers, FR.; Cale, TS. *Proc. 1st Int. Dielectric for ULSI Multilevel Interconnection Conf. (DUMIC)*, 1995, 204-210.

[341] International Technology Roadmap for Semiconductors, http://www.itrs.net

[342] Shah, N.; McVittie, J.; Shariff, N.; Gat, A. *Proc. Mat. Res. Soc. Symp.*, 1986, vol.52, 233-240.

[343] Cha, CL.; Vassiliev ,VYu.; Chor, EF.; See, AK. *Proc. 5th Int. Plasma-Induced Damage Conf. (P^2ID)*, 2000, 38-41.

[344] Navi, M.; Dunham, ST. *J. Electrochem. Soc.*, 1998, vol.145, no.7, 2545–2548.
[345] Farber, D.; Bae, S.; Okandan, M.; Reber, DM.; Kuzma, T.; Fonash, SJ. *J. Electrochem. Soc.*, 1999, vol.146, no.6, 2254–2257.
[346] Schnable, G.L. US Patent No 4582745, issued 15.04.1986.
[347] Planarization for ULSI Multilevel Interconnection. *Visual Booklet of Tutorial Course on the 5th Int. Conf. C.M.P. (CMP-MIC),* Wade, T., Ed., Santa Clara, 2000.

AUTHOR BIOGRAPHY

Vladislav Yu. Vasilyev received his MS in Chemistry from Novosibirsk State University in 1976, and his Ph.D. in physical chemistry and D.Sc. in solid state chemistry from Russian Academy of Sciences in 1990 and 2002, respectively. After university graduation, he joined the Novosibirsk Semiconductor Devices Manufacturing, Research and Development Division. From 1976 to 1996, he had positions ranging from an engineer to a chief technology engineer. His main area of interest is thin films and IC technology. In 1996-2000, he worked in the R&D department of Chartered Semiconductor Manufacturing, Ltd. as a principal engineer responsible for thin film development for various CMOS logic applications. From 2001-2005, he worked as a leading research scientist at the Institute of Semiconductor Physics, Siberian Branch of Russian Academy of Sciences. During the 2006-2008 period, he was an Invited Professor conducting ruthenium thin film research under the Korean Governmental Brain Pool Program at the Korea Polytechnic University, Department of Nano-Optical Engineering.

He has a total of over 30 years of experience in microelectronics technology. He is currently an IC technology advisor at the Novosibirsk Semiconductor Devices Manufacturing, a Deputy General Director of a sub-micron IC design center SibIS LLC, Novosibirsk, and a part-time Professor at the Novosibirsk State Technical University, Semiconductor Devices and Microelectronics Department.

He has published more than 120 scientific and technical papers, and made presentations at national and international conferences. He was granted with 15 Russian and US patents. He is an active member of The Electrochemical Society and a member of the American Nano Society.

INDEX

A

B

C

D

E

T

U

V

W

X

Y